Preface

The three volumes of **Science Matters** have been carefully designed to fulfill the requirements of the National Curriculum at Key Stage 4. The three volumes are entitled:

Science Matters – Materials
Science Matters – Physical Processes and
Science Matters – Life and Living

These three books provide a balanced science course for GCSE. This particular book is all about the science of **Materials** and the part that materials play in our lives, in industry and in society. It fully covers the requirements of attainment target 3 (AT 3) in the National Curriculum for Science and those for the new GCSE courses and examinations in chemistry. I hope you will find the book lively, colourful and interesting.

The book is divided into twelve major **sections**. These sections are divided into two-page units, each of which ends with several short **questions** to help you understand the topic you have just studied.

At the end of each section, there are **'Things to do'**. These pages contain **Activities** and **Study Questions**. Most of the study questions are like the questions on GCSE papers. Try to answer as many of these as possible because asking questions and answering them is a very good way to learn.

The **Activities** are related to different aspects of materials and chemistry. These activities include creative writing, text-related exercises, data response questions, surveys, role plays and problem-solving exercises. They will enable you to practise different skills. They will also help you to appreciate the wider social, environmental, economic and technological aspects of materials and their chemistry.

If you are looking for information on a particular topic, look it up in the contents list at the front of the book and in the index at the back.

Many people have influenced the planning and writing of **Science Matters – Materials.** I am particularly grateful to Caroline Evans, Vicki Smith, Gina Clarke and Eleanor Stillwell of Hodder and Stoughton and to my wife, Elizabeth. I have also appreciated the support and contributions of my co-authors Rosalind Jones, Nick England, John Sears and Sue Taylor.

Graham Hill

To Aunt Clara

10 99

SCIENCE MATTERS

Materials

Graham Hill

Hodder & Stoughton
LONDON SYDNEY AUCKLAND

CONTENTS

SECTION A

From raw materials to pure substances

Black gold – Crude oil is an important raw material. It is a source of chemicals and fuels

1 RAW MATERIALS

Mining iron ore

Iron ore can be made into steel and used to build structures like the Humber Bridge

Iron ore, like that in the photograph above, is useless. We can't eat it, grow things in it or build things with it. But if we heat it with limestone, coke and air we can turn it into iron, and from iron we can make steel.

- The study of materials and substances is called **chemistry**.
- Chemists and chemical engineers study materials and try to change useless materials like iron ore into useful materials like iron and steel.

These different materials like iron, iron ore, limestone, coke and air are called **substances**.

Raw materials from the Earth

Materials like iron ore, limestone, water and air, that occur naturally, are called **raw materials** or **naturally occurring materials**. They are found in the Earth, in the sea or in the air. On the other hand, materials like iron and steel do not occur naturally, but they can be made from iron ore. Because of this, iron and steel are called **man-made materials**. The chemical industry produces millions of tonnes of important man-made materials every year. These include metals, fuels, fertilisers and plastics.

Conserving and recycling

In our desire for more possessions and easier living we have taken more and more of the raw materials from the Earth. In doing this we have spoilt large areas of the Earth with mines, quarries, motorways and pylons. Large buildings have turned our cities into 'concrete jungles' and forests have been destroyed forever. As a result of this important raw materials, such as copper ore and oil, are being used up rapidly. Today there are 5000 million people on the Earth; there will be 7000 million by the year 2000. Unfortunately, *more* people need *more* food, *more* water, *more* raw materials and *more* land.

But *the Earth's resources will not last forever.* This fact has had some important effects.

- We are all more careful about the use of scarce resources. **Conservation** is essential.
- Waste materials, such as paper, plastics, glass, aluminium and steel, are being recovered and used again. This is called **recycling**.
- Scientists and engineers are producing cheaper **alternative materials** to those already in use.

Over 6 billion glass bottles and jars are used each year in Britain. Throwing these away produces 2 million tonnes of wasted glass

Bottle banks are a common sight in Britain. In future we shall have to recycle other materials like plastics, aluminium and steel if we are to make the most of the Earth's resources

QUESTIONS

1 How would you describe to a friend what chemistry is?
2 Make a list of 6 materials in your kitchen. If possible, give the chemical name of their main constituent (e.g. salt – sodium chloride). Say which materials occur naturally and which are man-made.
3 Coal is an important raw material. List 4 substances made or extracted from it.
4 (a) What does the word 'ore' mean?
(b) Which metals are obtained from the following ores: *bauxite*, *copper pyrites*, *haematite*, *galena*, *tinstone*?
5 Look at today's newspaper. What news or adverts does it contain about raw materials for the chemical industry?

2 MATERIALS AND THEIR PROPERTIES

Look at the variety of different materials in this room

Most of the outside of this building is glass. What are the advantages of using glass? What are the disadvantages?

Look around your home. Notice the variety of different materials and their uses.

- **Metals** are used to make knives and forks, pans and ornaments. Metals have been used for these items because they are hard, strong, shiny and can be made into different shapes.
- Curtains, carpets and clothes are made from **fibres** which are soft and comfortable. Some fibres like cotton and wool occur naturally as part of plants and animals, but others like polyester (Terylene) and Nylon are made from the chemicals in coal, crude oil and plants.
- **Plastics** such as polythene, perspex and PVC are used for articles like clingfilm, washing-up bowls, combs and records. Plastics are used because they are fairly cheap, flexible (bendy) and easily moulded into different shapes.
- Windows, jars and some ornaments are made of **glass**.
- Window frames, doors, chairs and tables are often made of **wood**.

1 Why is glass used for windows?
2 Why is wood used for doors and furniture?

If you have tried to answer the last two questions, you will see that we choose different materials for different uses because they have different **properties**.

> The properties of a material tell you what it is like and what it can be used for.

Classifying materials

Properties help us to see how one material is different from another. They also help us to put similar materials into groups or sets. For example, we can put drinks into different groups such as fizzy, milky, alcoholic and non-alcoholic. There is a special word for putting things into groups with similar properties. It is called classification. Scientists classify materials in several ways. One of the most useful ways is as **solids, liquids** and **gases**. These are sometimes called the three **states of matter**.

These photographs, and Table 1 below, show some important properties of solids, liquids and gases.

The rocks at Stonehenge show important properties of solids

This water flowing over the Niagara Falls shows important properties of liquids

The air inside this inflatable dinghy shows important properties of gases

Solids	**Liquids**	**Gases**
Solids have weight	Liquids have weight	Gases have weight
Solids keep their shape	Liquids take the shape of their container	Gases take the shape of their container
Solids keep the same volume	Liquids keep the same volume	Gases change their volume easily. Gases fill *all* of their container

Table 1 Important properties of solids, liquids and gases

QUESTIONS

1 Look at the clothes in the photo above.
(a) What fibres are these clothes made from?
(b) Which of these fibres are (i) naturally occurring (ii) man-made?

2 What properties of bricks and concrete make them important building materials?

3 At one time, gutters and drainpipes were made of iron. Today they are often made of plastics like PVC (polyvinyl chloride).
(a) What properties of iron made it useful for gutters and drainpipes?
(b) What were the disadvantages of iron?
(c) Why has iron been replaced by plastics?

4 Usually it is easy to tell whether a material is solid, liquid or gas. But sometimes it is more difficult.
(a) Is mud a liquid or a solid or does it contain both? Explain your answer.
(b) What state of matter is (i) smoke, (ii) jelly, (iii) custard, (iv) honey? Explain each of your answers.

5 (a) Make a table like the one below showing 4 different liquids that are used in a car and the jobs they are used for. The table has been started for you.
(b) Choose two of the liquids in your table. Say why they are used for their particular jobs.

Liquid used	**What is it used for?**
light oil . . .	brake fluid . . .

3 ELEMENTS – THE SIMPLEST MATERIALS

So far we know of 105 elements. Some like radium, have only been discovered in the last 100 years. But others, like gold, were known to ancient civilisations. This beautiful gold funeral mask belonged to Tutankhamun

The black material on badly burnt toast is carbon. No matter how this carbon is treated, it cannot be broken down into anything simpler

Robert Boyle first proposed the existence of elements in 1661. He described them as 'primitive and simple, or perfectly unmingled bodies'. *Elements could combine to make compounds, he said, and compounds could be divided into their elements*

When bread is well-toasted, the surface gets covered in a black solid which is carbon. Smoke, containing water vapour and carbon dioxide, also rises from the burnt toast.

Changes like this, that result in new materials, are called **chemical reactions**. The new substance is called the **product** of the reaction. The chemical reaction which produced carbon from bread is caused by heat from the toaster. Many foods are prepared by chemical reactions caused by heat.

We can summarise the reaction which takes place when bread is heated by writing a **word equation**:

$$\text{bread} \xrightarrow{heat} \text{carbon} + \text{carbon dioxide} + \text{water} + \text{toast}$$

No matter how the black carbon is treated, it cannot be broken down into a simpler substance.

> In 1661 Robert Boyle suggested the name **element** for a substance that cannot be broken down into a simpler substance. Elements are the simplest possible materials.

This is still a useful description of an element. Carbon is an element. But substances like water and carbon dioxide are not elements because they can be broken down into simpler substances. Substances like water and carbon dioxide which contain two or more elements are called **compounds**.

When electricity is passed through water containing a little sulphuric acid, it breaks up forming hydrogen and oxygen (Figure 1).

$$\text{water} \xrightarrow{\textit{electricity}} \text{hydrogen} + \text{oxygen}$$

The hydrogen and oxygen cannot be made any simpler: they are elements.

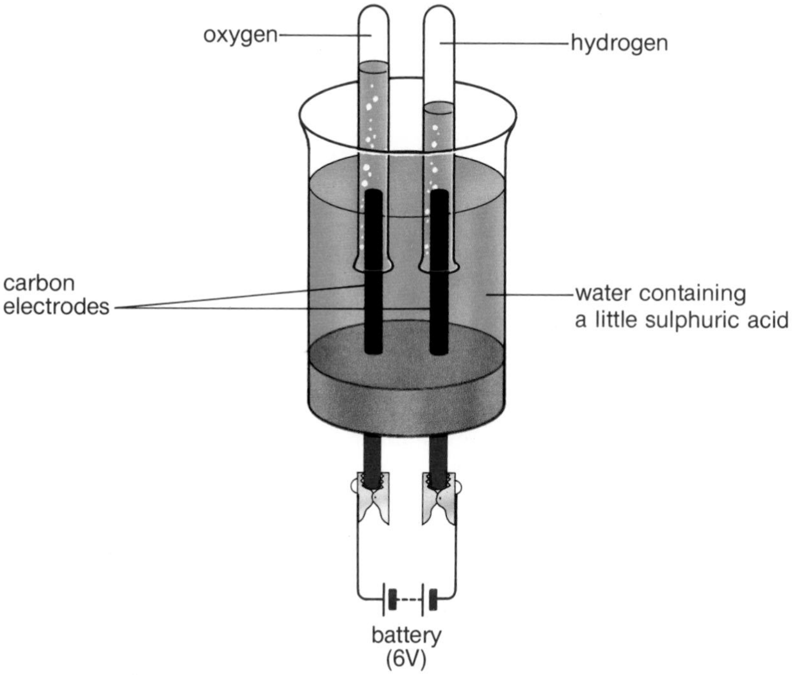

Figure 1
When electricity is passed through water containing sulphuric acid, the water splits up into hydrogen and oxygen

Everything in the universe is made up of elements. So far we know of 105 elements. These include aluminium, iron, copper, oxygen, nitrogen and carbon.

Although there are millions and millions of different materials in the universe, they can all be split up into one or more of the 105 known elements. For example, grass is made of carbon, hydrogen and oxygen. Sand is made of silicon and oxygen. So elements are *the building blocks for all substances*. In the same way, bricks are the building blocks for houses. Using a few different types of brick it is possible to build millions of different houses.

A reaction in which a compound splits up into two or more simpler substances is called **decomposition**. An example of this is the splitting of water into hydrogen and oxygen.

Water can be decomposed into hydrogen and oxygen, but if hydrogen is exploded with oxygen, these two simpler substances react to form water again. A reaction like this in which two or more substances join together to form a single product is called **synthesis**.

Notice that synthesis is the reverse of decomposition. Synthesis is a 'building up' whereas decomposition is a 'breaking down'.

$$\text{water} \underset{\textit{synthesis}}{\overset{\textit{decomposition}}{\rightleftarrows}} \text{hydrogen} + \text{oxygen}$$

A lot of toys can be made from a few different pieces of Lego. In the same way, all materials are made from a few different elements

QUESTIONS

1 Write a short article for 12 year olds explaining what elements are. Your article should include diagrams and photos if these will help.

2 Write down the names of all the elements mentioned in this unit.

3 A gas was produced when a solid was heated. How would you test to see if the gas is (i) carbon dioxide, (ii) water vapour, (iii) oxygen?

4 When red mercury oxide is heated, shiny beads of mercury are produced and oxygen gas is given off.
(a) Write a word equation for the reaction.
(b) Is this decomposition or synthesis? Explain your answer.

5 (a) Is it possible to decompose an element? Explain your answer.
(b) Is it possible to decompose a compound? Explain your answer.
(c) Which of the following can be decomposed?
sugar, zinc, hydrogen, salt.
(d) Explain why a synthesis reaction *always* produces a compound.

4 CLASSIFYING ELEMENTS

Why are metals ideal for making bells?

Element	Melting point /°C	Boiling point /°C	Electrical conductivity
Aluminium	660	2450	good
Carbon (graphite)	3730	4830	moderate
Copper	1083	2600	good
Gold	1063	2970	good
Iron	1540	3000	good
Oxygen	−219	−183	poor
Sulphur	119	445	poor

Table 1 The properties of some elements

Sulphur deposits around a volcano blowhole in Sumatra. Sulphur is an important non-metal. It is used to manufacture sulphuric acid

Look at the properties of the elements listed in table 1.

1 Which elements have a melting point above 500°C?
2 Which elements have a boiling point above 1000°C?
3 Which elements are good conductors of electricity?
4 Which of these elements can be polished to a shine?
5 Which of these elements can be bent or hammered into different shapes?
6 Which of these elements are metals?
7 Do the properties in Table 1 separate metals from non-metals?

Dividing elements into metals and non-metals is very useful. One of the best ways of checking whether an element is a metal or a non-metal is to see if it conducts electricity. This can be done by using the apparatus in Figure 1. The bulb lights when the element tested is a metal. When non-metals (except graphite) are tested, the bulb does not light because these are poor conductors.

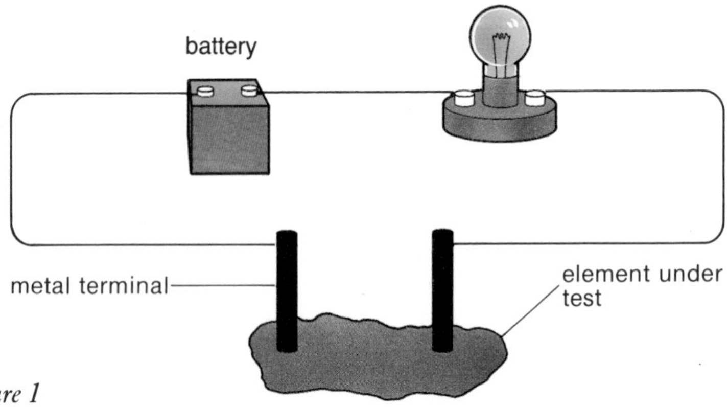

Figure 1

Although one or two elements, such as graphite, are difficult to classify, the classification of elements as metals and non-metals is very helpful. There are about 80 metals, about 20 non-metals and five or six elements that are difficult to place. Table 2 compares the properties of metals and non-metals. There is more about classifying elements in section F.

Marie Curie

Enrico Fermi

Dmitri Mendeléev

Property	Metals	Non-metals
State	usually solids at room temperature	solids, liquids or gases at room temperature
Melting point and boiling point	usually high	usually low
Density	usually high	usually low
Appearance	shiny	dull
Effect of hammering (malleability)	can be hammered into shape (malleable)	brittle (non-malleable) and soft when solid
Thermal and electrical conductivity	good	poor

Table 2 Comparing the properties of metals and non-metals

Albert Einstein

QUESTIONS

1 Look at the periodic table on page 126.
(a) Which elements have been named after the four scientists in these photos?
(b) Which other elements were named after famous scientists?
(c) Write a few paragraphs about the life and work of one of the scientists in the photos.

2 (a) Make a table showing the names of 20 metals and 10 non-metals.
(b) Write down the names of 2 elements that are difficult to classify as either metal or non-metal. Explain why they are difficult to classify.

3 How would you find out whether a chunk of a solid element is a metal or a non-metal?

4 A colourless gas *A* relights a glowing splint:
(i) *A* reacts with copper to give a black solid *B*;
(ii) *A* reacts with a hot dark grey solid *C* to give a colourless gas *D*;
(iii) *D* turns limewater milky;
(iv) *B* and *D* are both produced when the green powder *E* decomposes on heating.
(a) Identify the substances *A* to *E*.
(b) Write word equations for the reactions (i), (ii) and (iv).

5 USING ELEMENTS

Making a stained glass window using strips of lead

Aluminium is ideal for making greenhouse frames

Metals, like gold and silver, and the non-metal diamond, are attractive and unreactive. They are also very rare and precious compared to other materials. For these reasons, they are often used to make valuable ornaments and jewellery. This photo shows the Imperial State Crown

Look at the photographs on this page.

1 Why is aluminium used to make the frames of greenhouses?
2 Why is lead used to hold the pieces together in stained glass windows?

> The uses of an element depend on two things:
> - how easily it can be obtained
> - its properties

Aluminium is strong, it has a low density and it does not rust. These properties explain why it is used for the frames of windows and greenhouses, for tent poles and in aircraft construction. Aluminium is also a good conductor of heat and electricity. Because of these properties it is used for pans and other cooking vessels. Lead is a soft, malleable metal which does not rust. It is useful for roofing and for stained-glass windows. It also has a high density and because of this it is used to protect hospital workers from X-rays. Copper is a good electrical conductor. It is cheap compared to most metals and it can be drawn into wires. Because of these properties, it is used for electrical wires and cables.

Metals can be mixed with other elements to form **alloys**. Alloys have properties that are different from those of the elements they contain. So it is possible to make alloys with particular properties. The most common alloy is steel which is mainly iron with about 0.15% carbon. The carbon makes steel harder and tougher than pure iron. Stainless steel also contains chromium and nickel to prevent it from rusting. Iron, aluminium and copper are the most widely used metals. All three are used in large quantities in alloys. There is more about alloys in section E.

What about the uses of non-metals? Silicon and germanium conduct electricity but not as well as metals. Thus, these two elements are known as **semiconductors**. They are used to make 'silicon chips' for electronic circuits. Chlorine, a reactive gas which is poisonous in large quantities, is used at low concentrations to kill bacteria in drinking water and swimming pools.

The bands of rocks in the Grand Canyon are mainly sandstones and limestones. The main elements in sandstones are silicon and oxygen. The main elements in limestone are calcium, carbon and oxygen

Elements in the Earth's crust

Although there are 105 known elements, they occur in very different proportions on the Earth. Figure 1 shows the abundance of the commonest five elements in the Earth's crust. Notice that these five elements make up 91% by mass. The remaining 100 elements account for only 9%. The most abundant element in the Earth's crust is oxygen. This forms 23% by mass of the air and 89% of water. Most oxygen, however, is combined with other elements in rocks and soils. Sand and sandstone, for example, contain roughly equal proportions of oxygen and silicon, which is the second most abundant element. Clays also contain silicon and oxygen, together with aluminium. Aluminium is the third most abundant element and the most abundant metal in the Earth's crust. Iron is the fourth most abundant element. Most of it is combined with oxygen in iron ore (rust-coloured iron oxide). Chalk and limestone are made of calcium carbonate which is 40% calcium, 12% carbon and 48% oxygen. This accounts for most of the calcium in the Earth's crust.

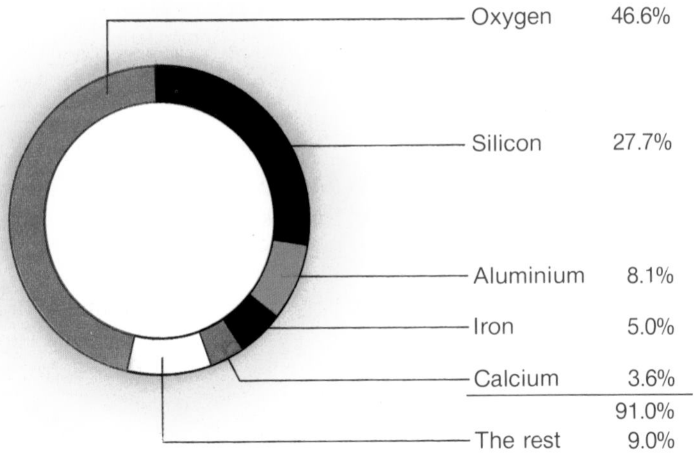

Figure 1

QUESTIONS

1 The percentages of different elements in your body are:

oxygen	65%
carbon	18%
hydrogen	10%
nitrogen	3%
calcium	2%
phosphorus	1%
other elements	1%

Represent this information as a pictogram or as a pie chart.

2 Prepare a four-minute talk on alloys. In your talk:
(a) explain what an alloy is,
(b) give some examples of alloys,
(c) show some uses of alloys,
(d) say why alloys are important.

3 Platinum is a rare, hard and unreactive metal with a high melting point. What uses does platinum have because of these properties?

4 Mercury is the only liquid metal at room temperature. It has a high boiling point, a high density and it is a good conductor of heat and electricity. What uses does mercury have on account of these properties?

5 (a) Use a textbook to make a list of the properties of helium.
(b) Why do you think helium is used to fill electric light bulbs?

6 COMPOUNDS AND MIXTURES

When aluminium pans and window frames are first made, they look shiny. Gradually, they become dull as a layer of white aluminium oxide forms on the surface. The aluminium and oxygen have not just *mixed* together. They have joined together to form a new substance called aluminium oxide. We say that the aluminium and oxygen have **reacted** or **combined** to form white aluminium oxide which is a compound. Salt and sugar are also compounds. Salt contains the elements sodium and chlorine chemically combined (Figure 1). Sugar is a compound of carbon, hydrogen and oxygen.

- A compound is a substance that contains two or more elements combined together.

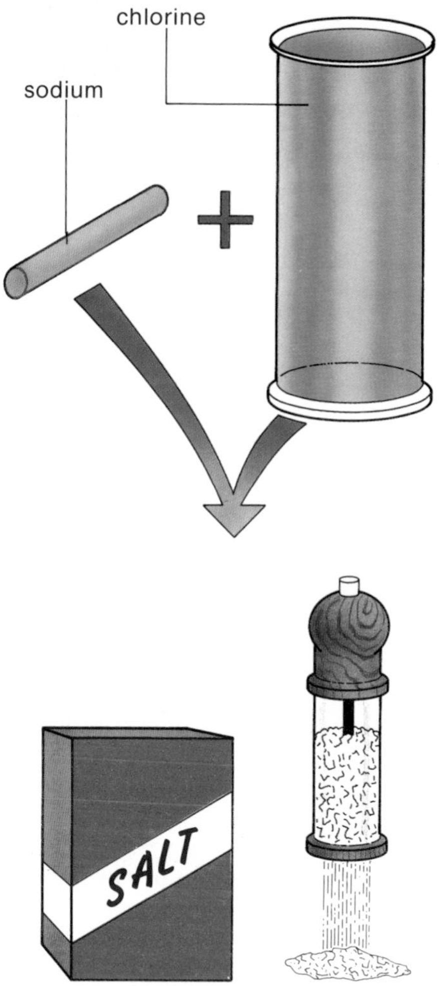

Figure 1
Sodium and chlorine are elements. They combine (react) to form salt (sodium chloride). Sodium chloride is a compound

Sugar is a compound containing the elements carbon, hydrogen and oxygen

When *two* elements form a compound, the name of the compound ends in **-ide**. For example,

aluminium + oxygen → aluminium ox**ide**
sodium + chlorine → sodium chlor**ide**
hydrogen + sulphur → hydrogen sulph**ide**

When a metal reacts with a non-metal, the non-metal forms the -ide part in the name of the compound. When two non-metals react, the more reactive non-metal forms the -ide part in the name of the compound.

Mixtures and compounds

The important differences between mixtures and compounds can be illustrated using iron filings and sulphur.

Making a mixture. *Mix some iron filings with powdered sulphur. In the mixture you can still see the yellow sulphur. You can separate iron from the mixture with a magnet*

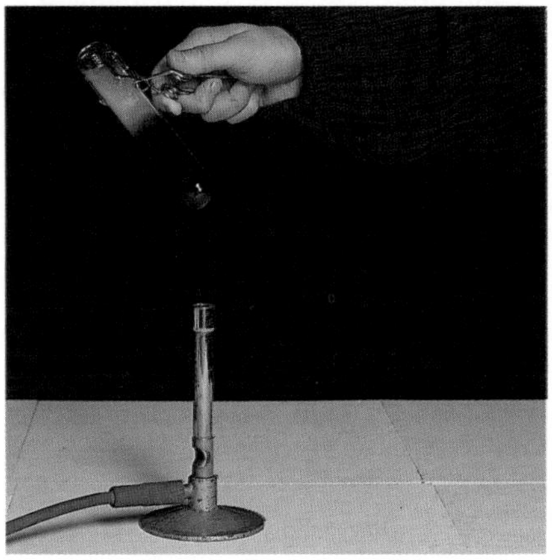

Making a compound. *Heat a mixture of iron filings and sulphur. The mixture gets red hot and iron sulphide forms*

Testing iron sulphide. *Iron sulphide contains sulphur, but you cannot see the yellow sulphur in it. You cannot separate iron from it with a magnet, either*

When a mixture forms:	When a compound forms:
The substances just mix together	The elements join together
No new substance forms	A new substance forms
No chemical reaction occurs	A chemical reaction occurs
The mixture has properties like the original substances	The compound has different properties from the elements in it
The original substances can be separated easily	The elements in it cannot be separated easily

Table 1 The differences between mixtures and pure compounds

QUESTIONS

1 Suppose your best friend has been absent from school during the lessons on elements, mixtures and compounds. Write down what you would say to your friend to explain what we mean by elements, mixtures, compounds and chemical reactions.

2 Write word equations for the reactions which occur when:
(a) coke (carbon) burns in air (oxygen),
(b) copper coins form a thin, dark layer of copper oxide,
(c) hydrogen explodes with air (oxygen).

3 When water is added to white (anhydrous) copper sulphate in a test tube, a blue substance is produced. At the same time, the test tube gets hot.
(a) Is the blue product a compound or a mixture?
(b) Give 3 pieces of evidence to support your answer.

7 SEPARATING MIXTURES

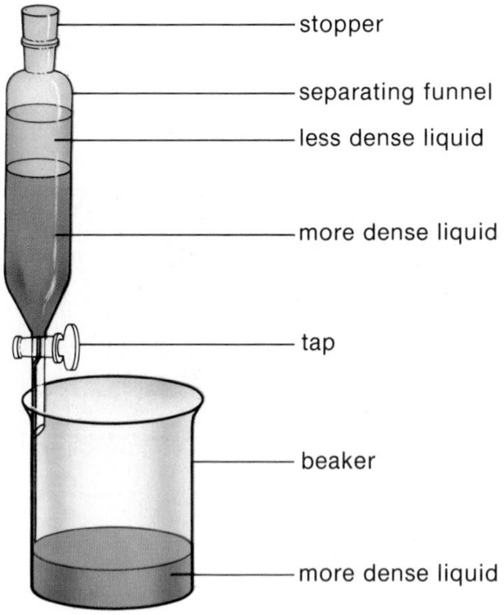

Figure 1

Most substances which occur naturally are mixtures. Very often these mixtures have to be separated before we can use them. Just imagine what would happen if we used untreated, muddy water for cooking and cleaning, or if we tried to run our cars on crude oil rather than petrol.

Muddy water contains particles of solid 'floating' in it. If the water is left for some time, the heavier particles of solid sink to the bottom as a **sediment**. The clearer water can then be poured off easily and separated from the sediment. This process is called **decantation**. Decantation is important in cooking. It is used to separate water from vegetables after cooking and also to separate two liquids which do not mix, like cream and milk. These liquids which do not mix are called **immiscible liquids**. Have you noticed how difficult it is to separate the cream from milk really well? The best way to separate two immiscible liquids is to use a separating funnel (Figure 1).

Another way to separate the solid sediments from muddy water is to use **filtration** (Figure 2). There are millions of tiny holes in filter paper. Liquids can pass through these holes, but solid particles are too large to do so. The solid that remains in the filter paper is called the **residue** and the liquid that trickles through is the **filtrate**.

Clean water for our homes

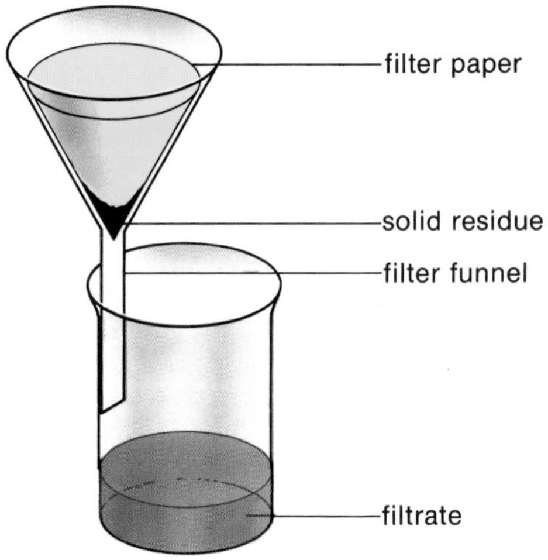

Figure 2

Filtration plays an important part in obtaining clean water for our homes. Figure 3 shows the main stages in the purification of river water.

The water is first stored in reservoirs where most of the solid particles can settle out. As the water is needed, it is filtered through clean sand and gravel which trap smaller particles of mud and suspended solids (Figure 4).

After filtering, the water is treated with small amounts of chlorine to kill harmful bacteria in the water. The purified water is then pumped to storage tanks and water towers from which it flows to our homes.

Filtration is also used to separate beer from its sediment (yeast) before bottling. The beer is filtered by forcing it through filter cloths to catch the sediment.

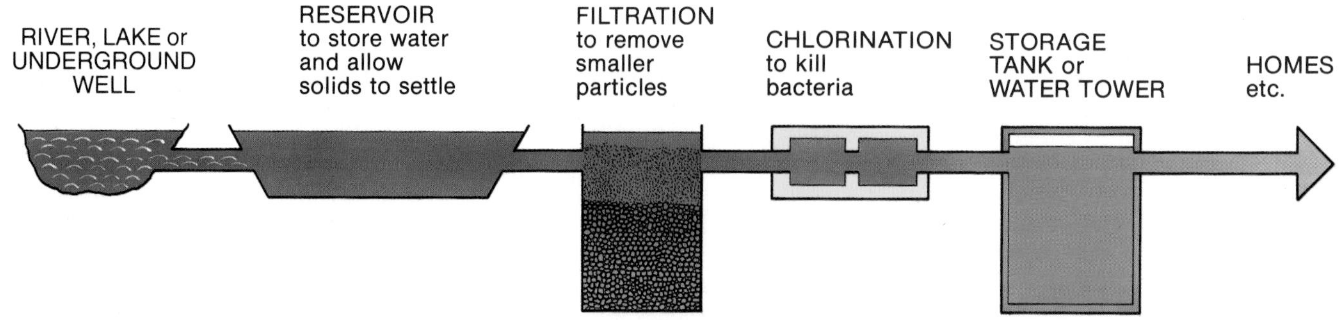

Figure 3

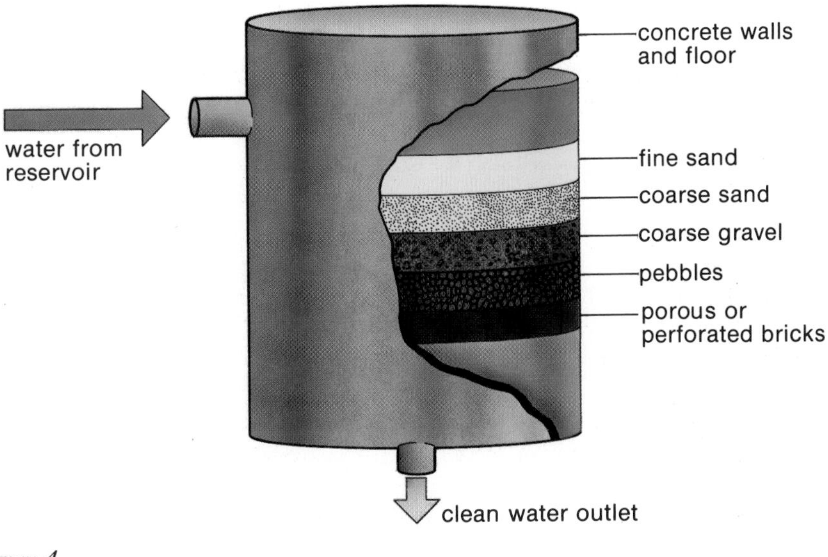

Figure 4

Some people like to decant the cream from the top of the milk for their breakfast cereal

Separating cream from milk

Another way to separate a solid from a liquid or to separate two immiscible liquids is to use a **centrifuge**. As the centrifuge spins round rapidly, a force pulls outwards and the heavier particles go to the bottom of the mixture. In the same way, a roundabout at the funfair drags you outwards. One of the most important uses of a centrifuge is in separating cream from milk to make skimmed milk. Whole milk is poured into a bank of spinning and sloping discs. The heavier milk is forced down and outwards and the cream rises up and inwards (Figure 5).

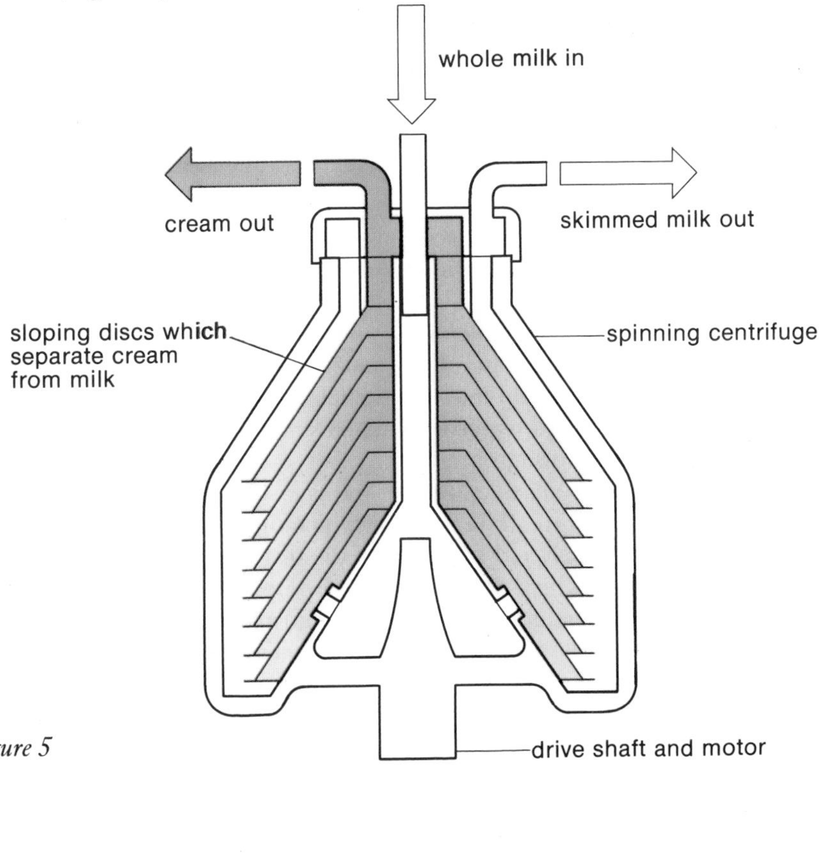

Figure 5

QUESTIONS

1 Draw a flow chart to summarise the main stages in the purification of river water for our homes.

2 Your younger brother and sister are playing at cooking. They decide to mix sugar with sand. Fortunately, you have a good chemistry set. Describe how you would separate dry sand and sugar crystals from their mixture.

3 You are given a yoghurt pot, some pebbles and some sand.

(a) Design a small-scale filtration plant which you could use to clean muddy river water. Draw a diagram of your design.

(b) How could you check how well your model works?

4 A company that separates cream from milk knows that there is an increased demand for cream during the summer. Assuming that the company can produce enough cream to satisfy the increased demand, what extra equipment will be needed in the summer?

8 SEPARATING SOLUTIONS

Sea-water is not pure – it contains salt and other dissolved substances essential for animals and plants that live in the water

Nail polish remover is a solvent for nail polish. The solvent is propanone (acetone)

Tap water is clean but *not* pure. It contains dissolved gases like oxygen and carbon dioxide and probably dissolved solids which make the water 'hard'. In hard water areas, the dissolved solids are left as scum after washing with soap. Clear sea water is another example of a liquid containing dissolved solids. You can't see the salt in clear sea water, but it must be there because you can taste it. The salt has been broken up into tiny particles which are too small to be seen even with a microscope. These particles are so small that they can pass through the holes in filter paper during filtration.

The mixture of dissolved salt and water forms a **solution**. The substance that dissolves is called the **solute**. The liquid in which the solute dissolves is the **solvent**. Solids such as salt and sugar which dissolve are described as **soluble**, and solids such as mud which do not dissolve are **insoluble**.

Although water is a very good solvent, many substances are insoluble in it. Oils and greases are insoluble in water but very soluble in other solvents, like trichloroethane. This liquid is used by dry cleaners to remove grease stains from clothes. Nail polish is insoluble in water but it dissolves easily in a liquid called propanone (acetone). Nail polish remover is made of propanone.

Evaporation and crystallisation

When sea water is left to dry in the sun, the water turns into vapour (evaporates) and leaves behind salt. Next time you are at the seaside, look for white rings of salt around the edges of rock pools.

This change of liquid to a gas or vapour is called **evaporation**. Evaporation can be used to separate a dissolved solid from its solvent. If the solvent evaporates slowly, then the dissolved solute is often left behind as well-shaped crystals. We say that it has **crystallised**. In hot countries, salt is obtained from sea water by allowing the water to evaporate, leaving the crystallised salt. Sugar is also obtained by a process of dissolving and evaporation. The sugar canes are crushed and then mixed with water to dissolve the sugar. The sugar solution is separated from the pulp and then water is evaporated off, leaving sugar crystals.

At room temperature, salt solution takes a long time to evaporate and form crystals. We can speed up the process by heating the solution gently (Figure 1). If the solution is evaporated quickly, the crystals are usually small. Larger crystals form if the solution is evaporated slowly. As the water evaporates, crystals begin to form at the edges of the solution. When this happens, it is best to stop heating and allow the solution to cool and crystallise slowly from then on. This gives larger, well-shaped crystals. If the crystals form in the solution, they can be dried between filter papers.

Evaporation is an essential process in drying wet clothes and in producing concentrated evaporated milk. As the solvent evaporates from the solution, the same amount of solid is left in less and less solvent. We say that a solution which contains only a little solute in a given amount of solvent is **dilute**. If a solution contains a lot of solute in a given amount of solvent we say it is **concentrated**. As more and more solvent evaporates, a solution becomes more and more concentrated. Eventually, the solution cannot become any more concentrated. If any more solvent evaporates, some solid solute will come out of solution and form crystals. A solvent that has dissolved all the solute it can is **saturated**. As solvent evaporates from the saturated solution, more and more solid crystallises out.

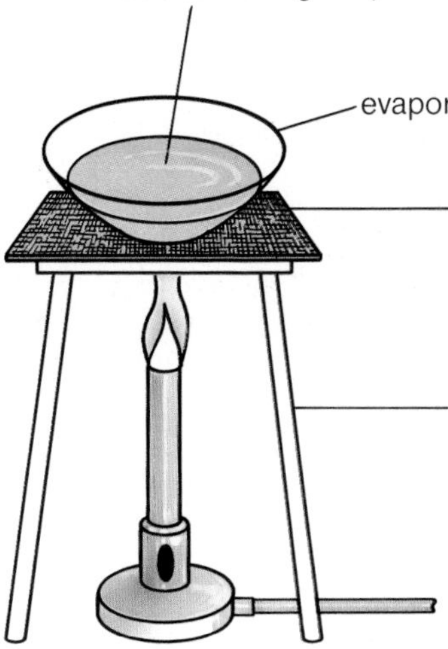

Figure 1

A workman harvesting salt from sea-water in Southern Spain

QUESTIONS

1 Which of the following are solutions?
ink, steam, tea, exhaust fumes, beer, orange juice

2 Which of the following are soluble in water?
sand, sugar, butter, aspirin, chalk, oil, vinegar

3 How would you obtain dry, well-shaped crystals from a solution of salt water? Describe your experiment.

4 Some people prefer to make coffee from ground filter coffee rather than use instant coffee. How would you prepare a jug of filter coffee? In your answer use the following words:
solution, solvent, solute, soluble, insoluble

5 Sea water contains dissolved solids including salt. Design an experiment to find the mass of solid dissolved in 10 cm^3 of sea water.

9 SEPARATING LIQUIDS

Some countries where there is very little rainfall rely on desalination plants like this one to obtain fresh water from sea-water

Pure water from sea water

We can get salt from sea water by evaporating off the water. But can we get *pure water* from sea water? When the sea water is evaporated, water vapour escapes into the atmosphere. If the water vapour is passed into a second container and cooled, it will turn back to water. All the solute (salt) stays in the solution, so the water which evaporates and cools is pure water.

When a vapour changes to a liquid the process is called **condensation**. The apparatus which cools the vapour to liquid is a **condenser**. Figure 1 shows how pure water can be obtained from sea water by evaporating the water and then condensing the water vapour. This process of evaporating a liquid and condensing the vapour is called **distillation**.

distillation = evaporation + condensation.

When the sea water in Figure 1 is heated, the water vapour which escapes passes through the inner tube of the condenser. This is cooled by the cold water flowing around it, so the water vapour condenses and drips from the lower end of the condenser into a receiver. The pure water which collects is called the **distillate**.

Distillation can be used to separate any solvent from its solution. In parts of the Middle East, where fuel is cheap, distillation is used on a large scale to get pure water from sea water. Pure water (distilled water) can also be obtained by distilling tap water. Most chemical laboratories have their own distillation apparatus called a **still** which produces distilled water.

Figure 1

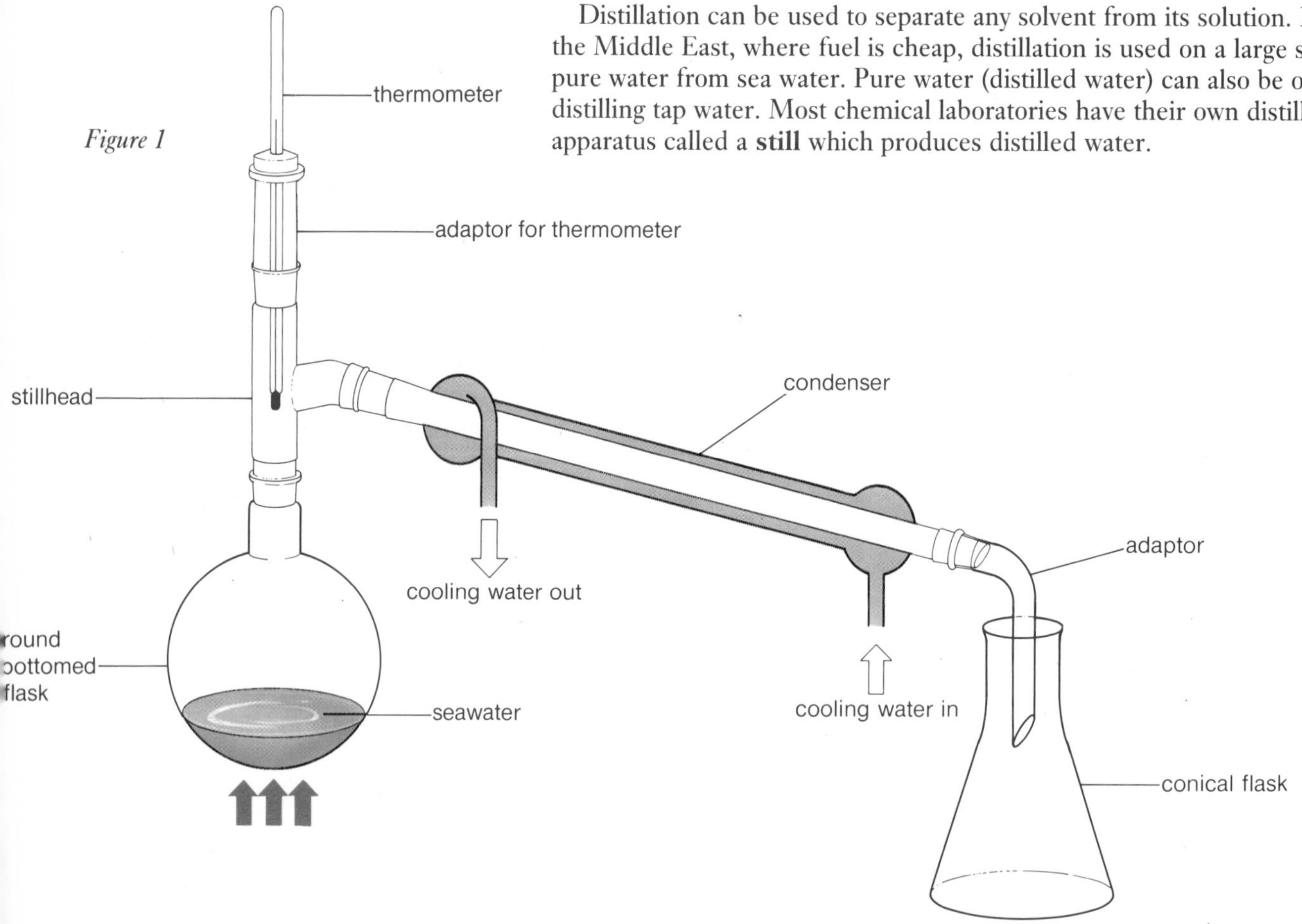

Petrol from crude oil

Crude oil is a mixture of many liquids and dissolved solids. Petrol and paraffin can be separated from it by distillation. When distillation is used to separate a mixture of liquids as in crude oil, it is called **fractional distillation** because it separates the mixture into two or more parts (fractions). Figure 2 shows the temperatures and products at different heights in an industrial fractionating column. Inside the column, there are horizontal trays with raised holes in them. The crude oil is heated by a furnace and the vapours then pass into the lower part of the fractionating column. As the vapours rise up the column through the holes in the trays, the temperature falls. Different vapours condense at different heights in the tower and are tapped off and used for different purposes. Liquids like petrol, which boil at low temperatures, condense high up in the tower. Liquids like lubricating oils, which boil at higher temperatures, condense lower down in the tower.

Fractional distillation is also used in industry to separate oxygen and nitrogen from the other gases in air (section B).

Camping Gaz provides a portable means of cooking by butane

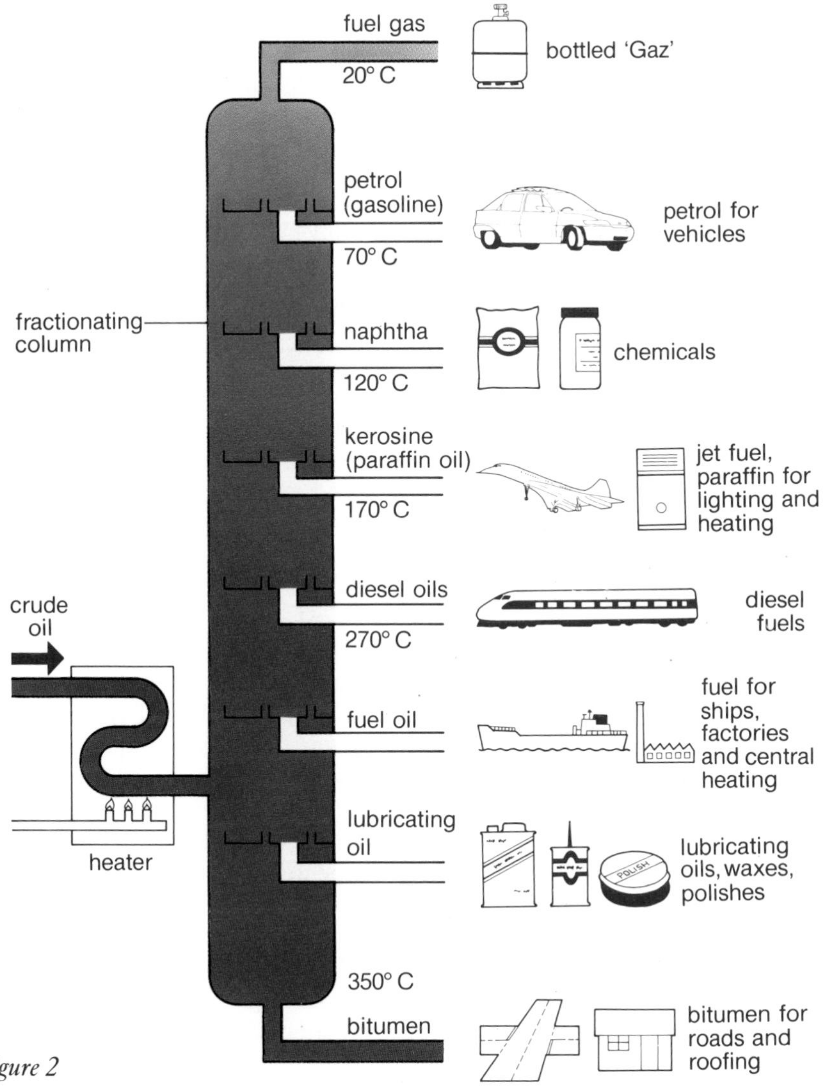

Figure 2

QUESTIONS

1 What is the difference between simple distillation and fractional distillation?

2 Two important processes that involve distillation are the manufacture of brandy and whisky. Brandy is made by distilling wine; whisky is made by distilling beer. Both brandy and whisky contain alcohol (boiling point 78°C) and water (b.p. 100°C).
(a) Which liquid boils at the lower temperature, alcohol or water?
(b) Which of these two liquids evaporates more easily?
(c) If beer is distilled, will the distillate contain a larger or smaller percentage of alcohol?
(d) Why is whisky more alcoholic than beer?

3 Evaporation is an essential process in drying wet clothes and in producing evaporated milk.
(a) What conditions (dry, damp, hot, cold, windy, calm) help evaporation of water from the clothes?
(b) What is evaporated from the milk?
(c) What is the difference between evaporated milk and ordinary milk?
(d) How do you think milk is evaporated industrially?

10 PURE SUBSTANCES

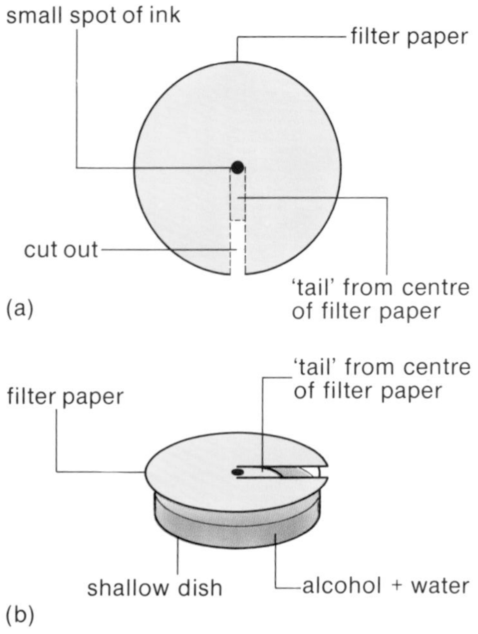

Figure 1

How pure is 'pure' honey?

The label in the photograph says that the honey is pure. This means that it has had nothing added to it or taken out of it. The so-called 'pure honey' contains many different substances. The proportions of these different substances gives each honey its own particular flavour.

> When chemists say that something is pure, they mean that it is a single substance and not a mixture of substances.

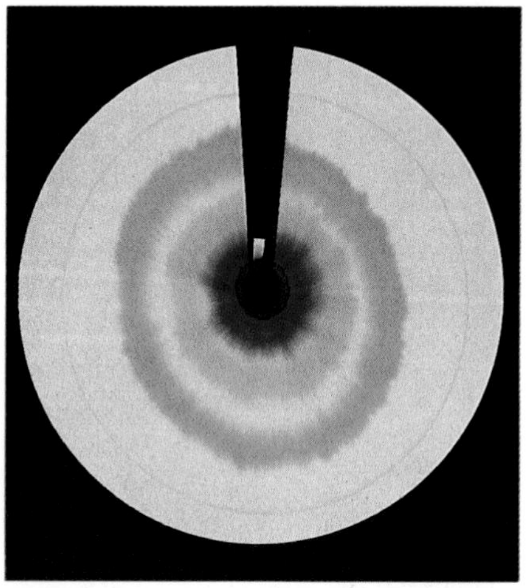

A chromatogram of black ink (The solvent was butan-1-ol: ethanol: 0.88 ammonia solution 3:1:1 by volume)

The honey is not pure to a chemist because it contains many substances. Most substances that occur naturally are mixtures, *not* pure substances. For example, air is a mixture of gases, mainly nitrogen and oxygen; petrol is a mixture of octane and other liquids and ink is a mixture of water and coloured dyes. If you have spilt ink on blotting paper you may have noticed that as the blot spreads out, it forms a series of different colours. Each of the coloured rings contains one of the dyes in the ink. Figure 1 shows how the dyes in ink can be separated more clearly.

Cut the filter paper as shown in Figure 1(a) and fold back the tail from the centre. Put a spot of ink at the centre of the filter paper and leave it to dry. Then, place the filter paper on the shallow dish so that the 'tail' dips into the mixture of equal parts water and alcohol (Figure 1(b)). As the liquid soaks up the 'tail' and through the filter paper, the different dyes in the ink spread out in a series of coloured rings (Figure 2). The green dye in grass and the various coloured dyes in flower petals can be separated by a similar process. The coloured substances are first removed from the grass or petals by grinding in a pestle and mortar with alcohol.

This method was first used to separate coloured substances. The process was therefore called **chromatography** from the Greek word *khroma* meaning colour. Nowadays, the substances being separated may be colourless. After chromatography has taken place, the paper is sprayed with a liquid which reacts with the colourless substances to produce coloured substances.

Chromatography can be used to separate mixtures and to decide whether a substance is pure. It is very important in medicine because it can be used to separate drugs and other constituents in blood and urine.

The colours in our foods and drinks can be enhanced by synthetic dyes and additives. Chemists can detect these dyes using chromatography and check that they are harmless.

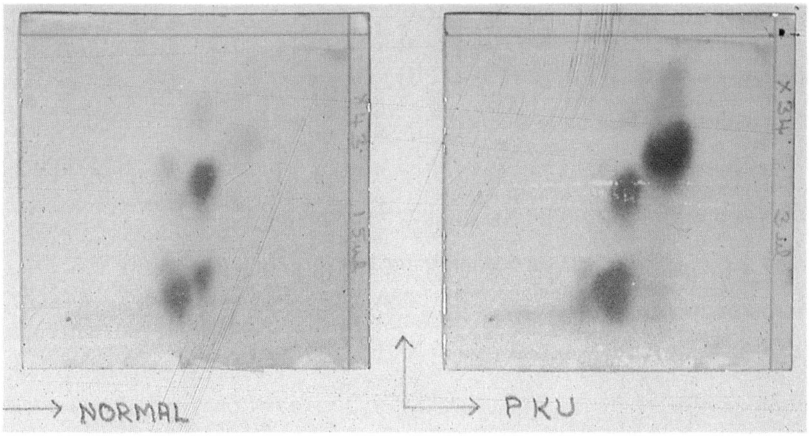

This photograph shows two chromatograms of substances called ketones in the urine. The one on the left is from a normal child. The one on the right labelled 'PKU' is from a child suffering from phenylketonuria. This causes the child to excrete ketones in the urine, which show up as larger spots on the chromatogram

Testing for pure substances

There are three ways of testing that a substance is pure:

- using chromatography
- checking its melting point
- checking its boiling point

If a substance is pure it will contain only one kind of material. So, if a coloured substance is pure, it will give only one coloured ring or band during chromatography.

The boiling point of a liquid changes as the atmospheric pressure rises and falls. On top of Mount Everest, where the pressure is much lower than at sea level, water boils at about 70°C. But, if the atmospheric pressure stays constant, a pure substance will always boil at the same temperature. Because of this, we must pick a standard pressure and measure all boiling points and melting points at this standard pressure. Scientists have chosen the pressure exerted by a column of mercury 760 mm high as standard atmospheric pressure. This pressure is called 1 atmosphere (atm). In SI units, 1 atm = 101 325 Nm^{-2}.

Pure water always boils at 100°C and pure ice always melts at 0°C at 1 atm pressure. The measurement of the melting point or the boiling point of a substance is the best way to decide if it is pure. If the substance is impure, its melting point and boiling point will differ from the values for the pure substance. For example, salty water boils above 100°C. As the water boils away and the solution gets more concentrated, the boiling point slowly rises.

QUESTIONS

1 Describe how you would use chromatography to test whether some red ink contains a single pure dye or a mixture of dyes.
2 How would you check whether a colourless liquid is pure water? Draw a diagram of the apparatus you would use.
3 Which of the following are pure substances and which are mixtures?
sea water, petrol, iron, steel, steam, soil, milk, salt
4 Find out how chromatography is used either in medicine, or in testing athletes for illegal drugs, or in checking the synthetic dyes used in foods.

SECTION A: THINGS TO DO

Activities

1 Making a cup of tea

Suppose you are going to make a cup of tea. Here are some rules that you should follow.

A Warm the teapot with boiling water and then pour the water away.
B Put one tea bag or one spoonful of tea in the pot for every two people.
C Pour freshly boiled water onto the tea.
D Let the tea infuse for five minutes and then stir.
E Serve the tea with milk and sugar to taste.

1 Why is it important to use boiling water?

2 What should the temperature of the water be when it is added to the teapot?

3 Why should the teapot be warmed before making the tea?

4 Why are tea leaves cut into small pieces before being used?

5 Some people prefer to use tea bags rather than loose tea leaves. Why is this?

6 What does the word 'infuse' mean in rule D?

7 Some people say that it makes no difference to the taste whether the tea is loose or whether it is in a bag. How could you test this? How will you make sure your test is fair?

8 Some people prefer to put milk in the cup before pouring in the tea. Other people put the tea in first and then add the milk. Plan an experiment to investigate whether pouring in the milk or the tea first makes any difference to the temperature.

9 If possible, try the experiments which you have suggested in parts **7** and **8**. What are your results?

SPARKS FLY IN SCHOOLGIRL HORROR!

Schoolgirl, Mandy Clarkson, aged 15, described how her hair had gone up in flames during a science lesson. Mandy said, "Suddenly my hair was in flames. I am sure my hair did *not* touch the Bunsen flame. I think it must have been caused by the gel on my hair."
Fortunately, Mandy is not badly burnt, but she will have to wear a wig for the next few months.

2 Sparks fly in schoolgirl horror!

Hair gels usually contain three ingredients

(i) a solvent which evaporates easily,
(ii) glycerine (propane-1,2,3-triol) or a polymer to gel the mixture,
(iii) perfume.

The solvent is probably flammable, but it is volatile and should evaporate quickly. Accidents like the one reported in the newspaper cutting have occurred two days after the gel was used.

The only ingredient which is likely to remain on the hair after some time is the glycerine or the polymer. Neither of these is very flammable. They are however spread over the hair to give a large surface area. Even so, it is difficult to see why the hair should catch fire so easily. One possible explanation is that the gelling stops the solvent evaporating.

Suppose you are taking an assembly at school. Your teacher has asked you to talk for three or four minutes about the hazards of hair gels. Remember that students may be at risk at home and in other lessons besides chemistry. Make this clear in your talk and suggest precautions that they should take when using hair gel. Now, write your talk.

3 Lemon tea and iced tea

Lemon tea and iced tea are two interesting ways of drinking tea that you may not have tried.

Lemon tea is pleasant all the year round.

(i) Prepare the tea in the usual way by pouring freshly boiled water onto the tea leaves.
(ii) Allow the tea to infuse for five minutes and then stir.
(iii) Use a tea strainer to collect the tea leaves and then serve the tea with a slice of lemon.
(iv) Sugar or sweetener may be added to taste.

Iced tea is pleasant on a hot summer's day.

(i) Prepare the tea in the usual way.
(ii) After the tea has been infused for five minutes, use a tea strainer to collect the tea leaves and pour the tea off into a jug.
(iii) Put the jug of tea into the fridge to cool.
(iv) Serve with ice cubes, lemon or orange slices if you wish.

Iced tea

Lemon tea

Copy the recipes for lemon tea and iced tea carefully.

1 Underline all the names of solutions in the recipes (e.g. lemon tea in the first line).

2 Put a circle around the names of any solvents.

3 Put a tick through the names of any solutes.

4 Put a cross through the names of any insoluble substances.

5 Draw a flow chart to show the stages in making a cup of lemon tea. Illustrate each stage with a diagram.

6 Why is lemon tea often recommended in slimming diets?

4 Word processing your own word quiz

1 Make a list of the important words in section A, units 1 to 5. After each key word, write down what it means. For example:

Chemistry – The study of materials and substances.
Substances – Materials like iron, iron ore, limestone, coke and air.
Raw materials – Materials that occur naturally.

In the diagram opposite, these three important words have been used to make a word quiz.

Can you see that the missing words is AIR?

2 Now make up a word quiz of your own using important words from the whole of section A. Write clues to find your missing word as in the example above.

3 Use a word processor to design and develop a neat version of your word quiz. Try to use all the facilities of your word processor including different sized letters, drawings (diagrams), shading, brushwork, etc.

4 Try out your word quiz on your friends.

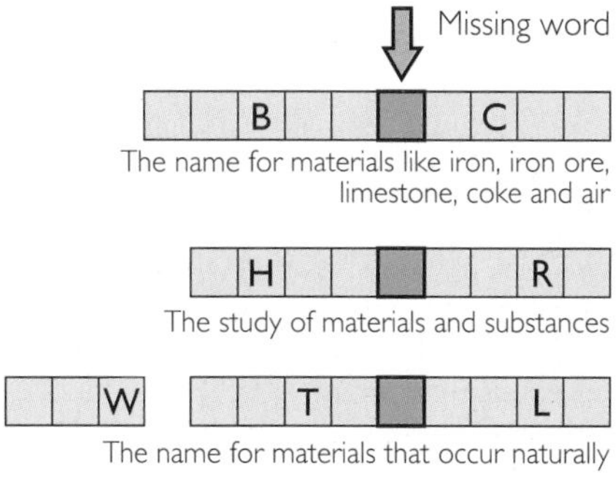

Study Questions

1 Some properties of six elements, *A* to *F*, are given below.

Element	Density /g cm^{-3}	Boiling point /°C	Electrical conductivity
A	3.12	58	Poor
B	8.65	765	Good
C	19.30	2970	Good
D	3.4×10^{-3}	−152	Poor
E	0.53	1330	Good
F	2.07	445	Poor

(a) Which of the six elements are metals?
(b) Which element will float on water as a solid at 20°C?
(c) Which element is a gas at room temperature?
(d) Which non-metal may be a solid at 60°C?
(e) Which metal is probably a liquid at 2000°C?
(f) What is the boiling point of the least dense metal?
(g) The elements in the list are cadmium, gold, lithium, bromine, krypton and sulphur. Which element is which?

2 What elements do the following contain?
octane (in petrol), marble (calcium carbonate), anhydrous copper sulphate, sodium chloride, sugar, bread, cheese, wood, a penny, alcohol, polythene, water

3 Seven processes used in chemistry to separate mixtures are:
A filtration, *B* distillation, *C* fractional distillation, *D* solution followed by filtration, *E* chromatography, *F* crystallisation, *G* evaporation to dryness.

Choose from *A* to *G* the method that would be used
(i) to obtain copper(II) sulphate crystals from copper(II) sulphate solution,
(ii) to obtain pure water from sodium chloride solution,
(iii) to isolate nitrogen from liquid air,
(iv) to obtain calcium carbonate from a mixture of calcium carbonate and water,
(v) to obtain sand from a mixture of salt and sand,
(vi) to separate a mixture of alcohol and water,
(vii) to find out how many coloured substances are present in a purple dye.

4 The following chromatogram was obtained in an experiment to analyse two mixtures.
(a) (i) Why was a pencil used to mark the chromatogram and not a pen?
(ii) Why must the base line, on which a small drop of each sample was placed, be above the level of the solvent at the start?
(b) From the chromatogram, which of the single substances were
(i) present in mixture 1?
(ii) present in mixture 2?
(iii) not found in either mixture?

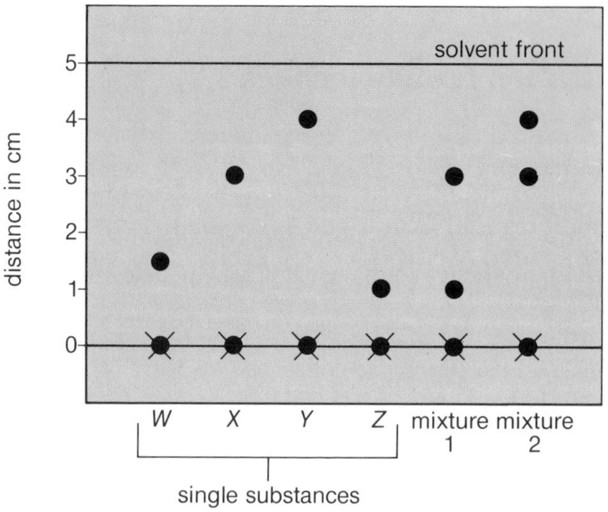

(c) The R_f value of a substance can be used to identify it by chromatography. The R_f value is defined as:

$$R_f = \frac{\text{distance moved by substance}}{\text{distance moved by solvent front in the same time}}$$

Calculate the R_f value of substance Y.

5 A sample of crude oil was distilled in the apparatus shown in figure 1. Four fractions of distillate were collected in the temperature ranges:

20°C – 70°C, 70°C – 120°C, 120°C – 170°C, 170°C – 220°C

(a) The crude oil is not heated alone but with mineral wool. Why is mineral wool used?
(b) Why is the receiving tube in a beaker of cold water?
(c) Why would it be better to use a condenser to condense the vapours?
(d) Which fraction will have condensed least efficiently?
(e) Which fraction of the distillate will collect first?
(f) Which fraction of distillate could be used as petrol?
(g) Which fraction of distillate is most flammable?
(h) Which fraction burns with the smokiest flame?
(i) Which fraction would be darkest in colour?
(j) Why is the mineral wool black at the end of the experiment?

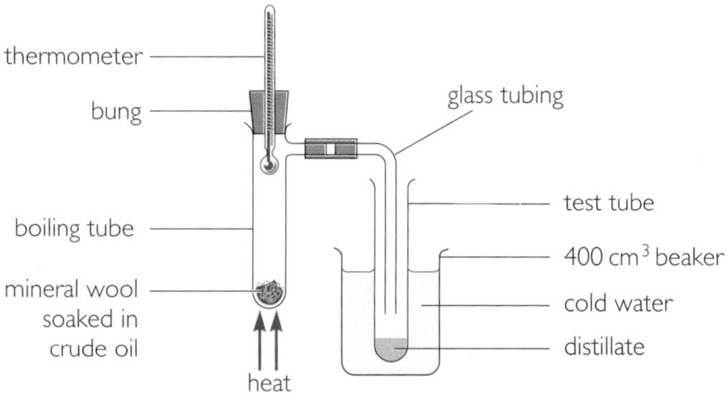

SECTION B

Our environment – air and water

The pollution around this lake is due to the chemicals used in paper making

1 RAW TO READY

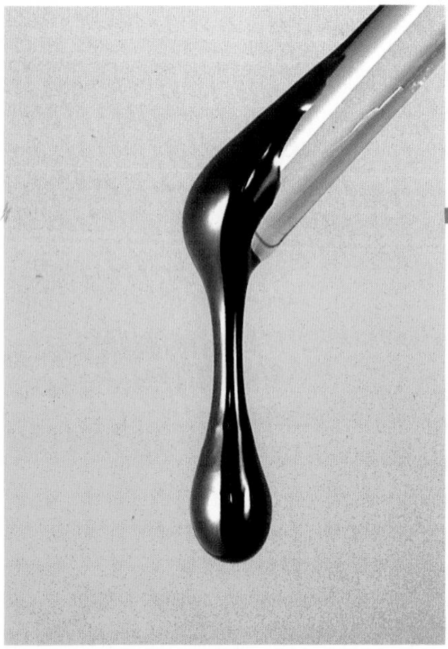

Crude oil is a thick, black, smelly liquid. It has very little use like this

Part of an oil refinery in Oregon, USA. At an oil refinery, crude oil is separated and converted into different products

Engine oil is a ready and useful product which can be manufactured from crude oil

Look at the photos above. They show how a dirty, smelly raw material like crude oil can be converted into a useful product like petrol. This example illustrates an important feature of the chemical industry and other manufacturing processes. They convert raw materials into ready and useful products.

raw materials e.g. crude oil	$\xrightarrow[\text{process}]{\text{manufacturing}}$	useful products e.g. petrol, engine oil

Table 1 shows the most important sources of raw materials for use in manufacture. It also lists some of the important materials and products we get from them.

Source of raw materials	Useful products obtained from the raw materials
Air	oxygen, nitrogen, argon
Water and the sea	drinking water, salt (sodium chloride), sodium, chlorine, magnesium, bromine
Rocks	metals (e.g. iron, copper, aluminium), limestone, sand, glass, cement, bricks, aggregate for building
Coal	fuels (e.g. coke), dyes and plastics from coal tar
Crude oil and natural gas	fuels, lubricating oils, waxes, plastics, dyes, perfumes, pesticides
Living things	timber, food (e.g. flour, sugar, cooking oil, meat, milk, eggs, fruit, vegetables), clothing (e.g. cotton, wool), rubber

Table 1 Important raw materials and the useful products we get from them

- When a chemical reaction occurs, a new substance is always produced.
- The new substance is called the product of the reaction.

Crude oil is one of the most important raw materials for the chemical industry. At an oil refinery, the crude oil is first separated into different fractions by distillation (section A, unit 9). The substances in these different fractions are then broken down and converted into useful materials like polythene, PVC and pesticides. These useful materials are very different from the starting materials in the crude oil. Processes like this which produce new materials are called chemical reactions.

QUESTIONS

1 Design a poster to show the useful products that we can obtain from either limestone or salt or plants.

2 Look at the photo above showing a potter working with clay.
(a) What does wet clay look like and feel like?
(b) How does the clay change as it dries out?
(c) How does the clay change after it is heated in a kiln?
(d) Why is clay a useful raw material?

3 Many scientific discoveries can be used in helpful ways or in harmful ways. Just think of explosives. Write two paragraphs on three of the following. One paragraph should show their helpful use. The other should show their harmful use.
fertilisers, insecticides, detergents, atomic energy, the contraceptive pill

4 Have a look at today's newspaper. What news or adverts does it contain which refer to
(a) important industrial products,
(b) our environment?

5 Apart from water, sodium chloride (salt) is the most abundant substance in the sea.
(a) How does it get into sea water?
(b) What is the simplest method of getting it from sea water in hot countries?
(c) Why is this method not used in Britain?
(d) In this country most salt is obtained from inland beds beneath the surface of the land in Cheshire. How did the salt come to be there?
(e) More than 25 million tonnes of salt are mined each year in Britain. What is this salt used for?

2 THE AIR – AN IMPORTANT RAW MATERIAL

River water and sea water contain dissolved oxygen. Fish take in water through their mouths and this flows over their gills. The gills extract oxygen from the water

The air is all around us. We need it to live. We need it to burn fuels and keep warm. We also use it to obtain useful products like oxygen, nitrogen and argon. Air is a mixture, but the substance in it that we need for both burning and breathing is **oxygen**.

What percentage of the air is oxygen?

Figure 1 shows an apparatus to find the volume of air used up when copper reacts with the oxygen in it.

copper + oxygen → copper oxide

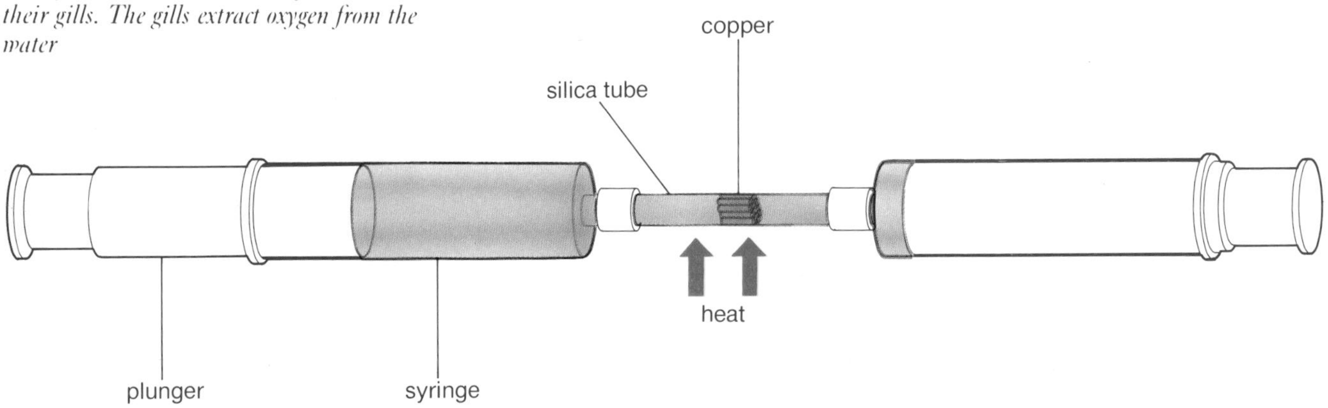

At the beginning, one syringe is empty and the other is filled with 100 cm^3 of air. The silica tube is heated strongly and the air is pushed to and fro several times so that all the oxygen in the air reacts with the copper. The tube is then allowed to cool and the volume of air in the syringe is measured. The heating and cooling are repeated until the volume of air which remains in the syringe is constant. Here are the results from one experiment.

Humans have no gills. They must carry a supply of air for underwater swimming

Volume of air in syringe before heating	100 cm^3
Volume of gas after first heating and cooling	82 cm^3
Volume of gas after second heating and cooling	79 cm^3
Volume of gas after third heating and cooling	79 cm^3

1 Has all the oxygen been used up after the first heating?
2 Has all the oxygen been used up after the second heating?
3 Why is the heating and cooling repeated three times?
4 How much oxygen did the copper remove?
5 What is the percentage of oxygen in the air?

Other gases in the air

Table 1 (below) gives some information about the gases in dry air.

Gas	% of air by volume	Boiling point /°C	Important uses	
Nitrogen	78.09	−196	making ammonia for nitric acid and fertilisers	
Oxygen	20.95	−182	steel making and welding (see section B, unit 3)	
Argon	0.93	−186	filling electric light bulbs (Argon is very unreactive. It will not even react with the white hot filament.)	
Carbon dioxide	0.03	−78	in fizzy drinks and fire extinguishers	

Notice that fresh air always contains about 0.03% carbon dioxide. This is because:

- carbon dioxide is produced when fuels burn and when plants and animals respire (see unit 4).
- carbon dioxide is used up when plants photosynthesise.

Although clean dry air contains only those gases in the table, ordinary air also contains water vapour and waste materials from industry (unit 5).

Obtaining gases from the air

The air is an important source of oxygen, nitrogen and argon. Figure 2 summarises how the gases are separated.

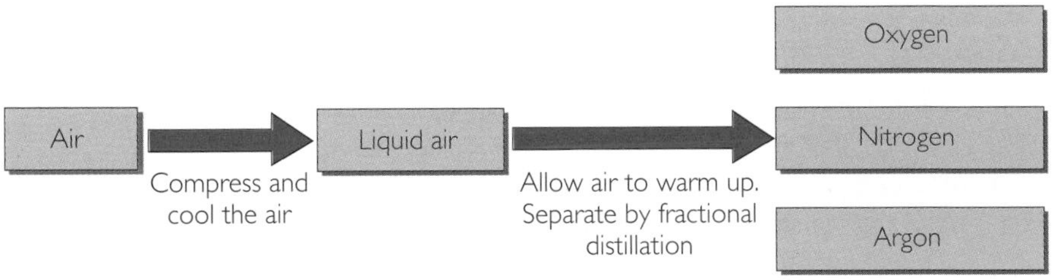

Look at the boiling points of the gases in Table 1.

1 Which gas forms a liquid first as the air is cooled?
2 Which gas boils off first when liquid air warms up?
3 Which gas boils off second when liquid air warms up?
4 Which gas boils off third when liquid air warms up?

QUESTIONS

1 In the syringe experiment to find the percentage of oxygen in the air, the hard-glass tube containing copper was weighed before and after the experiment.

Volume of oxygen in air	21 cm^3
Mass of tube + contents at start	11.365 g
Mass of tube + contents finally	11.393 g

(a) Why does the tube + contents increase in mass?
(b) What volume of oxygen does the copper react with?
(c) What mass of oxygen does the copper react with?
(d) What is the density of this oxygen (i.e. the mass of 1 cm^3 of it)?

2 Discuss the following questions with 2 or 3 others. The percentage by volume of carbon dioxide in ordinary air is 0.03%. Would you expect the percentage of carbon dioxide to be higher, lower or about the same as this
(a) in a crowded classroom?
(b) in a greenhouse full of plants on a sunny day?
(c) in the centre of a busy city?

3 OXYGEN

Metal sculptors use oxyacetylene torches to shape and cut their designs

The uses of oxygen

Oxygen is essential for burning and breathing, but the oxygen does not have to be pure. Pure oxygen is, however, required in large quantities for industrial and medical uses.

- **Manufacture of steel.** One tonne of pure oxygen is needed to produce every 10 tonnes of steel from impure iron ore. Oxygen is blown through the molten iron to remove carbon and sulphur impurities. The carbon and sulphur are converted to carbon dioxide and sulphur dioxide and then escape as gases.
- **Welding and cutting.** Pure oxygen is used in oxy-acetylene welding and cutting. When acetylene burns in oxygen, the temperature reaches 3200°C. This is hot enough to melt most metals which can then be cut or welded together.
- **Breathing apparatus.** Pure oxygen is used in life-support machines in hospitals. Oxygen helps the breathing of patients with lung diseases such as pneumonia. It is also mixed with anaesthetising gases during surgical operations.
  Mountaineers and deep sea divers also use oxygen mixed with other gases to breathe when supplies of air are not available.

Element	Reaction with oxygen	Product	Add water to product, then universal indicator
Carbon	glows red hot, reacts slowly	colourless gas (carbon dioxide)	dissolves pH = 5, acidic
Sulphur	burns readily with a blue flame	colourless gas (sulphur dioxide)	dissolves pH = 3, acidic
Sodium	bright yellow flame – white smoke and powder	white solid (sodium oxide)	dissolves pH = 11, alkaline
Magnesium	dazzling white flame – white clouds and powder	white solid (magnesium oxide)	dissolves slightly pH = 8, alkaline
Iron	glows red hot and burns with sparks	black-brown solid (iron oxide)	insoluble
Copper	does not burn, but the surface turns black	black solid (copper oxide)	insoluble

Table 1 Comparing the reactions of some elements with oxygen

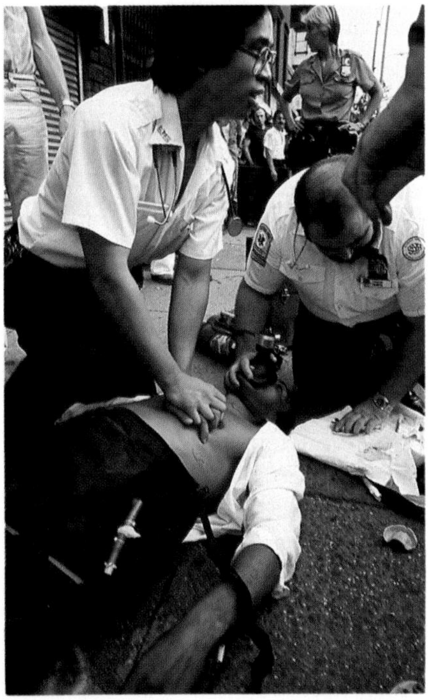

Oxygen being used to revive someone

How does oxygen react with elements?

Table 1 shows the results obtained when various elements are heated strongly in oxygen. All the elements burn better in oxygen than in air. The substances produced are **oxides**.

carbon + oxygen → carbon dioxide
sodium + oxygen → sodium oxide

- The metal oxides (sodium oxide, magnesium oxide, iron oxide and copper oxide) are all solids.
- The non-metal oxides (carbon dioxide and sulphur dioxide) are both gases.

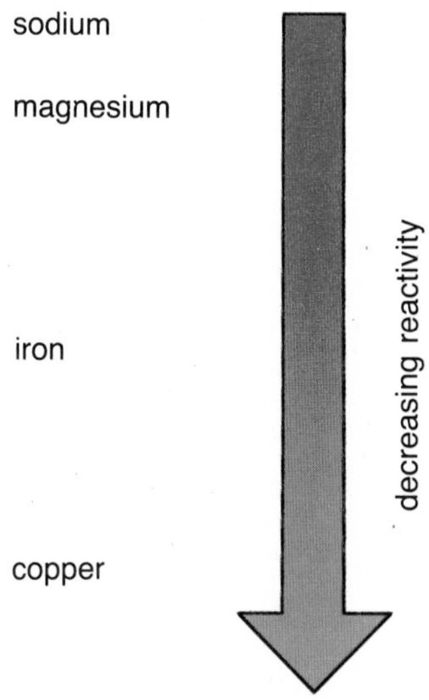

Figure 1

How do oxides react with water?

Each of the oxides was shaken with water. The solution produced was then tested with universal indicator. The results are given in Table 1.

Sodium oxide and magnesium oxide react with water to form alkaline solutions with a pH greater than 7. These alkaline solutions contain sodium hydroxide and magnesium hydroxide.

sodium oxide + water → sodium hydroxide
magnesium oxide + water → magnesium hydroxide

Iron oxide and copper oxide do not react with water or dissolve in it. So, they do not change the colour of the indicator.

The oxides of metals are called **basic oxides**. Most metal oxides are insoluble in water but a few, like sodium oxide and magnesium oxide, react with it to form alkaline solutions. These oxides are called **alkaline oxides**.

The non-metal oxides react with water to form acidic solutions with a pH less than 7. *These oxides of non-metals which give acids in water are called* **acidic oxides**.

carbon dioxide + water → carbonic acid
sulphur dioxide + water → sulphurous acid

What is the order of reactivity?

Some metals, such as sodium and magnesium, react very vigorously with oxygen producing a bright flame. Others, like iron, spark and glow. Some, like copper, react very slowly with no flame. From these results we can arrange the metals in an order of reactivity with the most reactive metal at the top and the least reactive at the bottom. This is called a **reactivity series (activity series)** for the metals. Figure 1 shows a reactivity series for the four metals that we used.

QUESTIONS

1 (a) What are the main uses of oxygen?
(b) Why do rockets carry liquid oxygen?
2 Design a rocket engine which is powered with liquid hydrogen and liquid oxygen.
3 What is meant by the following?
basic oxide, acidic oxide, alkaline oxide
4 Barium burns in air to form a solid oxide. This reacts with water to give an alkaline solution.
(a) Is barium a metal or a non-metal?
(b) Write word equations for the reactions described above.
(c) Which of the following elements would react like barium?
copper, sulphur, sodium, iron, nitrogen, calcium, carbon, potassium
5 Describe an experiment which you could do to find the percentage of oxygen, by volume, in the air you breathe out. Draw a labelled diagram of the apparatus you would use, describe the measurements you would make and explain how you would calculate a result.
6 Is air a mixture or a compound? Give three reasons to support your answer.

4 BURNING AND BREATHING

The Greeks believed that Prometheus took pity on humans and stole fire from the gods to give warmth to men and women

Some fuels produce sparks and flames when they burn

Why is burning so important?

Burning was the first chemical process used by humans. Our ancestors burnt fuels to keep warm, to cook food and to produce new materials like metals and clay pots. Burning is just as important to us. We burn fuels to keep warm, cook food, drive motor cars and generate electricity. Any substance which burns in air to produce heat is a **fuel**.

The most important fuels are coal, oil and natural gas. These fuels are mainly compounds containing carbon and hydrogen. During burning, these elements combine with oxygen in the air to produce carbon dioxide, water and heat.

$$\text{fuel} + \text{oxygen} \rightarrow \text{carbon dioxide} + \text{water} + \text{heat}$$

Reactions like this which give out heat are called **exothermic reactions**.

Sometimes, when a fuel burns, so much heat is produced that the products burst into flames. This happens in a Bunsen burner or a gas cooker when gases react with oxygen in the air to produce the flame.

Some fuels, like hydrogen, and explosives, like TNT and dynamite, react so fast that they cause explosions. Because of this *great care is needed in handling and using fuels.*

Antoine Lavoisier was a French chemist who lived in the 18th century. He showed that part of the air is used up when things burn, and said that this part of the air was oxygen

Breathing and respiration

When we breathe, air is taken into our lungs and then breathed out. But what happens to it in-between? One way to find out is to compare the air breathed in (**inhaled**) with the air breathed out (**exhaled**). The temperature and the percentages of oxygen, carbon dioxide and water vapour in ordinary air and in exhaled air are shown in Table 1.

	Ordinary air (inhaled air)	Exhaled air
% oxygen	20	15
% carbon dioxide	0.03	5
% water vapour	2	5
temperature/°C	15 (variable)	25

Table 1 Look at these results. How does inhaled air compare with exhaled air?

Once the air is in our lungs, oxygen can get into our bloodstream. The oxygen then passes to the rest of our bodies where it reacts with substances in our food. This process is called **respiration**. The results in Table 1 show that during respiration:

- oxygen is used up
- carbon dioxide is produced
- water is produced
- heat is produced

So respiration can be summarised as

$$\text{food} + \text{oxygen} \rightarrow \text{carbon dioxide} + \text{water} + \text{heat}$$

Notice the similarity between this word equation and that for fuels on the previous page. Foods are special kinds of fuels which are used in our bodies. They contain carbon and hydrogen in different compounds like fats, carbohydrates and proteins. Respiration is a kind of very slow burning. The same amount of heat is produced whether a substance burns in the air with flames or whether it is used up slowly in our bodies as a food.

Some athletes chew glucose sweets before a marathon. Glucose is a carbohydrate. It is a very good supply of fuel during a long race.

QUESTIONS

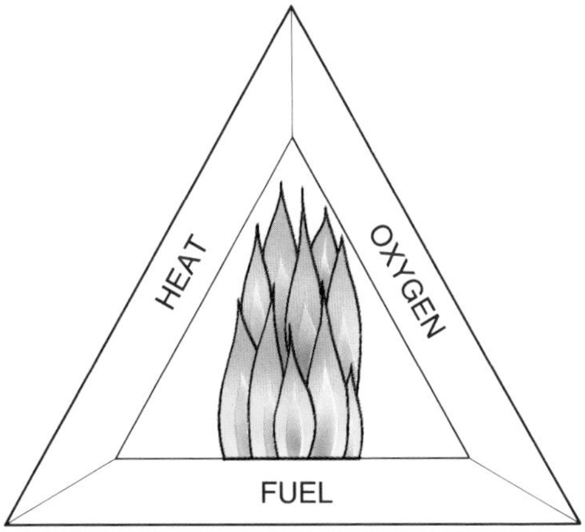

1 This diagram shows the fire triangle.
(a) Why are heat, oxygen and fuel shown along the sides?
(b) What happens if one of these is missing?
(c) What must you do to stop a fire?
(d) Have you ever needed to stop a fire? If so, what did you do and why?

2 How do you think the fire service would deal with
(i) a petrol fire, (ii) a forest fire, (iii) a chip-pan fire?
Say what they would do and why.

3 (a) What fuels are used for each of the following?
barbecueing, cooking in a tent, heating a school, getting a spaceship into orbit, lighting a housefire, running a car.
(b) For each of the uses in part (a) say why the particular fuel is chosen.

4 Look at the photograph of the French scientist Lavoisier on the last page. In May 1794, during the French Revolution, Lavoisier was arrested on a trumped-up charge and guillotined. The judge pushed aside his defence with the words '*The Republic has no need of men of science.*' A friend of Lavoisier is reported to have said '*Only a moment to cut off his head and a hundred years may not give us another like it.*'
(a) What is meant by (i) trumped-up charge, (ii) The Republic?
(b) Find out about Lavoisier's experiments.
(c) What important results did Lavoisier obtain?
(d) How did Lavoisier's results help scientists to understand burning?
(e) Pretend that you were Lavoisier's lawyer at his trial. How would you reply to the judge when he said 'The Republic has no need of men of science'?

5 AIR POLLUTION

Air pollution in a heavily industrialised area. Smoke like this causes windows, curtains and clothes to become dirty very quickly. Fumes in the smoke also attack the stonework on buildings and damage our health

The stonework of the Houses of Parliament had become very badly discoloured and damaged because of air pollution, as the (top) photo shows. But look at the difference after the buildings were cleaned in 1989!

Most air pollution is caused by burning fuels. The problems are therefore worst in industrialised areas. When a fuel burns, it reacts with oxygen to form oxides. If the fuel burns completely, then all the carbon in it is turned into **carbon dioxide** which is only slightly acidic. If there is not much air available, the carbon may form **soot** and **smoke** or it may be turned into **carbon monoxide**, which is a very poisonous gas. Carbon monoxide reacts with a substance in our blood called **haemaglobin**. This stops the blood carrying oxygen to the brain and other parts of the body. It is dangerous to run a car engine in a garage with the doors closed. This is because the lack of air may lead to the production of carbon monoxide. Some fuels, like coal and coke, contain small amounts of sulphur. When these fuels burn, **sulphur dioxide** is produced. This is a colourless, choking gas which irritates our eyes and lungs. The sulphur dioxide dissolves in water to form an acidic solution containing sulphurous acid.

$$\text{sulphur dioxide} + \text{water} \rightarrow \text{sulphurous acid}$$

When sulphur dioxide gets into rain, the rain water becomes very acidic. This **acid rain** harms plants and attacks the stonework of buildings. Acid rain has been blamed for the poor growth of trees in Scandinavia, and the death of fish in lakes and ponds. It is thought that the sulphur dioxide is produced in the industrial areas of northern England and southern Scotland. It is then carried by the prevailing winds across the North Sea to Scandinavia.

In car engines, petrol burns in the cylinders. A lead compound is added to petrol to help it burn smoothly. But waste lead compounds pass out with the exhaust gases. These lead compounds are poisonous because they affect the brain. In some countries, including the UK, there are strict laws to control the amounts of lead compounds which can be added to petrol. All new cars are manufactured so that they can run on **unleaded** petrol.

Car engines need air to burn the petrol. When the mixture is sparked, nitrogen and oxygen in the air combine to produce **nitrogen dioxide** which is an acidic gas. Nitrogen dioxide causes acid rain like sulphur dioxide.

The Tokyo Police are checking the exhaust gases of this car for pollutants. Tests on exhaust gases are now part of the MOT test in Britain

Acid rain has irreversibly damaged this statue on Wells Cathedral, Somerset

In the last few years, some scientists and politicians have become very concerned about the use of **CFCs (chlorofluorocarbons)** as refrigerator liquids and as the solvent in aerosol sprays. CFCs react with, and remove, ozone in the upper atmosphere. This allows more ultraviolet radiation in sunlight (normally absorbed by ozone) to get to the Earth. The extra ultraviolet radiation causes

1 damage to plants
2 greater risks of skin cancers.

CFCs, carbon dioxide and the other air pollutants also retain more heat than the gases in fresh air. This means that the temperature of the Earth is rising very slowly. This is called the **greenhouse effect**.

Air pollution shows that advances in modern technology have both advantages and disadvantages. We have the benefits of more fuels and more vehicles, but we also have the problems of more pollution. Fortunately, several countries now have strict laws to control air pollution. In 1956, the Clean Air Act in Britain made it an offence to pollute the air with soot and smoke from factories and homes. Local authorities were given the power to set up smoke control areas ('smokeless zones') to prevent air pollution. Only smokeless fuels, such as coke and anthracite, can be used in smokeless zones.

QUESTIONS

1 (a) What are the main substances that cause air pollution?
(b) How do they get into the air?
(c) Suggest 3 ways in which air pollution causes damage.
2 (a) What is acid rain and how is it caused?
(b) How does acid rain affect (i) lakes, (ii) forests?
3 What is (i) the Clean Air Act, (ii) a 'smokeless zone'?
4 (a) What further steps could be taken to reduce air pollution in heavily industrialised areas?
(b) What problems are there with stricter controls over air pollution?
5 Design a poster to show the dangers of CFCs.
6 It is the year 2030. You have just been elected as the first British Prime Minister to belong to the Green Party. Write down what you would say in your first interview on television about your policy on the environment.

6 WATER SUPPLIES

Water controls our lives. It affects our crops, our food, our health and the weather. 60% of the people in Asia and Africa cannot obtain clean water easily

Water is the most common liquid. We drink it, wash in it, swim in it and complain when it rains. In many ways water controls our lives.

- It determines where we can live.
- It determines whether we can grow crops and produce enough food.
- It determines which sports we can enjoy.
- It determines the weather we have.

All living things need water – people, animals and plants. About two thirds of your body is water. Every day you need about 2 litres (3½ pints) of water. This water may be part of your food or drinks. It replaces the water that you lose in urine, in sweat and when you breathe. Water is more important than food – most people can survive for 50 to 60 days without food, but only 5 to 10 days without water. For all these reasons, good management of our water supplies is essential. The treatment of our water supplies is covered in Section A, Unit 7.

How do we use water?

Most of the time, we take water for granted. It is only during a very dry summer that we need to use it carefully. In the UK, each person uses about 180 litres of water every day. In some parts of Africa, each person must survive on less than 10 litres per day. Every time you flush the toilet about 10 litres of water are used and taking a bath may use 100 litres. Figure 1 shows the amounts of water that we might use for different purposes each day.

Large amounts of water are also used in industry. Most of the water is used for cooling. The other large industrial uses of water are for washing and as a solvent. When all these uses are added together, it takes:

10 litres of water to make 1 litre of lemonade or beer;
200 litres of water to make 1 newspaper;
50 000 litres of water to make 1 average-sized car.

Where does water come from?

At the turn of a tap, we expect to get as much clean water as we need. But where does this treated water come from? If we trace the flow of water backwards, we find that it comes from

1 surface water in rivers and lakes,
2 ground water in underground wells and saturated rocks.

We can trace the water supply back even further and see that all surface water and underground water comes from rain. And there's a lot of it. It has been estimated that 380 000 km^3 of water falls as rain, hail or snow each year. When rain falls on the ground, some of it soaks deep into the earth as ground water. The rest runs off the land into rivers, lakes and oceans, as surface water.

This surface water covers about two thirds of the Earth. But where does the rain come from?

Figure 1 Average amounts of water used by one person each day in the UK

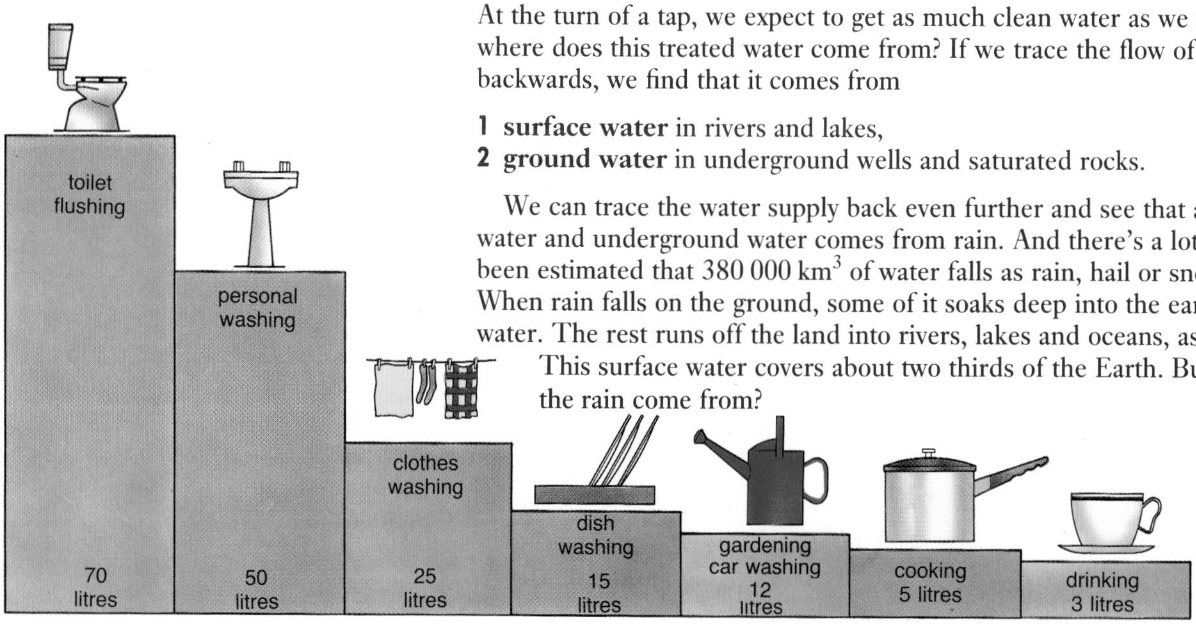

Figure 2 The water cycle

A large power station, like this one, uses 5 million litres of water every day for cooling. Because of this, power stations are often built near a large source of water, such as the sea, or a major river

Heat from the sun causes surface water to evaporate into the air. This water vapour collects in the air as clouds and rises. As the water vapour rises, it cools down and condenses, forming tiny drops of water. When these drops become larger they fall back to Earth as rain. This either soaks into the soil or joins rivers and oceans. The whole cycle then begins again. This continuous movement of water from the Earth's surface to clouds and then back to the earth as rain is called the **water cycle** (Figure 2).

QUESTIONS

1 Look at Figure 1.
(a) Which of the uses of water requires (i) the purest water, (ii) the least pure water?
(b) Write down the uses, in order, from those needing the purest water to those needing the least pure water.
(c) Which of the uses could be avoided in order to save water during a drought?
(d) Suppose you saved all the water from personal washing. To which of the other uses could this water then be put?

2 (a) Make a list of all the things that your family uses water for.
(b) Estimate the amount of water, in litres, which your family uses for each of these, in one week.
(c) Find out how much your family pays in water rates.
(d) Calculate how much your family pays for 1 litre of water.
(e) Do you think that you get value for money?

3 (a) Draw a flow diagram to show how your tap water originally comes from rain.
(b) Where is the water treatment plant nearest to your home?
(c) What other services does your water company provide besides supplying water?

7 WATER POLLUTION

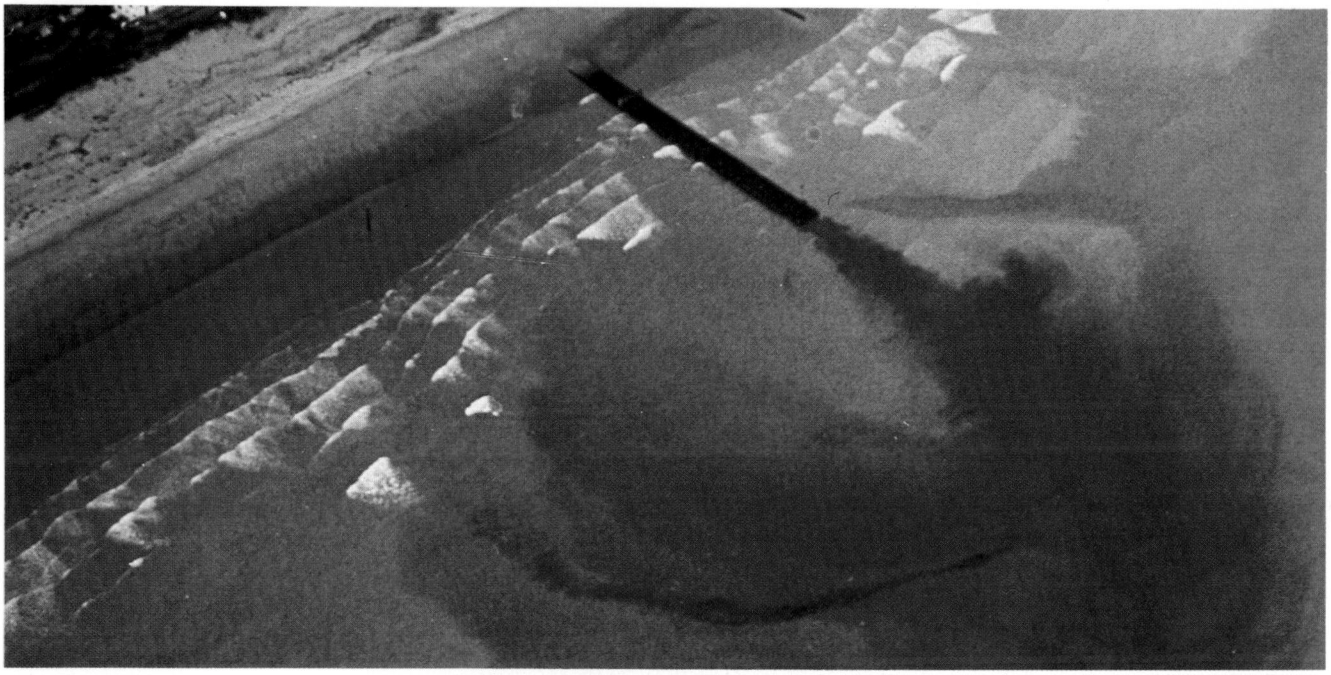

Sea-water is being badly polluted all over the world because sewage material is dumped into the oceans

Main sources of water pollution
Sewage
Fertilisers
Industrial chemicals
Pesticides
Oil
Detergents

Table 1 The main sources of water pollution

There are vast amounts of water in oceans, lakes and rivers, but it is easily polluted. The main sources of water pollution are listed in Table 1.

Sewage is the main cause of water pollution. At one time, all sewage was piped into rivers or into the sea. This caused health hazards and led to diseases such as cholera. Sewage also affects living things in the water. If the amount of sewage is small, then bacteria in the water can break it down to harmless materials like carbon dioxide, nitrates and water. But, if the amount of sewage is large, then the bacteria use up all the oxygen dissolved in the water as they feed on the sewage. If the oxygen concentration gets too low, most of the living organisms in the water (including the bacteria) die. The decaying materials make the water cloudy and smelly. In order to avoid this pollution, sewage plants treat the waste before it is returned to rivers and the sea. The sewage is first pumped into large tanks. Here it is mixed with air (aerated) so that it can be decomposed more rapidly by bacteria.

During the 1950s, the River Thames was almost lifeless between London Bridge and the sea. This was caused by the low oxygen concentration in the water. Since then, there have been improvements in pollution control. In 1974, the first salmon was caught since 1835 and now more than 70 species of fish have returned.

Table 2 summarises the sources and effects of other substances which cause water pollution.

Some people believe that heat is also a serious water pollutant. It is a particular problem near the cooling water outlets from large nuclear power stations where the water temperatures may be 10°C higher than normal. The higher water temperatures increase the corrosion of steel structures in the water and also reduce the amount of oxygen dissolved in the water. The reduced oxygen content can be harmful to some fish such as salmon and trout, although other species thrive in the warmer water.

Polluting substance	Source	Effect
Fertilisers	rain washes fertilisers into rivers and lakes	bacteria and algae grow faster, use up all the dissolved oxygen and then die
Industrial chemicals	oils, metal compounds, acids, alkalis, dyes, etc. from factories	poisonous to animals, plants and bacteria in the water
Pesticides	spraying of crops with chemicals	poisonous chemicals accumulate in the bodies of larger animals
Oil	oil from refineries and from ship-wrecked tankers	covers sea birds with oil; pollutes beaches
Detergents	factories, offices, homes	causes water to foam; poisonous to organisms in the water

Table 2 The sources and effects of various water pollutants

The guillemot and razorbills in this photo have all been killed by oil pollution at sea

This beach in the Gulf of Aqaba has been badly polluted with substances from a factory which processes phosphate rock (calcium phosphate)

Getting the balance right

It is easy to adopt a very simple view of pollution and say, 'All pollution is wrong'. But if we did this, and passed laws to protect our rivers and lakes, then industries would have to spend more money on avoiding pollution and getting rid of waste in some other way. These additional costs would be passed on to the customers and the goods from our factories would be more expensive. As a society, we have to decide between more pollution control and higher prices or less pollution control and lower prices. Remember too that *everyone* is responsible for pollution. We are all members of a society which consumes goods and materials and requires oil, fertilisers, pesticides and detergents.

QUESTIONS

1 (a) Why does sewage make fresh water murky and smelly?
(b) Why are bacteria mixed with sewage at some treatment plants?
(c) Why is solid sewage sometimes heated to 30°C and aerated regularly at the sewage plant?
2 Some people think that it is more important to improve the water supply and build water treatment plants in poor countries than to build hospitals.
(a) Discuss this view with 2 or 3 other people in your class.
(b) Do you agree with this view? Make a summary of your discussion so that you can report to the rest of your class.

8 PROPERTIES OF WATER

Living things, like this seal, can survive in water even in the Antarctic because ice floats on water

Water – an unusual substance

Water covers more than 70% of the Earth's surface in oceans, lakes and rivers. It freezes at 0°C and boils at 100°C. So it is a liquid at most places on the Earth. This is unusual. Can you think of another substance which occurs naturally as a liquid?

Water is also a very good solvent. Substances as different as salt, sugar, alcohol and oxygen will dissolve in it. Some substances are more soluble than others, as you can see in Table 1.

When the temperature falls below 0°C, water turns to ice. If water is trapped in pipes as it freezes, then the pipes may split open and cause a burst. The pipes split because water expands when it feeezes. This is another unusual property of water. Almost all other substances contract when they freeze. The expansion of water as it changes to ice means that ice is less dense than water. It causes rocks to crack and potholes to appear in the roads.

When ice forms on the surface of a pond, it acts as an insulator and it prevents the water below from freezing. Because of this, some living things can survive in water, even in Arctic conditions.

Water expands as it freezes. This can cause serious damage, such as burst pipes

Water as a solvent – solubility

When you stir a spoonful of instant coffee into a cup full of hot water, it dissolves very quickly. If you tried to make a cup of coffee with cold water, the coffee would not dissolve. In the same way, sugar dissolves better in hot drinks than in cold drinks. These everyday examples show that solids dissolve better in water as the temperature increases.

When sugar is stirred into coffee, it dissolves until a certain amount has been added. If more sugar is then added this does not dissolve, provided the volume of the liquid and its temperature do not change. The coffee is **saturated** with sugar. The extent to which a solute like sugar can saturate a solvent like coffee is expressed in terms of **solubility**.

Solute	Mass of solute which dissolves in 100 g of water at room temperature (20°C)
Sand	insoluble
Salt	36.0 g
Sugar	204 g
Alcohol	infinite
Oxygen	0.004 g
Carbon dioxide	0.014 g

Table 1 The solubility of some substances in water at 20°C

The solubility of a solute in a particular solvent is the mass of solute which saturates 100 g of solvent at a given temperature. Table 1 shows the solubilities of various substances in water at 20°C. The solubility of most solids increases with temperature.

Although most solids become more soluble in water as the temperature rises, gases become less soluble. Natural water in lakes and rivers contains dissolved gases from the air such as oxygen and carbon dioxide. More gas dissolves if the water is colder or if the water flows over waterfalls where it mixes freely with air.

Many water animals and plants depend on oxygen dissolved in the water. Fish die if there is less than 0.0004 g of dissolved oxygen per 100 g of water. This explains why fish caught in rivers cannot survive in fish tanks indoors where the water is warmer and cannot dissolve as much oxygen.

Notice in table 1 that carbon dioxide is more soluble than oxygen in water. This is because it reacts with water to form carbonic acid:

$$\text{carbon dioxide} + \text{water} \rightarrow \text{carbonic acid}$$

The carbonic acid which forms in natural waters reacts with chalk and limestone to form hard water. We shall return to this in a later unit.

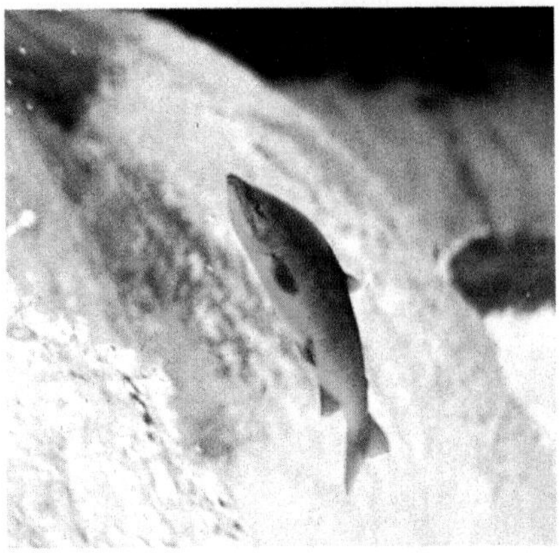

Some fish species require more dissolved oxygen in the water than others. Because of this they cannot survive at higher temperatures. Salmon require higher concentrations than most fish. They cannot survive if the water temperature is above 15°C

QUESTIONS

1 Why can water be described as 'an unusual substance'?

2 The solubilities of potassium nitrate and potassium chloride are given below at temperatures between 0°C and 60°C.

(a) Draw the solubility curves of the two solids by plotting solubility on the vertical axis and temperature on the horizontal axis.

(b) Which is the more soluble solid at 28°C?

(c) At what temperature are they equally soluble?

(d) How much potassium nitrate will crystallise from 100 g of water saturated with this substance at 50°C, if the temperature falls to 23°C?

Temperature /°C	0	10	20	30	40	50	60
Solubility of potassium nitrate /g per 100 g water	13	21	31	47	63	83	106
Solubility of potassium chloride /g per 100 g water	28	31	33	36	39	42	45

3 Discuss the following questions in groups of 3 or 4.

(a) Why is it necessary to give the temperature at which a solubility is measured?

(b) How could you make sure that a solution is saturated at a particular temperature?

(c) Why do you think that solids get more soluble in water as the temperature rises, but gases get less soluble?

(d) Why does the water in fish tanks need aeration (air bubbled in)?

4 Design an experiment to show that water expands when it freezes.

9 RUSTING

These steel cars are rusting badly because they have been out in the open

Articles made of iron and steel rust much faster if they are left outside in wet weather. A garden fork rusts more quickly when it is left out in the rain than when it is kept dry in a shed. This suggests that water plays a part in rusting. But what other substances are involved?

When aluminium is open to the air, it becomes coated with a layer of oxide. The aluminium has reacted with oxygen in the air. If rusting is similar to this, iron may also react with oxygen during rusting.

The apparatus in Figure 1 can be used to decide whether water and oxygen are involved in rusting. The test tubes are set up and left for several days. Tube 1 is the **control experiment.** It is the standard which we use to compare the results in other tubes. It contains iron nails in moist air. Tube 2 contains iron nails in dry air. The anhydrous calcium chloride absorbs water vapour. Tube 3 contains nails covered with boiled distilled water. The water has been boiled to remove any dissolved air. The layer of olive oil prevents air dissolving in the water.

The nails in tube 1 and tube 4 rust, but those in tubes 2 and 3 do not.

1 Does iron rust if there is (i) no air, (ii) no oxygen?
2 What conditions are necessary for rusting?

What happens during rusting?

Iron will only rust if both oxygen and water are present.

During rusting, iron reacts with oxygen to form brown iron oxide:

$$\text{iron} + \text{oxygen} \rightarrow \text{brown iron oxide}$$

At the same time, the iron oxide combines with water to form hydrated brown iron oxide. This is rust.

$$\text{iron oxide} + \text{water} \rightarrow \text{hydrated iron oxide (rust)}$$

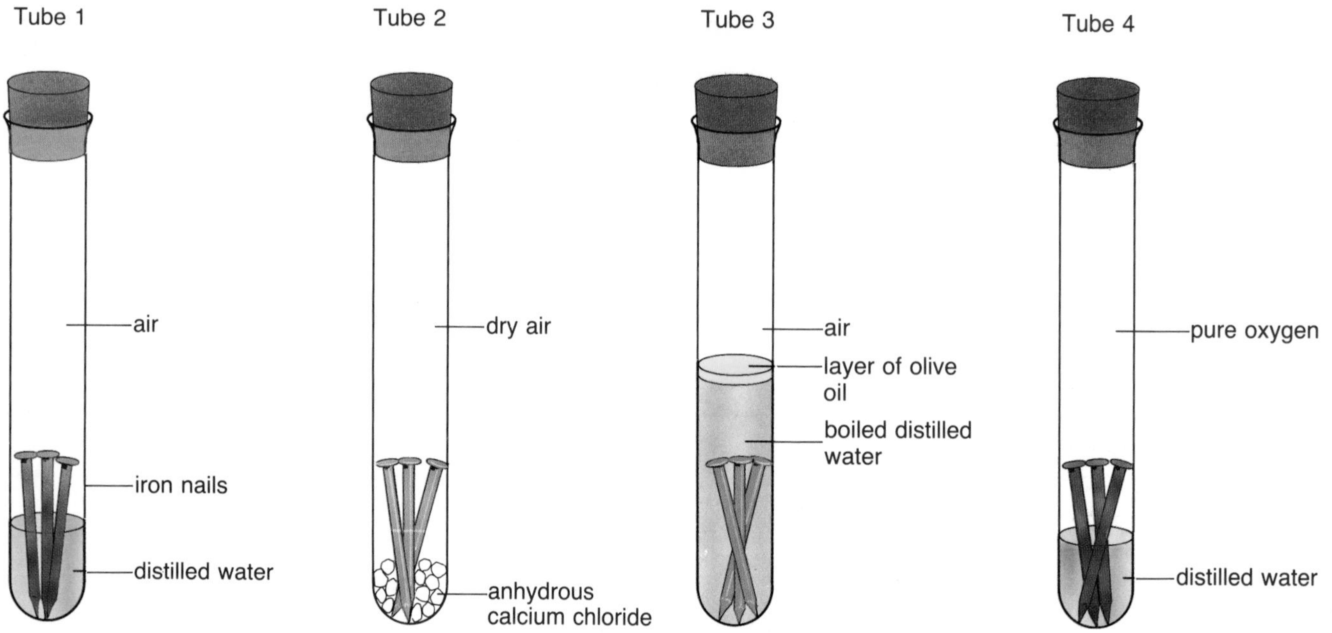

Figure 1

Substances, like rust, that have water as part of their structure, are described as **hydrated** and we call them **hydrates**. The water present in hydrates is called **water of crystallisation**.

Rusting costs millions of pounds every year because of

- the need to protect iron and steel objects,
- the replacement of rusted articles.

Although iron and steel rust more easily than several other metals, steel is used for ships, cars, bridges and other structures because it is cheaper and stronger than other building materials.

These car valves are being dipped in a solvent to prevent them from rusting

How is rusting prevented?

In order to stop iron and steel rusting, we must protect them from water and oxygen. The most important ways of doing this are:

- **Painting.** This is the usual method of preventing rusting in ships, vehicles and bridges.
- **Oiling.** The moving parts of machines cannot be protected by paint which would get scratched off. Instead they are oiled or greased (and this also helps lubrication).
- **Alloying.** Iron and steel can be mixed with other metals to form alloys. Stainless steel contains chromium, nickel and manganese mixed with iron.
- **Covering with a non-rusting material.** Buckets and dustbins are coated ('galvanised') with a layer of zinc. Other articles, like car bumpers, taps and kettles, are chromium plated. Some metal articles are covered with a layer of plastic.

Large structures such as the Forth Rail Bridge are being painted all the time. As soon as the painters have finished they begin again at the other end. It would take one man 72 years to paint the Forth Rail Bridge

QUESTIONS

1 Use the experiment on rusting, and its results, to explain the following: *control experiment, hydrates, anhydrous, water of crystallisation*

2 Explain the following statements.
(a) Iron objects rust away completely in time.
(b) Iron railings rust more quickly at the bottom.
(c) Iron on shipwrecks in deep sea-water rusts very slowly.

3 What experiments would you do to find out whether
(i) iron rusts more quickly in sea water or in distilled water,
(ii) steel rusts more quickly than iron?
Draw a labelled diagram of the apparatus you would use in each case. What results would you expect?

4 Aluminium and iron are the two most widely used metals.
(a) Why are these two metals used so much?
(b) Iron is used in spite of the fact that it rusts easily. Mention three ways in which rusting can be prevented.
(c) What advantages are there in using iron rather than aluminium?
(d) Iron can be magnetised. It can also be attracted by a magnet. What uses does iron have because of these properties?
(e) What advantages are there in using aluminium for certain purposes rather than iron?

10 IRON FOR INDUSTRY

The main raw material for making iron is iron ore (haematite). This ore contains iron(III) oxide. The largest deposits of iron ore in the UK are in Northamptonshire and Lincolnshire, but most of this is of poor quality. The best quality ores are found in Scandinavia, America, Australia, North Africa and Russia.

Iron ore is converted to iron in a special furnace called a **blast furnace**. This furnace is a tapered cylindrical tower about twenty metres tall. Figure 1 shows a diagram of a blast furnace with a summary of the reactions to produce iron.

This is a modern blast furnace. The ancient Egyptians were the first people to extract iron from iron ore on a large scale. Today the world production of iron is about 750 million tonnes per year

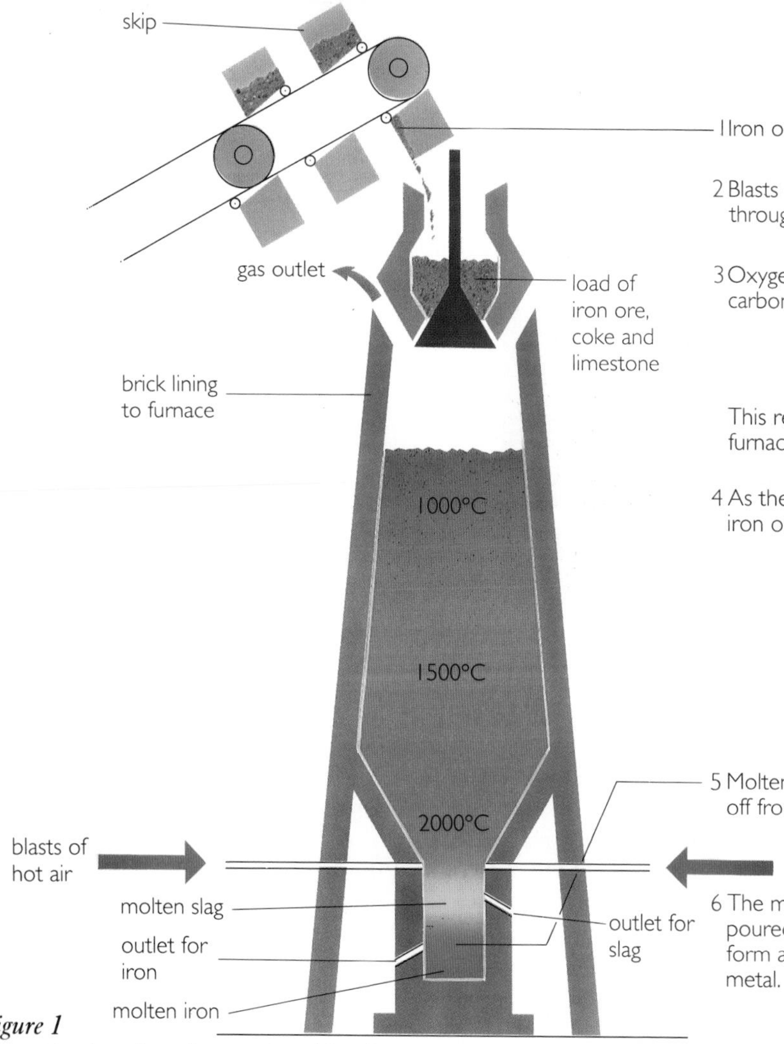

1 Iron ore, coke and limestone are added at the top of the furnace

2 Blasts of hot air (which give the furnace its name) are blown in through small holes near the bottom

3 Oxygen in the blasts of air reacts with coke (carbon) to form carbon monoxide

carbon + oxygen ⟶ carbon monoxide

$$2C + O_2 \longrightarrow 2CO$$

This reaction is very exothermic and the temperature in the furnace gets to 2000°C.

4 As the carbon monoxide rises up the furnace, it reacts with the iron ore (iron(III) oxide) to form iron.

iron(III) oxide + carbon monoxide ⟶ iron + carbon dioxide

$$Fe_2O_3 + 3CO \longrightarrow 2Fe + 3CO_2$$

5 Molten iron runs to the bottom of the furnace where it is tapped off from time to time

6 The molten iron is either used immediately to make steel or poured into moulds to solidify. The large chunks of iron which form are called 'pigs'. So, the name 'pig-iron' is used for the metal.

Figure 1
Extracting iron from iron ore in a blast furnace

Why is limestone used in the furnace?

Iron ore usually contains impurities like earth and sand (silicon dioxide, SiO_2). Limestone helps to remove these impurities. The limestone decomposes at the high temperatures inside the furnace. It forms calcium oxide and carbon dioxide.

$$\text{calcium carbonate} \rightarrow \text{calcium oxide} + \text{carbon dioxide}$$
$$CaCO_3 \rightarrow CaO + CO_2$$

The calcium oxide then reacts with sand (silicon dioxide) and other substances in the impurities to form **slag** (calcium silicate).

$$\text{calcium oxide} + \text{silicon dioxide} \rightarrow \text{calcium silicate}$$
$$CaO + SiO_2 \rightarrow CaSiO_3$$

The slag falls to the bottom of the furnace and floats on the molten iron. The slag is used for building materials and cement manufacture.

From pig-iron to steel

Pig-iron contains about 8% of impurities. The main impurities are carbon (3–5%) and sulphur (1–2%). These impurities make pig-iron very hard and brittle compared to iron and steel.

In order to make pig-iron into steel, the percentages of carbon and sulphur must be reduced to about 0.1%. This is done by blowing oxygen onto the hot, molten pig-iron. The oxygen converts carbon and sulphur to carbon dioxide and sulphur dioxide which escape as gases.

QUESTIONS

1 The three stages involved in extracting a metal from its ore are
(a) *mining* and concentrating the ore,
(b) *reducing* the ore to the metal,
(c) *purifying* the metal.
Draw a flow diagram relating to these 3 stages for the manufacture of steel from iron ore.

2 (a) Why are furnaces used to make iron called *blast* furnaces?
(b) Why are blast furnaces usually built near coal fields?
(c) What materials are added to a blast furnace?
(d) What materials come out of a blast furnace?

3 (a) The prices of metals are given in the 'commodities' section of the more serious daily papers. Look up the prices for iron (steel), aluminium, copper and zinc. How does the price of iron (steel) compare with other metals?
(b) Suggest 3 reasons why iron is the cheapest of all metals.
(c) Suggest 3 reasons why iron is used in greater quantities than any other metal.

11 OXIDATION AND REDUCTION

What substances are being oxidised and reduced in this photo?

When food goes bad, substances in the food combine with oxygen in the air. The process involves oxidation. Carbon dioxide and water are produced

Many everyday reactions involve substances combining with oxygen to form oxides. Burning, breathing and rusting are three important examples.

During burning, fuels containing carbon and hydrogen react with oxygen to form carbon dioxide and water (hydrogen oxide):

$$\text{fuel} + \text{oxygen} \rightarrow \text{carbon dioxide} + \text{water}$$

During respiration, foods containing carbon and hydrogen react with oxygen to form carbon dioxide and water:

$$\text{food} + \text{oxygen} \rightarrow \text{carbon dioxide} + \text{water}$$

During rusting, iron reacts with oxygen and water to form hydrated iron oxide:

$$\text{iron} + \text{oxygen} + \text{water} \rightarrow \text{hydrated iron oxide}$$

Chemists use a special word for reactions in which substances combine with oxygen. They call the reaction **oxidation** and the substance is said to be **oxidised**. But if one substance combines with oxygen, another substance (possibly oxygen itself) must lose oxygen. We say that substances which lose oxygen in chemical reactions are **reduced** and we call the process **reduction**.

Oxidation and reduction always happen together. If one substance combines with oxygen and is oxidised, another substance must lose oxygen and be reduced. We call the combined process **redox** (REDuction + OXidation).

Redox in industrial processes

A redox reaction is sometimes used in welding steel. A mixture of powdered aluminium and iron oxide (known as 'thermit') is used. Aluminium is more reactive than iron, so it removes oxygen from iron oxide to form aluminium oxide and iron. (Figure 1)

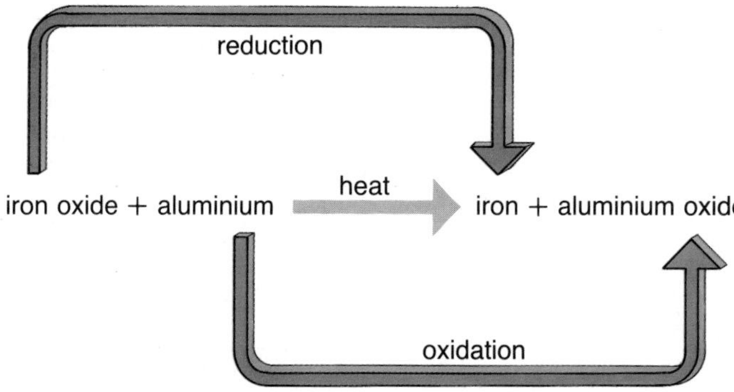

Figure 1

Aluminium gains oxygen and is oxidised. Iron oxide loses oxygen and is reduced. The whole process involves both oxidation and reduction. It is an example of redox.

This reaction between aluminium and iron oxide is very exothermic. The iron is produced in a molten state. As the iron solidifies, it will weld two pieces of steel together. Welding together Inter-City railway lines using this method has increased the average 'life' of the track from 23 to 30 years. The photo (right) shows the process in action.

The redox reaction between aluminium and iron oxide can be used to weld railway lines together. Molten iron from the reaction runs into a mould around the rails to be joined. When the iron has cooled the mould is removed and excess metal trimmed off

In a blast furnace, iron oxide can be reduced with coke (carbon). Why do you think aluminium has to be used in this method?

Redox reactions also occur in the manufacture of metals. Aluminium is manufactured by passing electricity through aluminium oxide. The electricity decomposes the aluminium oxide into aluminium plus oxygen. In this process the aluminium oxide is reduced to aluminium.

$$\text{aluminium oxide} \xrightarrow{\textit{electricity}} \text{aluminium + oxygen}$$

The overall reaction in the manufacture of iron from iron ore is:

$$\text{iron oxide + carbon} \rightarrow \text{iron + carbon dioxide}$$

Which substance is oxidised and which substance is reduced in this process?

QUESTIONS

1 Look at the food going bad in the photo (on page 46).
(a) What chemical processes occur when food decays?
(b) What methods and precautions do we use at home to stop food going off?
(c) How do these methods work?

2 Which substance is oxidised and which is reduced in each of the redox reactions below?
(a) aluminium + water → aluminium oxide + hydrogen
(b) hydrogen + oxygen → water
(c) copper oxide + hydrogen → copper + water

3 (a) Will copper oxide and magnesium react on heating? Explain your answer.
(b) Will magnesium oxide and copper react on heating? Explain your answer.
(c) Element *W* reacts with the oxide of element *X* but not with the oxide of element *Y*. Write *W*, *X*, and *Y* in order of reactivity (most reactive first).

4 Find out how either chromium or titanium is manufactured. How is redox involved in these processes?

5 A sample of dry rust was heated strongly in a test tube. Droplets of a liquid, *A*, formed on the cooler parts of the tube. *A* turned anhydrous copper sulphate blue. The residue left in the tube, *B*, was heated strongly in a stream of gas, *C*. A grey-black solid, *D*, was formed (which conducted electricity) and a colourless liquid, *E*, condensed on the cooler parts of the apparatus. *E* turned anhydrous copper sulphate blue.
(a) Identify the substances, *A–E*.
(b) One of the reactions described above involves redox.
(i) Which is it?
(ii) Write a word equation for the reaction involved.
(iii) Explain which substance is oxidised and which is reduced.

SECTION B: THINGS TO DO

Activities

1 Fuels in the home

1 Make a survey of the methods of central heating and the fuels used in the homes of the people in your class. You will need to find the following information:
(i) the number of homes in your survey,
(ii) *central heating* – how many homes use:

- gas-fired central heating
- oil-fired central heating
- solid-fuel (coke, coal) central heating
- electricity for central heating (i.e. electric storage heaters)

(iii) *other fuels* – how many homes use:

- electric fires
- fan heaters
- gas fires
- wood
- coal fires
- paraffin heaters
- smokeless fuel
- solar heaters

2 Draw a bar chart showing the number of homes using the different forms of central heating listed in part (ii).

3 Draw a bar chart showing the number of homes using the other fuels listed in part (iii).

4 How do you think the number of homes using the different forms of central heating has changed in the last 20 years?

5 How do you think the number of homes using the different forms of central heating will change in the next 20 years?

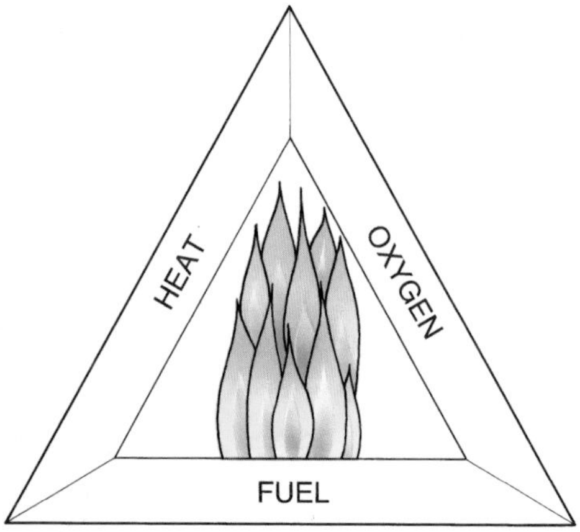

The fire triangle

2 Fire fighting

The fire triangle shown in Figure 1 can be used to explain how fires start and how they can be put out.

1 Why does the fire triangle have labels for fuel, oxygen and heat?

2 When the fire service are called to a fire, they try to remove either the fuel, the oxygen, the heat or more than one of these three. Why is this?

3 Which one of fuel, oxygen or heat is being removed from the fires described below?
(a) When the fat in the chip pan suddenly caught fire, Nicola grabbed a damp cloth and covered the chip pan.
(b) When the forest fire started, the foresters cleared the fire breaks (i.e. the wide tracks in the forest) just ahead of the fire.
(c) The firemen sprayed water on the burning beams of the warehouse.
(d) Julian was sitting close to the fire. Suddenly his coat was in flames. Shakila acted quickly and rolled Julian in the rug on which he was sitting.

Fire-fighters continue to damp down a smouldering car to ensure that it is no longer a fire hazard

3 Pollution of the River Rhine and the North Sea

The reports

In November 1986, *The Times* newspaper published reports concerning pollution of the River Rhine. One report is reproduced on the right. Read the report on the right, and the article below.

Safety standards questioned

The Sandoz chemical plant in Basle has been accused of insufficient safety standards after a fire which led to serious pollution of the Rhine.

A report suggests that the storage building where the fire broke out had no trays to catch leaking chemicals, no automatic sprinklers and no smoke-warning or heat warning systems.

The building stored a powerful fungicide containing mercury. Only 10 metres from this building, there was another store containing sodium. If water, which was used to fight the fire, had reacted with the sodium a serious explosion might have resulted.

An investigation into the cause of the fire will probably take weeks to prepare. Meanwhile, many German towns along the Rhine face a water shortage because their wells have been closed.

In Basle, demonstrators marched through the city to the Sandoz Chemical Company. They carried banners saying, ' We don't want to be tomorrow's fish.'

The demonstrators were angry about reports that at least 200 gallons of a poisonous liquid containing mercury had leaked into the River Rhine. The West Geman government may demand damages for the accident and press for more co-ordination between countries over inland waterway disasters.

The Rhine Fishing Federation claimed that plants and animals would be killed in 155 miles of the Rhine between Basle and Manheim. Salvage workers had recently removed several hundred kilograms of dead eels from the river.

Pollution alert in North Sea

By Pearce Wright

Ships which monitor North Sea pollution and radioactivity levels for the Ministry of Agriculture and Fisheries have been alerted to track a flood of mercury due to be discharged from the Rhine later this week.

Their measurement of how this lethal plume is dispersing will be relayed to experts on the protection of coastal and offshore North Sea fisheries.

Aquatic life in the Rhine was destroyed as an estimated 30 tons of mercury and other chemicals

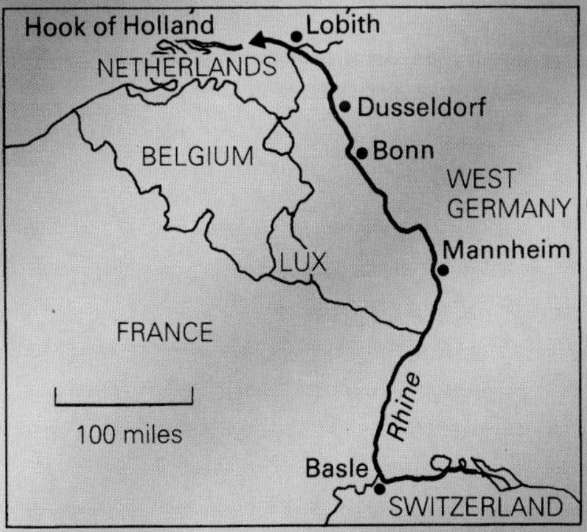

used in the manufacture of pesticides drifted down the river from Basle in Switzerland, through Germany, and to the Netherlands at the weekend.

The mercury, the key ingredient of a powerful fungicide, was washed into the river when firemen fought a blaze at the Sandoz chemical plant, near Basle.

It undid efforts of 10 years to clean up the Rhine, which had been criticized as "becoming Europe's sewer" because of the effluent from rapid industrial growth in Germany, Switzerland and France.

Mercury is one of the heavy metals that are mined for various industrial purposes. Others include lead, arsenic and cadmium. They are also called the toxic metals because very small concentrations are required to poison animals and plants.

A similar incident with mercury 25 years ago, though spread over a longer period, gave the first hints of the extreme toxicity of the heavy metals.

It happened in the small Japanese coastal town of Minamata, where mercury poisoning originating in wastes from a chemical factory spread from fish to fishermen and their families.

The disaster occurred because the mercury was transformed into a highly biologically active form of organic mercury compound after it was discharged. That anxiety will exist over the mercury pouring into the North Sea.

The enquiry

Imagine that an enquiry has been set up to look into the accident. Some people have been asked to prepare statements for the enquiry team. These people include:
(i) a scientist who is particularly interested in the toxic (poisonous) properties of mercury and its compounds,
(ii) a safety officer with particular interest in safety standards in industrial plants,
(iii) the Mayor of Basle.

The statements

Make a group with two other students. Decide who will act as the scientist, who will act as the safety officer and who will act as the Mayor of Basle. Discuss the issues in your group. Then, write down the statements you will each make to the enquiry team (about 300 words). Practise your talks to each other. (Your teacher may ask you to read your statements to the rest of the class.)

Study Questions

1 Many fuels contain small amounts of sulphur. This forms sulphur dioxide when the fuel is burned. The sulphur concentration in lichens was measured at various points in a town. The map across shows points (*A–F*) where measurements were made. The table below shows the results.

Point	Sulphur concentration/ parts per million
A	0.02
B	3.30
C	2.10
D	1.50
E	0.01
F	0.02

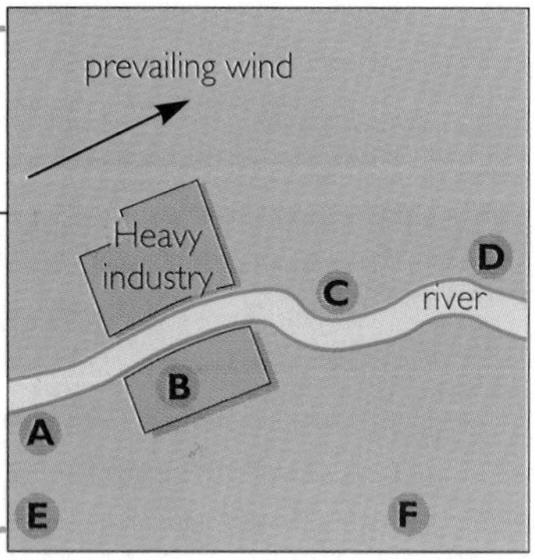

(a) Write an equation to show the formation of sulphur dioxide when a fuel is burnt.
(b) Why did the lichens contain traces of sulphur?
(c) How does sulphur dioxide cause acid rain?
(d) Study the map and the results. What conclusions do you make?
(e) Explain the difference in sulphur concentration between points *A* and *C*.
(f) Since 1955, the number of species of flowering plants within 1 kilometre of the town centre was monitored, together with the output of sulphur dioxide.

Year	Millions of tonnes of sulphur dioxide	Numbers of species of flowering plants
1955	5.6	442
1965	6.5	330
1970	5.8	322
1975	5.4	320
1980	5.4	317
1985	5.1	317

(i) How does the sulphur dioxide seem to be affecting the flowering plants?
(ii) Why do you think the number of species of flowering plants has fallen?
(iii) What steps should be taken to enable the number of species of flowering plants to increase?

2 The apparatus above can be used to find the composition of the air. 100 cm^3 of air were placed in syringe *A* with syringe *B* empty. The copper was heated strongly and the air was passed to and fro between syringes *A* and *B* over the hot copper, and finally returned to syringe *A*.

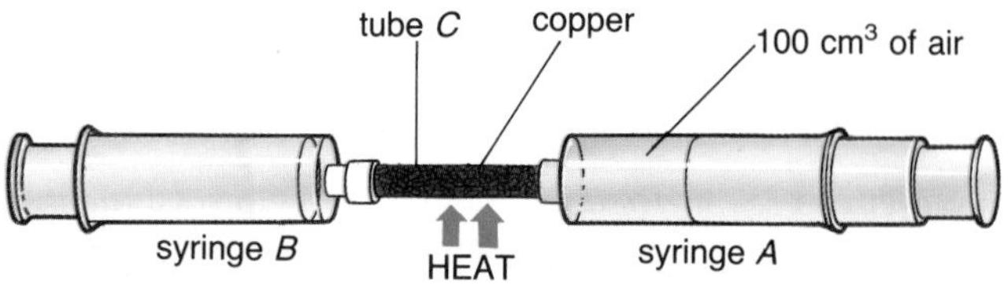

(a) What compound is formed in tube *C*?
(b) Which gas does the copper remove from the air?
(c) What volume of gas is left in syringe *A* at the end of the experiment?
(d) What is the main gas remaining in syringe *A* at the end of the experiment?

3 The Clean Air Act of 1956 made the emission of dark smoke from a chimney illegal. A definition of dark smoke was made in terms of a white card with a scale composed of areas with differing degrees of black shading. New industrial plants burning solid fuel at rates of 10 tonnes per hour or more must install efficient devices to trap the grit and dust produced.
(a) How would you construct a shaded card to use for testing the darkness of smoke emitted from a chimney?
(b) How could you represent the extent of darkness of the smoke using this shaded card?
(c) What devices could industrial plants use to trap grit and dust?
(d) How would you design and locate (i) industrial furnaces, (ii) industrial chimneys, to reduce smoke pollution?

4 (a) What processes cause smoke pollution in the atmosphere?
(b) Why do industrial plants get rid of their smoke through tall chimneys?
(c) What effect does smoke pollution of the atmosphere have on health? Which diseases are most linked to smoke pollution?
(d) Devise experiments (i) to show that city air is polluted with smoke particles, (ii) to obtain clean air.

5 Since the year 1900, the percentages of carbon dioxide and lead compounds in the atmosphere have increased. Suggest reasons for the increase in (i) carbon dioxide, (ii) lead compounds.

6 Steam is passed over red-hot coke and a mixture of two gases (*A* and *B*) is produced. This mixture is then burned in pure oxygen and, when the products of the reaction are cooled to room temperature, a colourless liquid, *C*, and a colourless gas, *D*, remain. *D* reacts with a dilute solution of calcium hydroxide (lime water) to form a white precipitate, *E*. *D* allows magnesium to burn in it, forming two solids, *F* and *G*. When *D* is passed over red-hot coke, the gas *A* is formed.
(a) Identify the substances *A–G*.
(b) Write word equations for all the reactions involved.

SECTION C

Particles

A painting of John Dalton collecting marsh gas. Marsh gas comes fom the rotting materials at the bottom of a pond. John Dalton was the first scientist to use the word 'atom' for the smallest particle of an element

1 EVIDENCE FOR PARTICLES

Look at the photos below and try to answer the questions in the captions.

What happens to ice when it melts? How does the ice manage to change into water?

When sugar is added to tea or coffee, it dissolves quite quickly. The sugar seems to disappear. What happens to the sugar when it dissolves? Where does the sugar go?

How does the liquid get through the filter paper when filter coffee is made? The coffee grains don't get through the paper. Why is this?

In order to answer these questions you will need to use the idea that

> All materials are made up of particles.

If possible, get some sugar and stir it into warm water. Watch the sugar disappear as you stir the water. We can explain how the sugar dissolves and disappears using the idea of particles. Both sugar and water are made up of very small particles. The particles are much too small to see, even under a microscope. When sugar dissolves, tiny particles break off each solid granule. These sugar particles mix with the water particles in the liquid. So, the solution can taste sweet even though you cannot see the sugar (Figure 1).

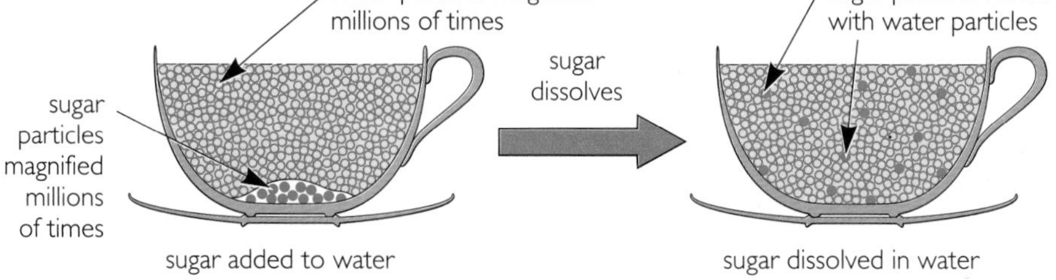

Figure 1
Explaining what happens when sugar dissolves in water

Only a small amount of curry powder gave this meal its hot taste

How large are the particles of materials?

Anyone who cooks knows that a small amount of pepper, ginger or curry powder will give something a really strong taste. Too much spice can spoil the whole meal. This suggests that tiny particles in the spice can spread throughout the whole dish. Figure 2 shows an experiment which will help you to get some idea about the size of particles.

Dissolve 1 g of dark purple potassium manganate(VII) crystals in 1000 cm^3 of water. Take 100 cm^3 of this solution and dilute it to 1000 cm^3 with water. Now take 100 cm^3 of the once-diluted solution and dilute this to 1000 cm^3 with water. Repeat the dilution again and again until you get a solution in which you can *just* see the pink colour. It is possible to make 6 dilutions before the pink colour is so faint that it is only just noticeable. When the potassium manganate (VII) dissolves, its particles spread throughout the water making a dark purple solution. When this solution is diluted, the particles spread further apart.

This experiment shows that the tiny particles in only 1 g of potassium manganate(VII) can colour 1 000 000 000 cm^3 of water. This suggests that there must be millions and millions of tiny particles in only 1 g of potassium manganate(VII).

Similar experiments show that the particles in all substances are extremely small. For example, there are more air particles in a thimble than grains of sand on a large beach.

When oil is spilt on water or on the road, it covers the surface in a thin, brightly-coloured film

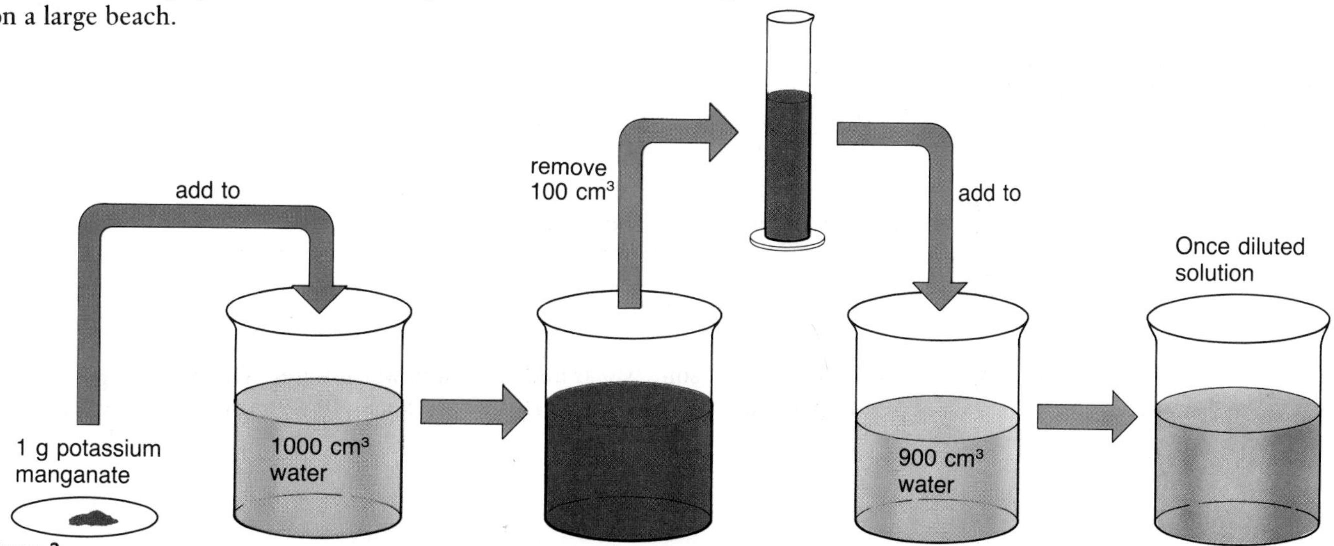

Figure 2

QUESTIONS

1 Krisnan and Christine were talking about dissolving sugar in tea. Krisnan thought that the sugar would weigh less when it was dissolved because it would be floating in the tea.
(a) What do you think happens to the mass of a substance when it dissolves?
(b) Plan an experiment to test your suggestions in part (a).
(c) If possible, carry out your suggested experiment. Explain your results.

2 Get into groups of two or three. Use the idea of particles to discuss and explain what happens when:
(a) water in a kettle boils to produce steam,
(b) you add water to a clay flower pot and the outside of the pot becomes wet,
(c) puddles disappear on a fine day,
(d) tightly tied balloons go down after some time.

3 Look back at Figure 2 and the experiment involved. Suppose that 1 g of potassium manganate(VII) has a volume of 1 cm^3 and that there is one particle of potassium manganate(VII) in every drop of the final 1 000 000 000 cm^3 of faint pink solution.
(a) Estimate the number of drops in 1 cm^3 of the faint pink solution.
(b) How many particles of potassium manganate(VII) are there in 1 000 000 000 cm^3 of the faint pink solution?
(c) Calculate the volume of one particle of potassium manganate(VII) in the crystal.

4 A goldsmith used 1.93 g of gold (density 19.3 g/cm^3) to make an extremely thin sheet of gold, 100 cm^2 in area.
(a) What is the volume of the gold foil? (Use the density to obtain this.)
(b) What is the thickness of the gold foil?
(c) What is the largest possible size of gold particles?

PARTICLES IN MOTION

The best evidence that particles of matter are constantly moving comes from the study of *diffusion* and *Brownian motion*.

Why is it possible to smell perfume from several metres away?

- **Diffusion.** Fish and chips have a delicious smell. You can smell them a long way from where they are being fried. How does the smell get from the fish and chips to your nose?

Particles of gas are released from the fish and chips. The particles mix with air particles and move away from the chips. *This movement and mixing of particles is called* **diffusion**.

Gases diffuse to fill all the space available to them – even heavy gases like bromine (Figure 1). Gases consist of tiny particles, moving around haphazardly and colliding with each other and the walls of their container. The gas particles don't care where they go, so sooner or later they will spread into all the space available.

Diffusion also occurs in liquids, although it takes place more slowly than in gases (Figure 2). This means that liquid particles move around more slowly than gas particles. Diffusion does not happen in solids.

Diffusion is very important in living things. It explains how the food you eat gets to different parts of your body. After a meal, food passes into the stomach and through the intestines. Large particles are broken down into smaller ones. The smaller particles can then diffuse through the walls of the intestines into the bloodstream.

Diffusion is also important in getting oxygen from your lungs into your blood.

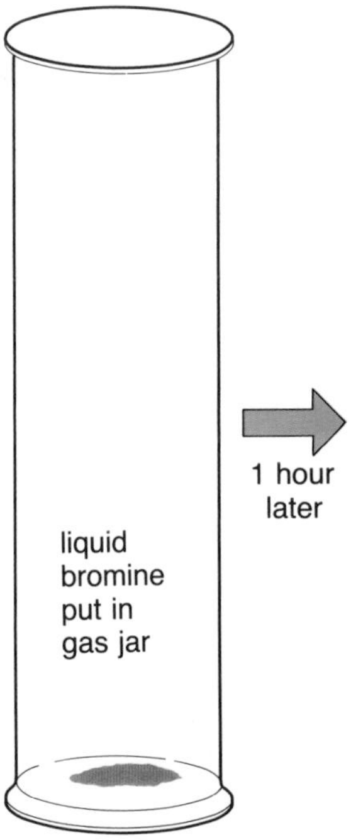

Figure 1
Liquid bromine evaporates and the vapour diffuses. Care: bromine is very poisonous. You should never handle it yourself

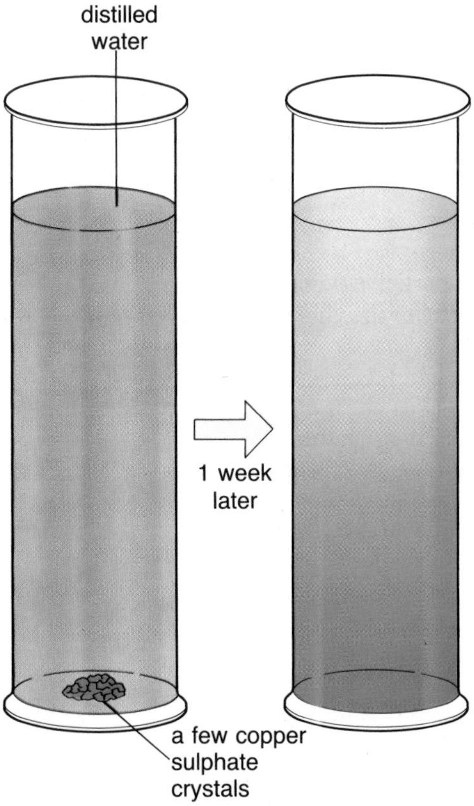

Figure 2
Why is the blue colour darker near the bottom of the gas jar? Will all the solution eventually be the same colour? Explain your answer

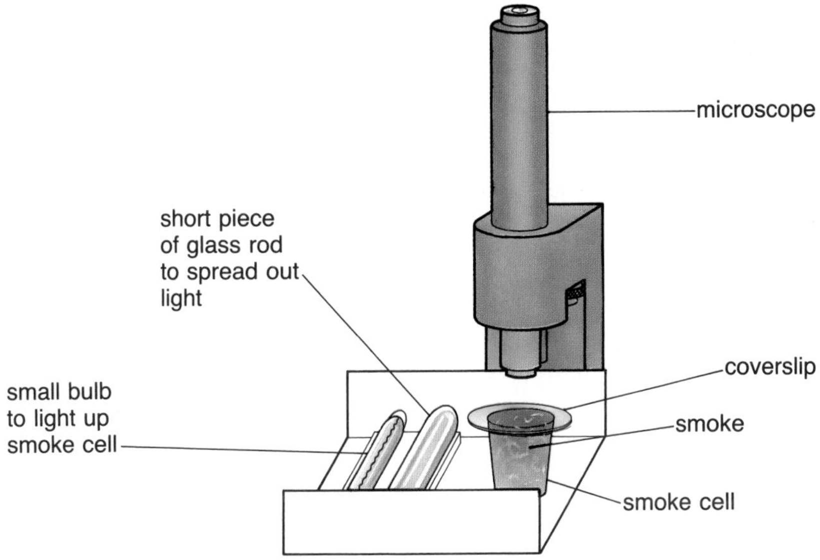

Figure 3

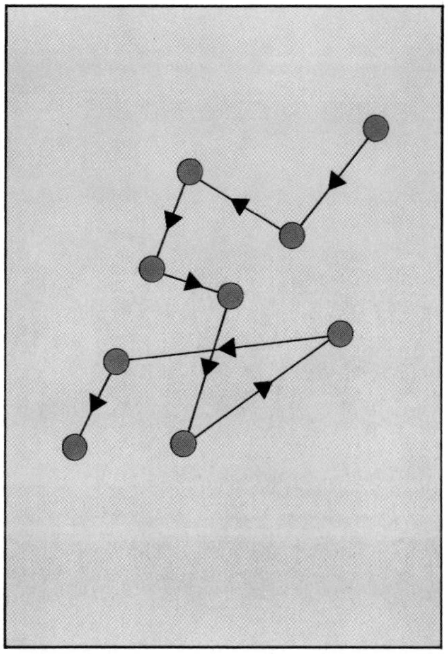

Figure 4
Path of a particle during Brownian motion

• **Brownian motion.** In 1827 a biologist called Robert Brown was using a microscope to look at pollen grains in water. To his surprise, the pollen grains kept moving and jittering about randomly. Similar random movements can be seen when you look at smoke particles through a microscope (Figure 3). *This movement of tiny particles in a gas or liquid is called* **Brownian motion.**

Smoke from a smouldering piece of string is injected into the smoke cell using a teat pipette. Under the microscope, the smoke particles look like tiny pinpoints of light which jitter about (Figure 4).

The movement of the smoke particles is caused by the random motion of oxygen and nitrogen particles in the air around them. The particles of smoke are small, but they are much larger than air particles. Through the microscope we can see smoke particles, but air particles are much too small to be seen. These air particles move very fast and hit the smoke particles at random. The smoke particles are therefore knocked first this way and then that way so they appear to jitter about (Figure 5). As the temperature rises, particles have more energy and they move about faster. This means that gases and liquids diffuse faster when the temperature rises. Particles undergoing Brownian motion also jitter about faster as the temperature rises because they are being bumped more often by the small particles in the gas or the liquid.

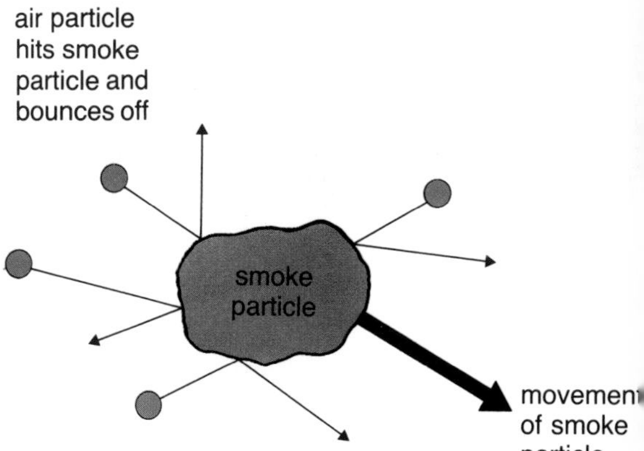

Figure 5

QUESTIONS

1 Imagine that you are a particle of sugar in a lump of sugar which someone has just added to their tea. Describe what happens as
(a) you are stirred into the tea,
(b) you dissolve and mix with the tea and milk,
(c) you are sipped and then swallowed.

2 (a) Why is it important to switch off the engine when filling a car's petrol tank?
(b) Why is it possible to smell some cheeses even when they are wrapped in cling film?

3 You can smell hot, sizzling bacon several metres away, yet you have to be near cold bacon to smell it. Explain this difference in terms of particles.

4 (a) Why does bromine vapour eventually fill the gas jar in Figure 1?
(b) How would the bromine diffuse if all the air was removed from the gas jar?

3 PARTICLES IN MIXTURES

When salt is mixed with water, a clear **solution** is produced. This clear solution is a **mixture**. Tiny particles of salt are spread evenly throughout water in the solution. It is impossible to see the particles of salt in the solution because they are so small.

When mud is mixed with water, a cloudy mixture is produced. If the water is decanted from its sediment of mud, it still looks cloudy. In this case, particles of mud remain 'floating' in the water. The mud particles are much larger than the salt particles in salt solution. The mud particles cannot dissolve in the water, so a cloudy liquid is formed. This cloudy liquid is an example of a **colloid**. A colloid is made from two substances which cannot dissolve in each other or mix properly with each other. It contains bits or globules or particles of one substance spread through the other. These particles are too large to dissolve, but too small to settle out.

The photographs below show some examples of important colloids.

Most dairy products are colloids. They consist of tiny drops of liquid spread throughout another liquid or a solid. These are called **emulsions**. *Butter and cheese contain tiny globules of water spread throughout a fatty solid. Milk and yoghurts contain tiny globules of fat spread throughout a watery liquid*

Mist, fog and clouds consist of tiny droplets of water suspended in the air. Colloids with tiny particles of solid or liquid suspended in a gas are called **aerosols**

Shaving cream and soap suds consist of small bubbles of air spread throughout soap. Colloids with tiny bubbles of gas in a solid or liquid are called **foams**. *Expanded polystyrene is an example of a solid foam*

Emulsion paint is another important example of a colloid. This contains three different kinds of particles suspended in water. These are particles of a white pigment (titanium oxide), particles of latex and particles of china clay.

The particles in emulsion paint are not visible to the naked eye

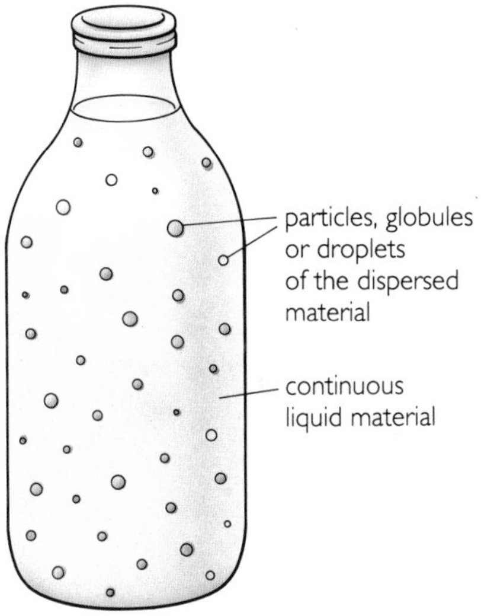

Figure 1
The continuous and dispersed materials in a colloid

In colloids, the particles, droplets or globules of one material are mixed up with the other, but not dissolved in it. The particles are scattered about or *dispersed* in a random manner. In fact, every colloid has two parts. One part is the *continuous* material like water in milk or air in mist. The other part is the *dispersed* material like fat globules in milk or water droplets in mist (Figure 1).

QUESTIONS

1 There are eight different types of colloid. These are listed in the table below.

Continuous material	Dispersed material	Type of colloid
gas	liquid	aerosol
gas	solid	aerosol
liquid	gas	foam
liquid	liquid	emulsion
liuqid	solid	sol
solid	gas	solid foam
solid	liquid	solid emulsion
solid	solid	solid sol

(a) Emulsion paint is not really an emulsion. What is it?
(b) Why is it not possible to make a colloid from two gases?
(c) What type of colloid are the following?
pumice, ice cream, bread, whipped cream, smoke, deodorant spray, currant cake, expanded polystyrene, lipstick, suntan lotion, gloss paint, clouds

2 Jelly is made by mixing gelatine with warm water. Gelatine is a protein which consists of very large molecules. The gelatine disperses in the water. It does not dissolve. As the mixture of gelatine in water cools, a **gel** is formed. In this gel, water droplets are trapped in a continuous network of gelatine.

(a) What type of colloid is (i) gelatine in warm water, (ii) the jelly when it has set?
(b) Draw a diagram to illustrate the colloid of gelatine in warm water.
(c) Draw a diagram to illustrate the colloid in the jelly when it has set.

3 (a) Measure the diameter of the solid particles in the emulsion paint in the photograph above.
(b) Use the magnification given in the caption to calculate the *actual* diameter of the particles.
(d) How many times are these particles larger than single atoms? (The diameter of an atom is about $\frac{1}{100\,000\,000}$ cm = 10^{-8} cm.)

CHANGE OF STATE

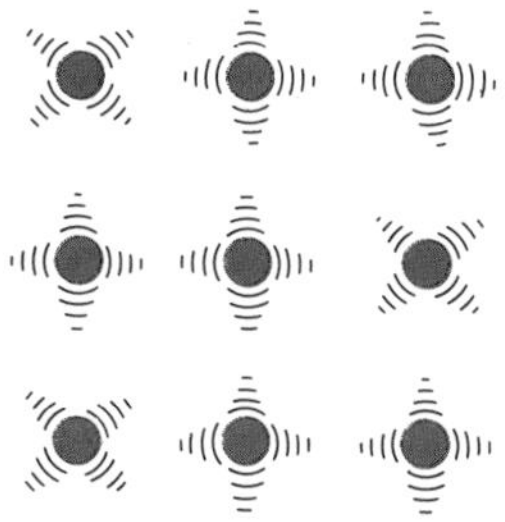

Figure 1

Figure 2

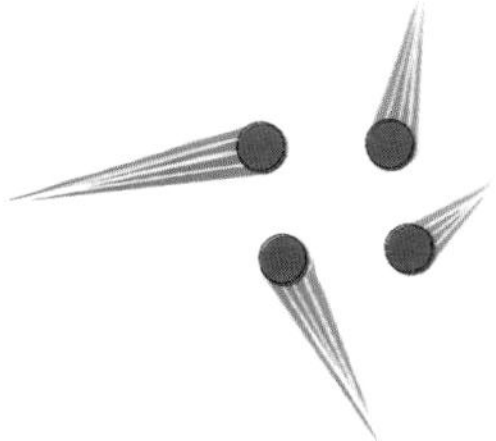

Figure 3

The kinetic theory of matter

The idea that matter is made of moving particles is called the **kinetic theory**. (The word 'kinetic' comes from a Greek word *kineo* which means 'I move'.) The main points of the theory are as follows:

1 All matter is made up of tiny, invisible, moving particles.
2 Particles of different substances have different sizes.
3 Small particles move faster than heavier ones at a given temperature.
4 As the temperature rises, the particles move faster.
5 In a solid, the particles are very close and they can only vibrate about fixed positions (Figure 1).
6 In a liquid, the particles are a little further apart. They have more energy and they can move around each other (Figure 2).
7 In a gas, the particles are far apart. They move rapidly and randomly in all the space they can find (Figure 3).

The movement of particles in solids, liquids and gases can be compared to the movement of children in school. During lessons, the pupils stay in their seats but they are not still. They write notes, answer questions and wriggle (*vibrate*) about in their seats. They are behaving like the particles of a *solid*. When the lesson ends, the pupils pack up their things and move in a group to another part of the school for the next lesson. The pupils stay together, but *move around each other*. They resemble the particles in a *liquid*. When the bell rings at break, the children spill out into the playground where they *move around freely and randomly*. The pupils now resemble a *gas*.

During lessons pupils resemble solid particles

Between lessons pupils resemble liquid particles

In the playground pupils resemble gas particles

Changes of state

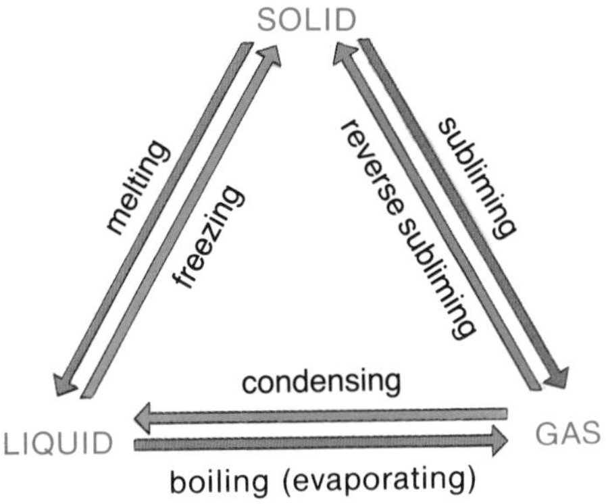

Figure 4

Solids, liquids and gases are sometimes called the three **states of matter**. The kinetic theory can be used to explain how a substance changes from one state to another. A summary of the different changes of state is shown in Figure 4. These changes are usually caused by heating or cooling a substance.

- **Melting and freezing.** When a solid is heated, the particles gain energy and vibrate faster and faster. Eventually, they break free from their fixed position and begin to move round each other. The solid melts to form a liquid. *The temperature at which the solid melts is the* **melting point**.

The temperature at which a solid melts tells us how strongly its particles are held together. Substances with high melting points have strong forces between their particles; substances with low melting points have weak forces between their particles. Metals and alloys, like iron and steel, have high melting points. This suggests that there are strong forces between their particles. This is why metals can be used as girders and supports.

The forces between metal particles in steel are so strong that steel can be used in thin cables to lift heavy loads

- **Evaporating and boiling.** When a liquid is heated, the particles gain energy and move around each other faster and faster. Some particles near the surface of the liquid have enough energy to escape from those around them into the air, and some of the liquid evaporates to form a gas.

Eventually, a temperature is reached at which the particles are trying to escape from the liquid so rapidly that bubbles of gas actually start to form inside the liquid. *The temperature at which this evaporation begins to occur within the bulk of the liquid is the* **boiling point**. Liquids which evaporate at low temperatures are described as *volatile.*

The temperature at which a liquid boils tells us how strongly the particles are held together in the liquid. Liquids with high boiling points have stronger forces between their particles than liquids with low boiling points.

QUESTIONS

1 How do the particles move in (i) a solid, (ii) a liquid, (iii) a gas?
2 What do you understand by the following?
kinetic theory, states of matter, melting point, boiling point
3 What happens to the particles of a liquid (i) as it cools down, (ii) as it freezes?
4 Get into a group with 2 or 3 others in your class. Discuss the following points together. Try to explain each point using the kinetic theory.
(a) Gases exert a pressure on the walls of their container.
(b) Solids have a fixed size and shape.
(c) Liquids have a fixed size but not a fixed shape. They take the shape of their container.
(d) Solid blocks of air freshener used in toilets can disappear without leaving any solid.
(e) Many solids are hard and difficult to bend.

5 ENERGY AND CHANGE OF STATE

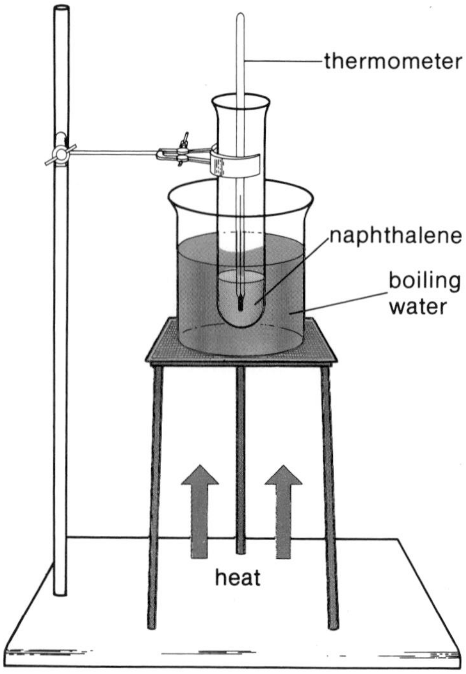

Figure 1

Melting a solid

When a solid melts, it changes state and forms a liquid. This is an endothermic process. Heat is needed to melt the solid. Figure 1 shows an experiment to investigate how the temperature of naphthalene changes when it is surrounded by boiling water. The temperature of the naphthalene is plotted against time in Figure 2.

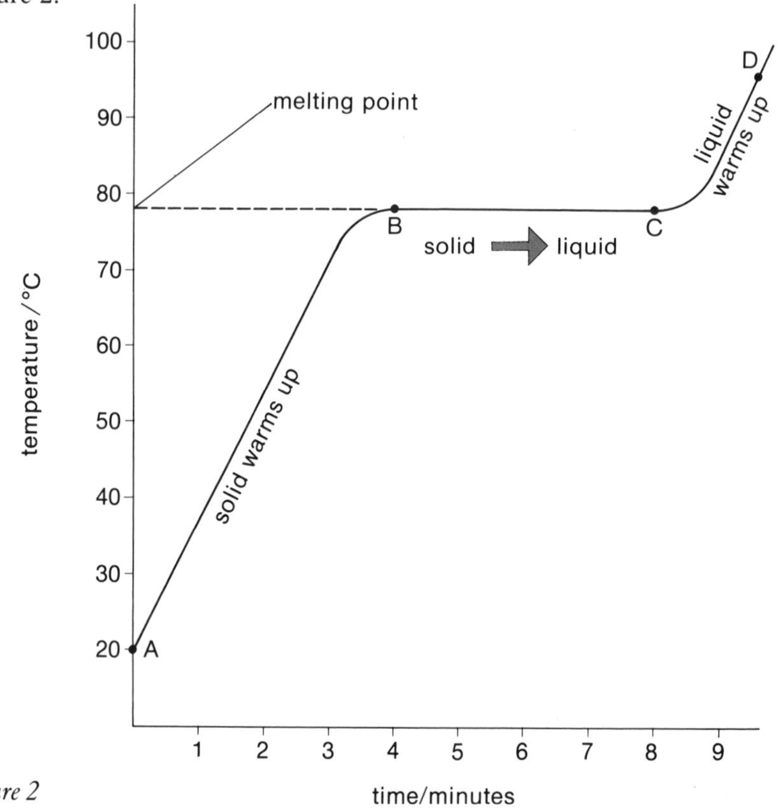

Figure 2

When the solid is heated, its temperature rises along the curve *AB* in Figure 2. As heating continues, the temperature stays constant along *BC*. There is no rise in temperature even though energy is being supplied to the naphthalene. *Along BC the heat supplied does not raise the temperature. It is needed to break the particles away from their fixed positions in the solid crystals until they can move around each other in the liquid.* The naphthalene is melting along *BC* and the graph shows that the melting point is 78°C. At *C*, all the naphthalene has melted. On further heating, the temperature rises again along *CD* as the heat supplied warms up the liquid naphthalene. Figure 3 shows what happens to the naphthalene particles as the solid melts between *B* and *C*.

Molten lava is rock which has become so hot near the Earth's core that it has become liquid. As it cools on the surface, it becomes solid once more

Vaporising a liquid

When a liquid is heated, the temperature rises until its boiling point is reached. Once the liquid is boiling, the temperature stays constant because the energy being supplied is needed to separate liquid particles so they can move around at high speed in the gaseous state (Figure 3).

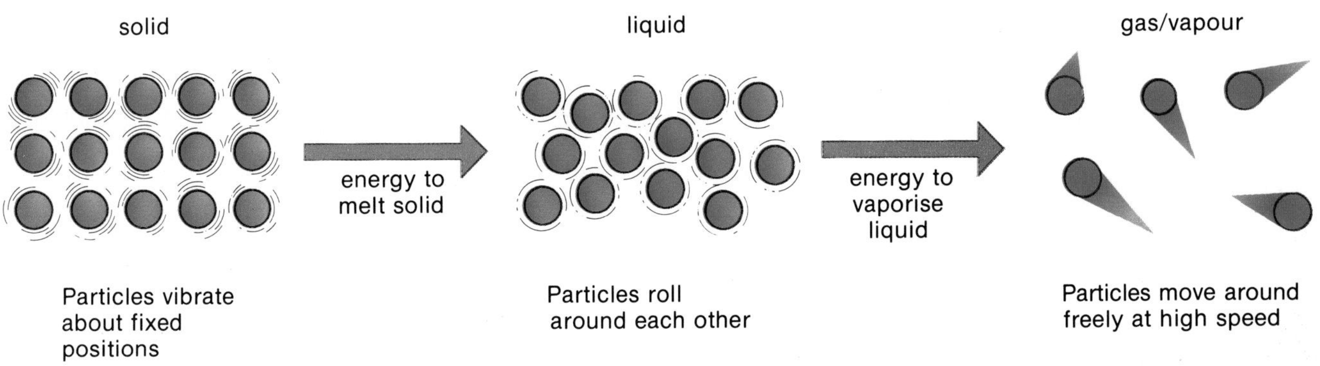

Figure 3

When a substance warms up or melts or vaporises, it absorbs energy. When a substance cools, freezes or condenses, it loses energy. The energy changes between a liquid and its vapour are used in refrigerators (Figure 4). The liquids used in refrigerators are called *freons* or *CFCs* (unit B5). They are non-poisonous compounds of carbon, chlorine and fluorine with a low boiling point. Liquid freon vaporises in the coiled pipes around the ice box. This takes heat from the refrigerator and from the food inside it, so they cool down. The freon vapour is then pumped into the pipes and fins at the back of the machine. Here the pressure rises and the freon vapour turns to liquid. As it condenses, the freon vapour gives out heat to the pipes and the air around them. Finally the liquid freon returns to the pipes around the ice box and the process is repeated.

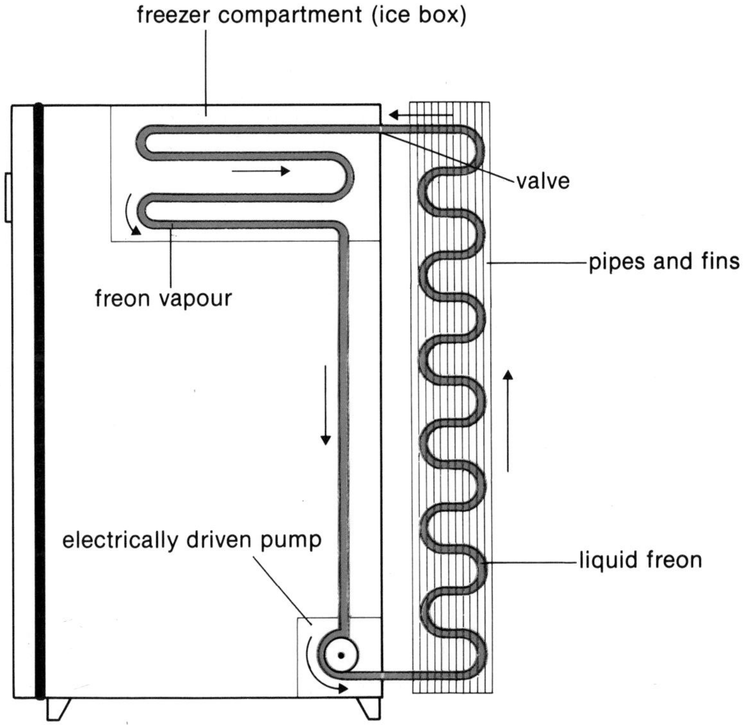

Figure 4
A household refrigerator

QUESTIONS

1 (a) How does sweating help us to keep cool in hot weather?
(b) Dogs with thick coats cannot sweat like humans. How do they keep cool?
2 Why should the liquid used in a refrigerator be (i) volatile, (ii) non-poisonous, (iii) non-flammable?
3 Draw a sketch graph of temperature against time to show what happens when ice is heated from −20°C until the water boils.
4 Find out how pressure cookers work and why they cook food more quickly than normal saucepans. Design a leaflet explaining how a pressure cooker works to someone who has just bought one.

6 GAS PRESSURE

Gas pressure is important. It helps us to inflate balloons and tyres (Figure 1). Gas pressure from the air (*atmospheric pressure*) dictates our weather. When atmospheric pressure is high, clouds are driven away and the weather is fine and dry. When atmospheric pressure is low, the weather is often cloudy and rainy.

When atmospheric pressure is high, clouds are driven away and the weather is fine and dry

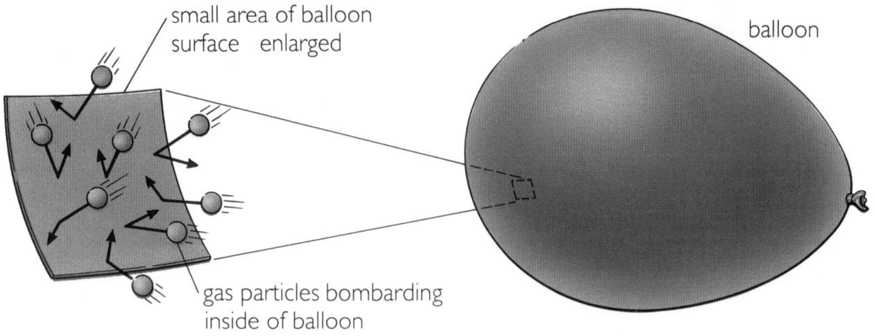

Figure 1
Gas particles move around randomly inside the balloon. These collide with each other and with the walls of the balloon. Constant bombardment of the inside of the balloon by millions of gas particles every second causes the gas pressure. Actually, the balloon has gas pressure on its outside and *its inside. The pressure inside is greater than that outside. This extra pressure inside the balloon helps to keep it inflated*

Pressure and volume

Figure 2 shows what happens to the particles in a gas if you hold your finger over the end of a bicycle pump and push in the handle. The gas particles have been pushed into a smaller volume. This means that the particles bombard the sides of the pump, and your finger, more often. So the pressure increases. You can feel the increased pressure on your finger. This simple experiment shows that *if you decrease the volume of a gas, its pressure will increase.*

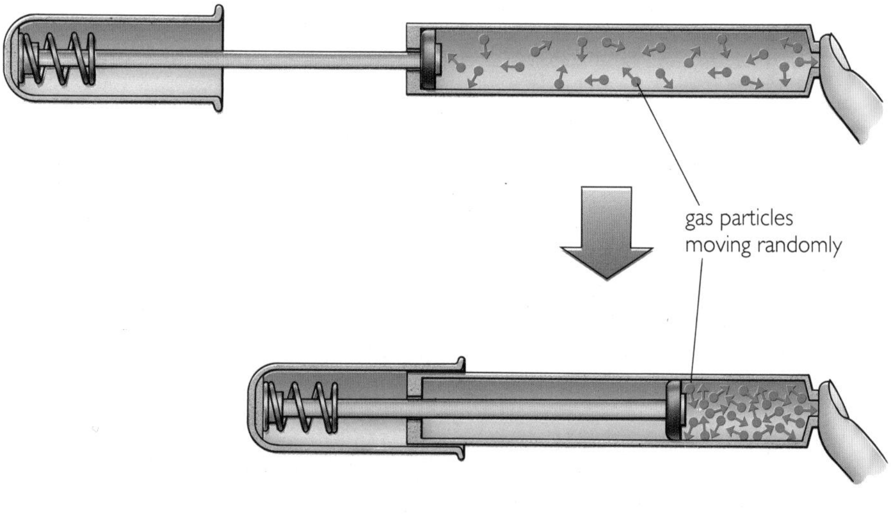

Figure 2

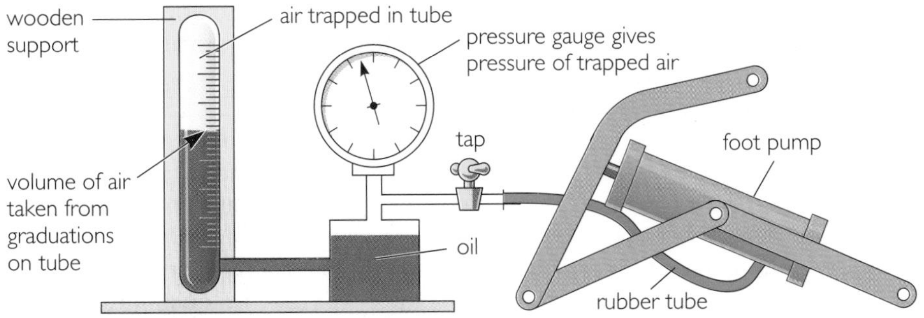

Figure 3

Figure 3 shows the apparatus we can use to see how the volume (V) of air changes as the pressure (P) increases. (The volume of a gas will also change if the temperature changes. So the temperature *must* be kept constant during this experiment.) The footpump is used to increase the pressure of the air trapped in the tube.

Table 1 shows the results of one experiment.

Notice that:

- the volume is *halved* from 100 cm^3 to 50 cm^3 when the pressure *doubles* from 100 000 Pa to 200 000 Pa.
- the volume goes down to *one quarter* (25 cm^3) when the pressure *quadruples* to 400 000 Pa.

Pressure /Pa	Volume /cm^3
100 000	100
150 000	67
200 000	50
250 000	40
300 000	33
350 000	29
400 000	25

Table 1 The volume of some trapped air at different pressures (Pressure is measured in Pascals, Pa. 1 Pa = 1 newton per square metre. Atmospheric pressure is about 100 000 Pa)

So, as the volume decreases, the pressure increases by the same proportion. This is an **inverse** (upside-down) relationship and we say that **the volume of a gas is inversely proportional to the pressure**.

Using symbols we can write this as

$$V \propto \frac{1}{P} \quad \text{or} \quad V = \frac{\text{constant}}{P}$$

$$\therefore \quad PV = \text{constant}$$

If pairs of readings are taken, we can say that

$$P_1V_1 = P_2V_2$$

This relationship between P and V was first discovered by Robert Boyle in 1662. It is usually called **Boyle's Law**. This says that:

> The volume of a fixed mass of gas is inversely proportional to its pressure, if the temperature stays constant.

QUESTIONS

1 (a) Use the results in Table 1 to plot a graph of volume (vertically) against pressure (horizontally).
(b) How would you describe the shape of the graph?
(c) Work out the value of $\frac{1}{P}$ for each pressure.
(d) Now plot a graph of volume (vertically) against $\frac{1}{P}$ (horizontally).
(e) What is the shape of this graph? What can you conclude from this graph?
(f) Use the graph to predict the volume of the gas when the pressure is 500 000 Pa.

2 A diver on the sea bed releases an air bubble with a volume of 2 cm^3. When the bubble reaches the surface, its volume is 8 cm^3. If the atmospheric pressure is 100 000 Pa, what is the pressure on the sea bed?

3 Pressure in the main gas grid pipelines is 800 kPa. The pressure in gas storage vessels is 100 kPa. What volume of gas from the grid pipelines is required to fill a gas storage vessel of volume 160 000 m^3?

7 THE GAS LAWS

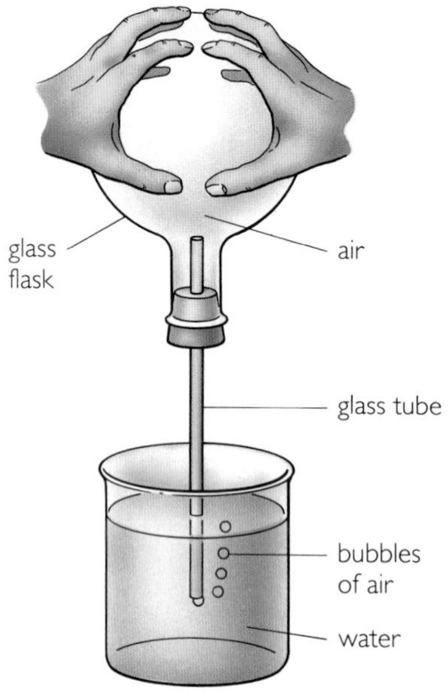

Figure 1

When a gas is heated, it expands. Figure 1 shows a simple experiment which you could do to illustrate this. Heat from your hands warms the air inside the flask. As the air expands, bubbles escape from the tube into the water. This simple experiment shows that *the temperature of a gas can affect its volume.*

If you are motoring on a hot summer's day, the pressure of the air inside your car's tyres may rise by about 20 kPa. In this case, the air in the tyres cannot escape, unlike the air in the flask in Figure 1. But this example shows that *the temperature of a gas can affect its pressure.*

Notice that temperature can affect both the volume and the pressure of a gas. So if we want to study the effect of temperature on these two quantities, we must keep one fixed (**constant**), while we see how the other changes.

How temperature affects volume

Figure 2 shows the apparatus we can use to investigate the effect of temperature on the volume of a gas, *at constant pressure*. The gas being tested is air. This is trapped in a capillary tube by a short thread of mercury. The pressure of the trapped air is constant during the experiment because the top end of the tube is open. The capillary tube is attached to a centimetre scale so that the length of the trapped air column can be measured. (The length of the air column will be a measure of its volume.)

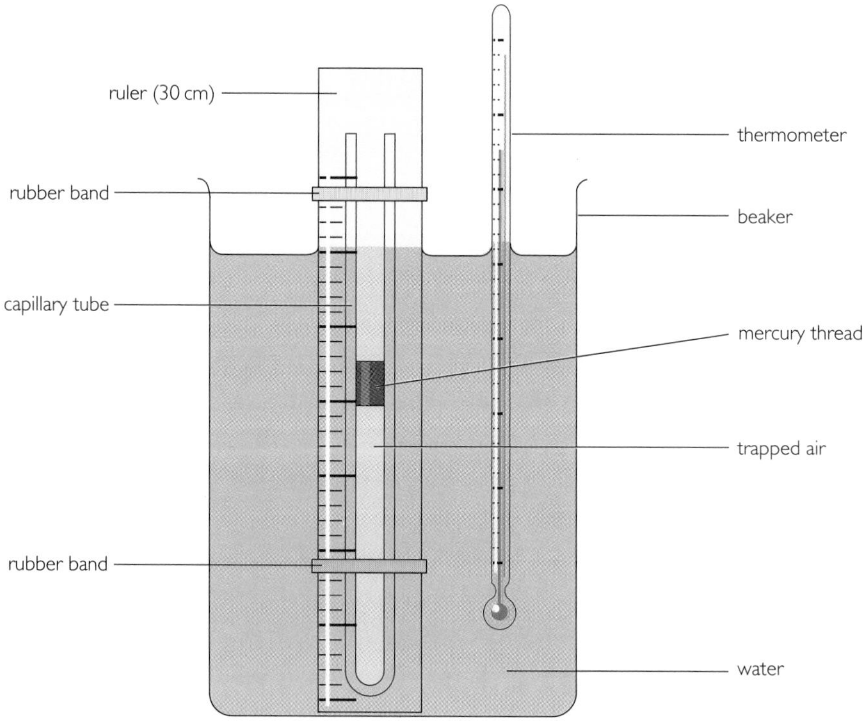

Figure 2

The length of the air column and the temperature of the water are first noted. The water is then heated until its temperature has increased by about 10°C. When the temperature of the trapped air is the same as that of the water, the length of the air column is taken. This procedure is repeated at 4 or 5 other temperatures.

Figure 3 shows the graph obtained when the length of the air column (proportional to its volume) is plotted against temperature. This shows that the volume of the gas increases steadily with temperature. If the graph is drawn backwards (**extrapolated**), it hits the temperature axis at −273°C. This suggests that the volume of the gas has contracted to zero at −273°C.

When the experiment is repeated with other gases, we get similar results to these. The volumes of all gases seem to contract to zero at −273°C. As we cannot imagine any volume less than zero, scientists have assumed that −273°C is the lowest possible temperature. It is sometimes called **absolute zero**.

Sometimes it is useful to have a temperature scale whose zero is absolute zero. This is called the **absolute temperature scale** or the **Kelvin scale**. Units on this scale are called Kelvins (K). So −273°C is 0 K and 0°C is 273 K. Notice how easy it is to convert from °C to K. You just add 273 (Figure 4).

The graph in Figure 3 is a straight line through the origin of the absolute temperature scale. So we can say:

> The volume of a fixed mass of gas is proportional to its absolute temperature, if the pressure stays constant.

This result is sometimes called **Charles' Law** after Jacques Charles who first discovered it in 1787.

Using symbols we can write Charles' Law as:

$$V \propto T \quad \text{or} \quad \frac{V}{T} = \text{constant}$$

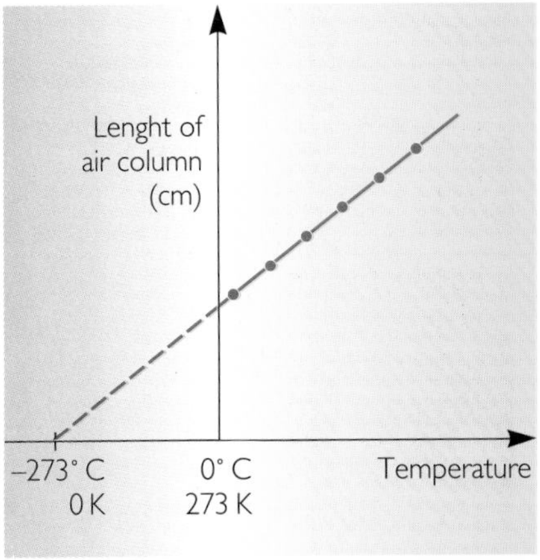

Figure 3

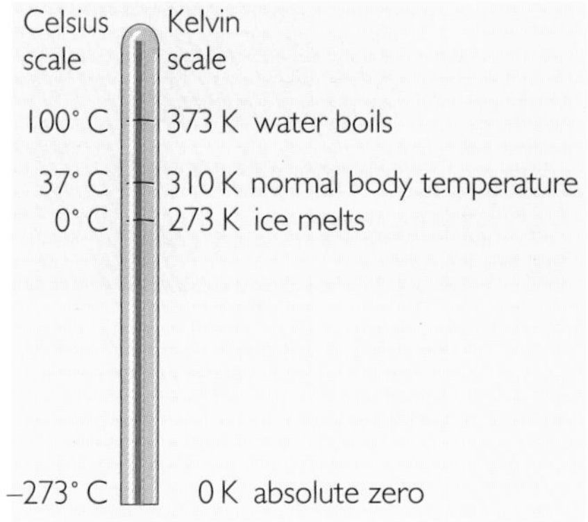

Figure 4
The Celsius and Kelvin temperature scales

Combining Boyle's Law and Charles' Law

Boyle's Law and Charles' Law are sometimes called Gas Laws.

We can combine Boyle's Law, $PV = \text{constant}$

and Charles' Law, $\frac{V}{T} = \text{constant}$,

into a single equation, known as the **Gas equation**:

$$\frac{PV}{T} = \text{constant}$$

If P_1, V_1, T_1 and P_2, V_2, T_2 are the pressure, volume and temperature of the same mass of gas under two different conditions, we can write

$$\frac{P_1V_1}{T_1} = \frac{P_2V_2}{T_2}$$

This equation is very useful in working out the volume of a gas when the pressure and temperature change.

QUESTIONS

1 A bicycle pump contains 58 cm^3 of air at 17°C and a pressure of 1 atmosphere. What will the pressure be when the air is compressed to 10 cm^3 and its temperature rises to 27°C? (Hint: In your calculations, temperatures should be in K.)

2 Liam checked the tyre pressure in his car early one morning when the temperature was 7°C. It was 160 kPa. After a long journey the pressure had risen to 180 kPa. What was the temperature of the air in the tyres at the end of his journey?

3 Use the idea of moving particles to explain why
(a) the pressure of the air inside a car tyre increases if the temperature rises,
(b) the volume of a balloon decreases if the temperature falls.

8 ATOMS – THE SMALLEST PARTICLES

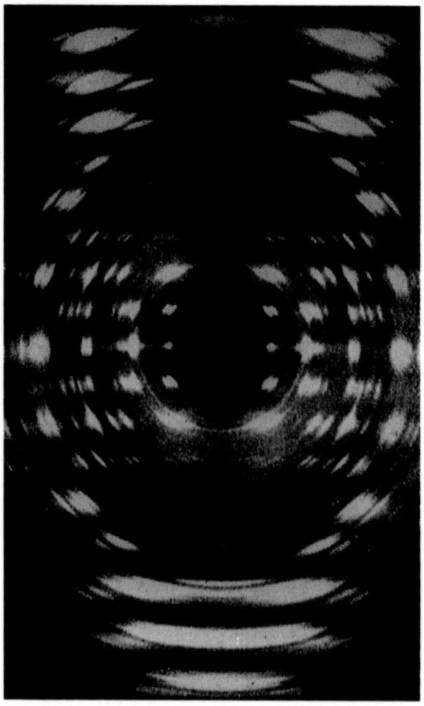

This X-ray photograph of DNA has bright spots caused by X-rays being reflected by particles in the crystal

Dalton's theory of atoms

Experiments involving diffusion and Brownian motion (unit 2 of this section) show that substances are made up of small particles. Photographs taken by X-rays and electron microscopes also give us good evidence for particles. X-ray photographs of crystals, such as the one on the left, show white spots on a black background. The white spots are caused by X-rays being reflected by regularly spaced particles in the crystal.

Electron microscopes can magnify objects more than a million times. In 1958, scientists in the USSR observed individual particles of barium and oxygen using an electron microscope. Figure 1 shows an electron microscope photo of a manganese compound. The magnification is 127 000 times. The dark lines are rows of manganese particles in the crystal. These particles are the smallest possible particles of manganese – they are manganese *atoms.*

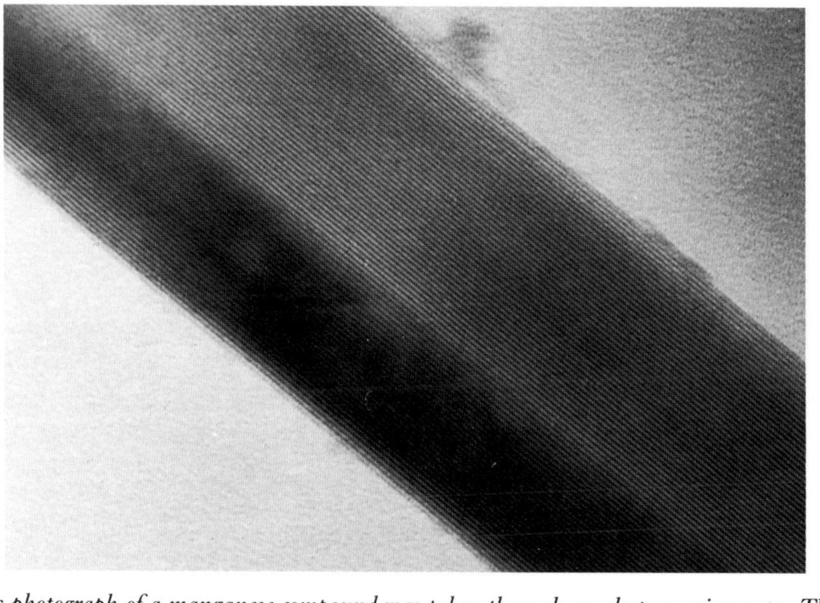

This photograph of a manganese compound was taken through an electron microscope. The dark lines are rows of manganese atoms

John Dalton – the first scientist to use the name 'atom' for the smallest particle of an element. Dalton was born in 1766 in the remote village of Eaglesfield in Cumbria. His father was a weaver. For most of his life, Dalton taught at the Presbyterian College in Manchester

> The smallest particle of any element is called an **atom**

The word 'atom' was first used by John Dalton in 1807, when he put forward his **Atomic Theory of Matter**. The main points in Dalton's theory are:

1 All matter is made up of tiny particles called atoms.
2 Atoms cannot be made or destroyed.
3 Atoms of the same element are exactly alike, with the same mass, colour, etc.
4 Atoms of different elements have different masses, colours, etc.
5 Atoms can join together to form larger particles in compounds.

Nowadays, we call these larger particles **molecules.** Dalton's theory about atoms and molecules is still very useful today.

Representing atoms with symbols

Dalton also suggested a method of representing atoms with symbols. Figure 2 shows how he represented the atoms of some common elements. The modern symbols which we use for different elements are based on his suggestions.

Table 1 gives a list of the symbols of some of the common elements. (A longer list of symbols appears on page 269.) Notice that most elements have two letters in their symbol; the first letter is a capital, the second letter is *always* small. These symbols come from either the English name (O for oxygen, C for carbon) or the Latin name (Au for gold – Latin: *aurum*; and Cu for copper – Latin: *cuprum*).

Element	Symbol	Element	Symbol	Element	Symbol
Aluminium	Al	Helium	He	Oxygen	O
Argon	Ar	Hydrogen	H	Phosphorus	P
Bromine	Br	Iodine	I	Potassium	K
Calcium	Ca	Iron	Fe	Silicon	Si
Carbon	C	Lead	Pb	Silver	Ag
Chlorine	Cl	Magnesium	Mg	Sodium	Na
Chromium	Cr	Mercury	Hg	Sulphur	S
Cobalt	Co	Neon	Ne	Tin	Sn
Copper	Cu	Nickel	Ni	Uranium	U
Gold	Au	Nitrogen	N	Zinc	Zn

Table 1 The symbols for some elements

Using symbols, we can represent compounds as well as elements. For example, water is represented as H_2O because the smallest particle of water (a molecule) contains two hydrogen atoms and one oxygen atom. Carbon dioxide is written as CO_2 – one carbon and two oxygen atoms. H_2O and CO_2 are called **formulas**. Formulas show the relative numbers of atoms of each element in a compound.

Chemists use symbols in the same way that typists use shorthand. Using symbols, they can represent substances and chemical changes quickly. For example, the formula for potassium manganate(VII) is $KMnO_4$. If you have to name this substance many times, how would you prefer to write it – using symbols or using the long name!

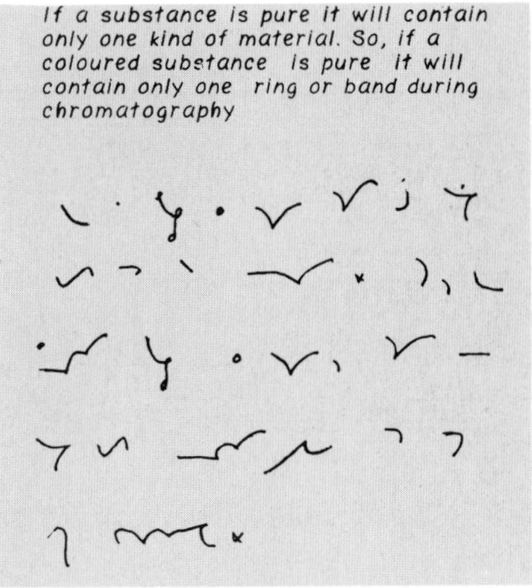

By using shorthand you can make notes more quickly. Chemists use chemical symbols as their shorthand

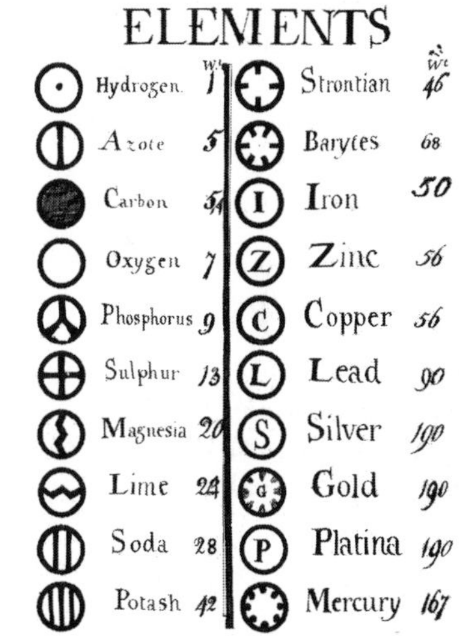

Dalton's symbols for some common elements

QUESTIONS

1 Look at the electron microscope photograph of a manganese compound, on the opposite page.
(a) Estimate the thickness of one of the dark lines in the photo. This dark line is a row of manganese atoms.
(b) Calculate the actual diameter of a manganese atom. (Assume the magnification is 127 000.)

2 Find out about the life and work of John Dalton. Prepare a short talk about Dalton for the rest of your class. If possible, illustrate your talk with drawings, diagrams and photos.

3 How many atoms of the different elements are there in one molecule of (i) methane (natural gas), CH_4, (ii) sulphuric acid, H_2SO_4, (iii) sugar, $C_{12}H_{22}O_{11}$, (iv) chloroform, $CHCl_3$?

4 Look at Dalton's symbols for the elements in the photo above.
(a) What do we call 'Azote' and 'Platina' today?
(b) Six of the substances in Dalton's list are compounds and *not* elements. Pick out two of these compounds and write their correct chemical names.
(c) Which one of Dalton's symbols do you think is the most appropriate? Why do you think it is appropriate?
(d) Look at the numbers at the right-hand side of the elements in Dalton's list. What do you think these numbers represent?

9 MEASURING ATOMS

How large are atoms?

Experiments with thin films of oil on water show that olive oil particles are about $\frac{1}{10\,000\,000}$ cm thick. But olive oil particles are large molecules containing more than 50 atoms. If we estimate that one molecule of olive oil is about 10 atoms thick, how big is a single atom?

Atoms are about $\frac{1}{100\,000\,000}$ *of a centimetre across*. This means that if you put 100 million of them side-by-side, they will measure only 1 cm. It is very difficult to imagine anything as small as this. You will get some idea of the size of atoms from Figure 1. If atoms were magnified to the size of marbles, then on the same scale, marbles would have a diameter of 1500 km – one third of the distance between New York and London.

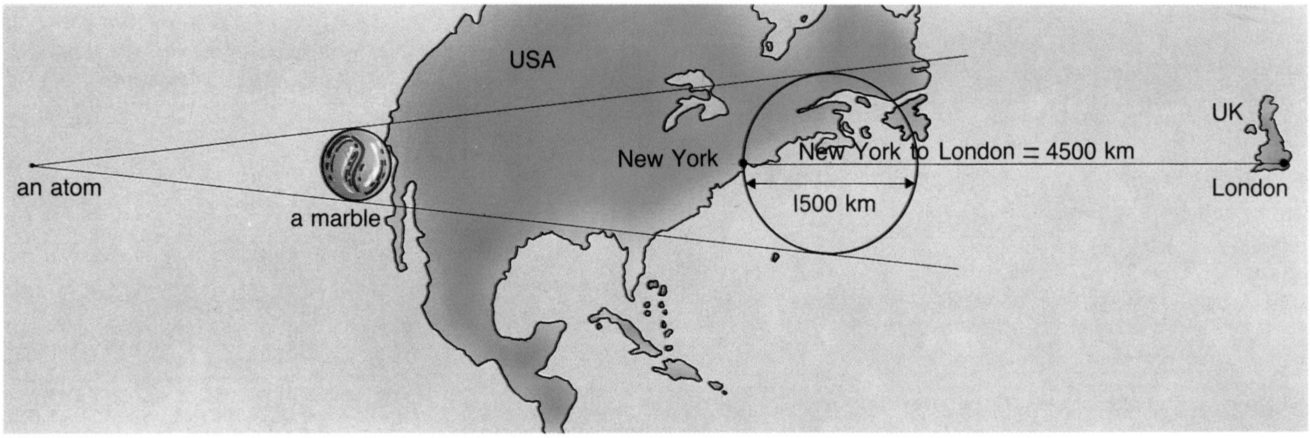

Figure 1
If atoms are magnified to the size of marbles, on the same scale a marble would have a diameter of 1500 km

Figure 2 will also help you to realise just how small atoms are. It shows a step-by-step decrease in size from 1 cm to $\frac{1}{100\,000\,000}$ cm. Each object is one hundred times smaller than the one before it. The dice on the left is about 1 cm wide. In the next square, the grain of sand is about $\frac{1}{100}$ cm across. The bacterium in the middle is 100 times smaller again – about $\frac{1}{10\,000}$ (10^{-4}) cm from end to end. In the next square, the molecule of haemoglobin is 100 times smaller than this – about $\frac{1}{1000\,000}$ (10^{-6}) cm in diameter. Finally, in the right-hand square, the atom is one hundredth of the size of the haemoglobin molecule – about $\frac{1}{100\,000\,000}$ (10^{-8}) cm in diameter.

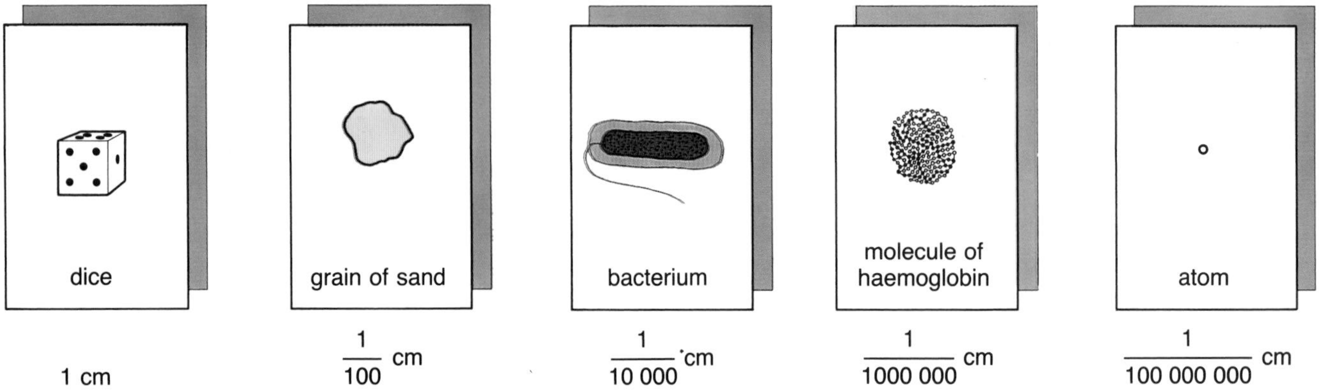

Figure 2
Step-by-step to the size of atoms

How heavy are atoms?

A single atom is so small that it cannot be weighed on a balance. However, *the mass of one atom can be compared with that of another atom using an instrument called a* **mass spectrometer** (Figure 3).

In a mass spectrometer, atoms are passed along a tube and focused into a thin beam. This beam of particles passes through an electric field (which speeds them up) and then through a magnetic field where they are deflected. The extent to which an atom is deflected depends on its mass – the greater the mass, the smaller the deflection. Using the amount of deflection, it is possible to compare the masses of different atoms and make a list of their relative masses.

Relative atomic masses

Chemists compare the masses of different atoms using **relative atomic masses** (RAMs). The symbol A_r is sometimes used for relative atomic mass. The element carbon is chosen as the standard against which the masses of other atoms are compared. Carbon atoms are given a relative atomic mass of 12 and the relative masses of other atoms are obtained by comparison with this. A few relative atomic masses are listed in the table. Other relative atomic masses are given on page 269. From Table 1 you will see the carbon atoms are 12 times as heavy as hydrogen atoms; iron atoms are 55.8 times as heavy as hydrogen atoms.

Element	Symbol	Relative atomic mass
Carbon	C	12.0
Hydrogen	H	1.0
Oxygen	O	16.0
Iron	Fe	55.8
Copper	Cu	63.5
Gold	Au	197.0

Table 1 The relative atomic masses of a few elements

Figure 3
A simple diagram of a mass spectrometer

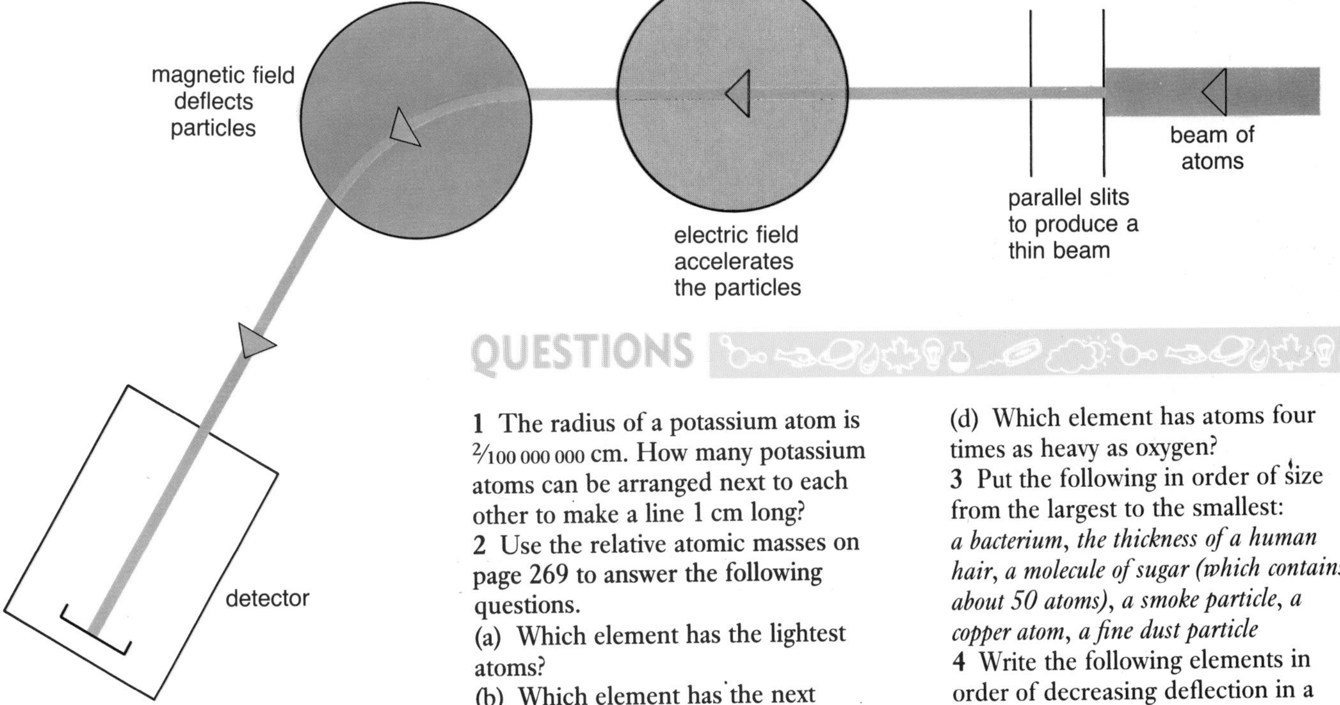

QUESTIONS

1 The radius of a potassium atom is 2/100 000 000 cm. How many potassium atoms can be arranged next to each other to make a line 1 cm long?

2 Use the relative atomic masses on page 269 to answer the following questions.

(a) Which element has the lightest atoms?

(b) Which element has the next lightest atoms?

(c) How many times heavier are carbon atoms than hydrogen atoms?

(d) Which element has atoms four times as heavy as oxygen?

3 Put the following in order of size from the largest to the smallest: *a bacterium, the thickness of a human hair, a molecule of sugar (which contains about 50 atoms), a smoke particle, a copper atom, a fine dust particle*

4 Write the following elements in order of decreasing deflection in a mass spectrometer (put the element which is deflected the most first): *copper, calcium, carbon, cobalt, chlorine*

10 USING RELATIVE ATOMIC MASSES

This photograph shows an aluminium colander (mass 108 g), a copper bracelet (mass 16 g), some iron nails (mass 5.58 g) and some barbecue charcoal (mass 150 g). How many moles of each element do the objects contain? (Al = 27, Cu = 64, Fe = 55.8, C = 12)

Counting atoms

Relative atomic masses show that one atom of carbon is 12 times as heavy as one atom of hydrogen. Therefore, 12 g of carbon will contain the same number of atoms as 1 g of hydrogen. An atom of oxygen is 16 times as heavy as an atom of hydrogen, so 16 g of oxygen will also contain the same number of atoms as 1 g of hydrogen. In fact, *the relative atomic mass (in grams) of every element* (1 g of hydrogen, 12 g carbon, 16 g oxygen, etc.) *will contain the same number of atoms.* This number is called **Avogadro's constant** in honour of the Italian scientist Amedeo Avogadro. *The relative atomic mass in grams is known as one* **mole** *of the element.* So, 12 g of carbon is 1 mole of carbon and 1 g of hydrogen is also 1 mole. 24 g of carbon is 2 moles and 240 g of carbon is 20 moles. Notice that:

$$\text{The number of moles} = \frac{\text{mass}}{\text{RAM}}$$

Experiments show that Avogadro's constant is 6×10^{23}. Written out in full this is 600 000 000 000 000 000 000 000. Thus *1 mole of an element always contains 6×10^{23} atoms.*

We can use the mole idea to count (calculate) the number of atoms in a sample of element. For example:

12 g (1 mole) of carbon contains 6×10^{23} atoms
so 1 g ($\frac{1}{12}$ mole) of carbon contains $\frac{1}{12} \times 6 \times 10^{23}$ atoms
=>10 g ($\frac{10}{12}$ mole) of carbon contains $\frac{10}{12} \times 6 \times 10^{23} = 5 \times 10^{23}$ atoms

$$\text{i.e. number of atoms} = \text{number of moles} \times 6 \times 10^{23}$$

This bank clerk is counting money by weighing the coins. In the same way, chemists count atoms by weighing

Chemists are not the only people who 'count by weighing'. Bank clerks use the same idea when they count coins by weighing them. For example, 100 1p coins weigh 356 g; so it is quicker to take 100 1p coins by weighing out 356 g of them than by counting.

Chemists often need to count atoms. In industry, nitrogen is reacted with hydrogen to form ammonia, NH_3, which is then used to make fertilisers. In a molecule of ammonia, there is one nitrogen atom and three hydrogen atoms. In order to make ammonia, chemists must therefore react:

1 mole of nitrogen + 3 moles of hydrogen
(14 g of nitrogen) (3×1 g = 3 g of hydrogen)

not 1 g of nitrogen and 3 g of hydrogen.

To get the right quantities, chemists must measure in moles *not* in grams. Thus the mole is the chemist's counting unit.

Finding formulas

We have used some formulas already, but how are they obtained? How do we know that the formula of water is H_2O?

All formulas are obtained by doing experiments. When water is decomposed into hydrogen and oxygen, results show that

$$\begin{aligned} 18\text{ g of water gives } 2\text{ g of hydrogen} &+ 16\text{ g of oxygen} \\ = 2\text{ moles of hydrogen} &+ 1\text{ mole of oxygen} \\ = 2 \times 6 \times 10^{23}\text{ atoms of hydrogen} &+ 6 \times 10^{23}\text{ atoms of oxygen} \end{aligned}$$

Since 12×10^{23} hydrogen atoms combine with 6×10^{23} atoms of oxygen, it means that 2 hydrogen atoms combine with 1 oxygen atom. Therefore, the formula must be H_2O. These results are set out in Table 1. By finding the masses of the elements which form a compound, we can use relative atomic masses to calculate the number of moles of atoms that are present and this gives us the formula.

	H	O
Masses present	2 g	16 g
Mass of 1 mole	1 g	16 g
∴ moles present	2	1
Ratio of atoms	2	1
=> Formula	H_2O	

Table 1 Finding the formula of water

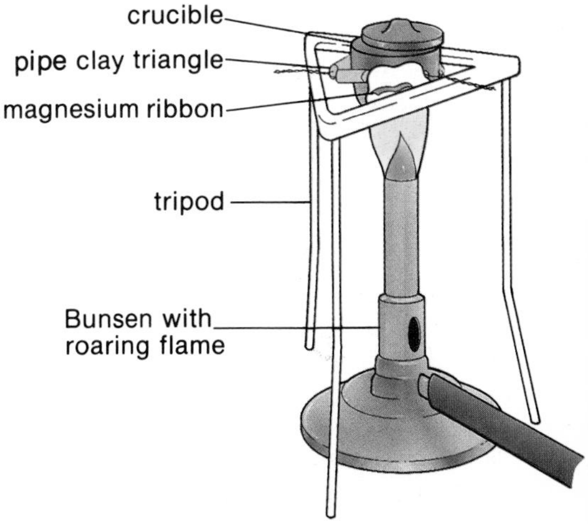

Finding the formula of magnesium oxide

When magnesium ribbon is heated, it burns with a very bright flame to form white, powdery magnesium oxide:

$$\text{magnesium} + \text{oxygen} \rightarrow \text{magnesium oxide}$$

Weigh 0.24 g of clean magnesium ribbon. Heat this strongly in a crucible until all of it forms magnesium oxide (Figure 1). Put a lid on the crucible to stop magnesium oxide escaping, but keep a small gap so that air can enter. When the magnesium has finished reacting, reweigh the crucible + lid + magnesium oxide. 0.40 g of magnesium oxide should have formed.

$$\begin{aligned} &0.24\text{ g magnesium gives } 0.40\text{ g magnesium oxide} \\ &\Rightarrow \text{mass of oxygen reacting} = 0.40 - 0.24 = 0.16\text{ g} \end{aligned}$$

Table 2 shows how to obtain the formula of magnesium oxide from these results.

	Mg	O
Masses reacting	0.24 g	0.16 g
Mass of 1 mole	24 g	16 g
∴ moles present	0.01	0.01
Ratio of moles	1	1
∴ ratio of atoms	1	1
=> Formula	MgO	

Table 2 Finding the formula of magnesium oxide

QUESTIONS

1 How many atoms are there in (i) 52 g chromium (Cr = 52), (ii) 2.8 g nitrogen (N = 14), (iii) 0.36 g carbon (C = 12), (iv) 20 g bromine (Br = 80)?

2 What is the mass of (i) 3 moles of bromine (Br = 80), (ii) ¼ mole of calcium (Ca = 40), (iii) 0.1 mole of sodium (Na = 23)?

3 Methane in natural gas is found to contain 75% carbon and 25% hydrogen. Calculate the formula of methane using a method like that in Table 1.

4 What are the formulas of the following compounds?
(a) a compound in which 10.4 g chromium (Cr = 52) combines with 48 g bromine (Br = 80)
(b) a nitride of chromium in which 0.26 g chromium forms 0.33 g of chromium nitride (N = 14).

5 Plan an experiment to find the formula of black copper oxide. (Hint: the copper oxide can be reduced to copper by heating strongly in a current of natural gas.)
Describe
(a) what you would do,
(b) the weighings you would make,
(c) any precautions you would take,
(d) how you would calculate your result.

11 PARTICLES IN ELEMENTS

Look at the properties of iron and oxygen in Table 1. How do the properties of metals, like iron, differ from those of non-metals, like oxygen? We can use our ideas about particles in elements and in solids, liquids and gases to explain the properties of iron and oxygen. This will help you to understand the properties of metals and non-metals.

Property	Iron	Oxygen
State at room temperature	solid	gas
Melting point	high	very low
Boiling point	high	very low
Density	high	very low
Electrical conductivity	good	poor
Squashiness (compressibility)	very hard	easy to squash (very compressible)

Table 1 The properties of iron and oxygen

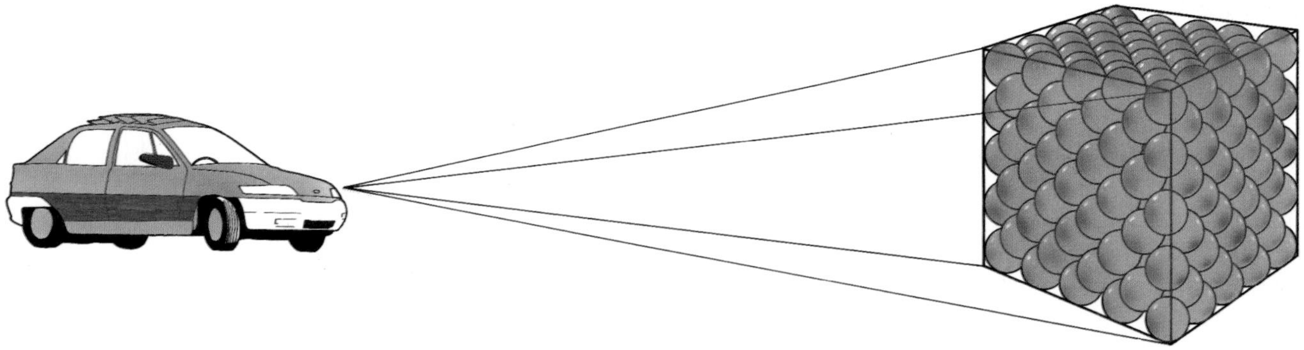

Figure 1

Iron and other metals

In iron and other metals, the particles are single atoms. These atoms are packed together very closely and in a regular pattern (Figure 1). The regular arrangement of particles builds up to give a solid with a regular shape called a **crystal**. The atoms are packed tight and there are strong forces between them. This gives the metal a high density and makes it hard and strong. The strong forces hold the atoms together very tightly and restrict their movement. This results in a high melting point and a high boiling point.

All atoms have a small positively-charged core or **nucleus** surrounded by tiny negative particles called **electrons**. In metals, the electrons are fairly free to move around and away from the nucleus. They can easily be attracted to the positive terminal of a battery and form an electric current. This is why metals conduct electricity. In non-metals the electrons cannot move so easily. So non-metals are usually poor conductors.

The strong forces between metal atoms allow iron to be used in bridges like the one above at Coalbrookedale

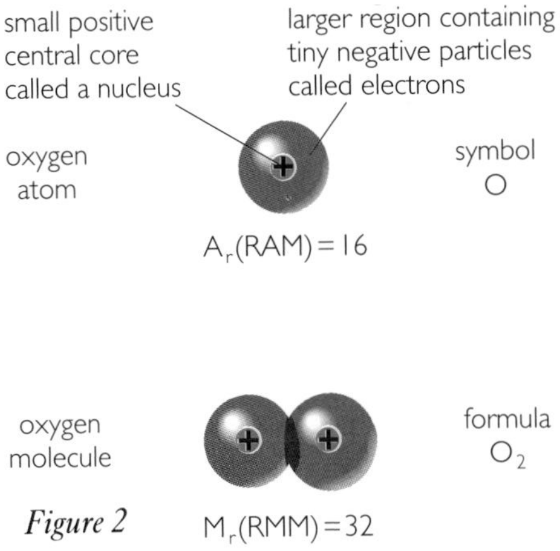

Figure 2

Oxygen and other non-metals

The relative atomic mass of oxygen is 16, but experiments show that particles of oxygen in the air have a relative mass of 32 (Figure 2). This shows that oxygen consists of molecules containing two atoms joined together (Figure 3). The formula of oxygen gas is therefore O_2. We say that O_2 has a **relative molecular mass (RMM)**, or a **relative formula mass**, of 32.

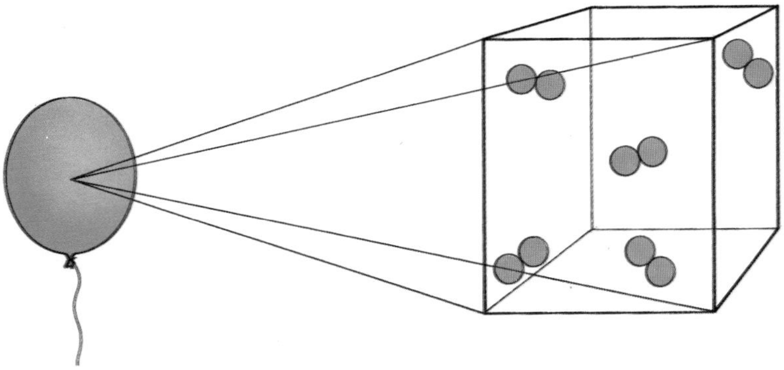

Figure 3

Most non-metals are like oxygen. They have a small number of atoms joined tightly together to form a molecule (Table 2). All the common gaseous elements have molecules containing two atoms. They are often described as **diatomic molecules**. There are strong forces between the two atoms in one molecule of oxygen, but very weak forces between one molecule and another. So, the O_2 molecules are easy to separate and oxygen has a very low melting point and boiling point. The molecules are also widely-spaced at room temperature so the density is very low. Since the particles are widely spaced, it is possible to squeeze them into a much smaller volume. This explains why gases can be easily compressed in bagpipes, in balloons, in tyres and in inflatable toys.

Element	Formula
Oxygen	O_2
Hydrogen	H_2
Nitrogen	N_2
Chlorine	Cl_2
Sulphur	S_8
Argon	Ar

Table 2 The formulas of some non-metals

QUESTIONS

1 The formula for nitrogen is N_2. What does this mean?
2 How many atoms are there in one molecule of sulphur?
3 Why do metals have a high density?
4 Why does sulphur melt at a much lower temperature than copper?
5 Suppose your best friend has missed the last few lessons. Write down how you would explain to him or her what we mean by (i) an atom, (ii) the difference between an atom and a molecule, (iii) symbols, (iv) the difference between symbols and formulas, (v) relative atomic mass, (vi) relative molecular mass.

12 PARTICLES IN REACTIONS

The original balance with which Dalton studied the amounts of substances which react together. Dalton suggested that atoms were simply rearranged in chemical reactions

Writing equations

So far, we have used word equations to show what happens when substances react. When magnesium reacts with oxygen, the product is magnesium oxide. We can summarise this in a word equation as:

magnesium + oxygen → magnesium oxide

It would be easier to use formulas instead of names when we write equations. We could write magnesium as Mg, oxygen as O_2 and magnesium oxide as MgO. This gives

$$Mg + O_2 \rightarrow MgO$$

But notice that this does not balance. There are two oxygen atoms in the O_2 molecule on the left and only one oxygen atom in MgO on the right. So, MgO on the right must be doubled to give

$$Mg + O_2 \rightarrow 2MgO$$

Unfortunately, the equation still does not balance. There are two Mg atoms on the right in 2MgO, but only one on the left in Mg. This can easily be corrected by writing 2Mg on the left, i.e.

$$2Mg + O_2 \rightarrow 2MgO$$

This example shows the three stages in writing an equation:

1 *Write a word equation for the reaction,*

e.g. hydrogen + oxygen → water

2 *Write formulas for the reactants and products,*

e.g. $H_2 + O_2 \rightarrow H_2O$

3 *Balance the equation by making the number of atoms of each element the same on both sides,*

e.g. $2H_2 + O_2 \rightarrow 2H_2O$

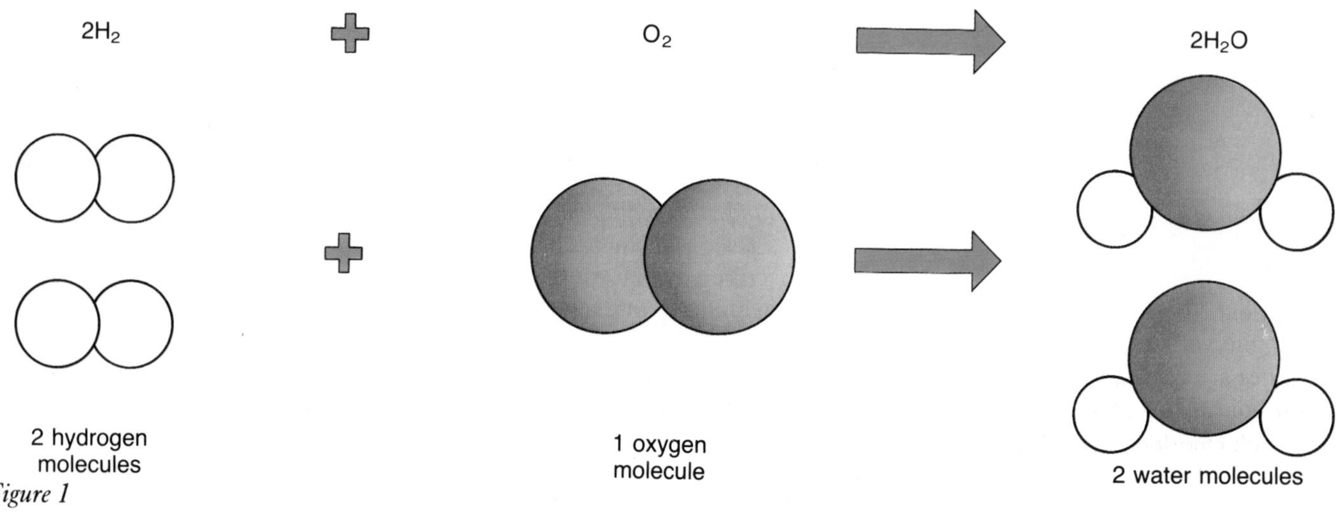

Figure 1

An equation is a summary of the starting substances and the products in a chemical reaction. Equations with formulas also help us to see how the atoms are rearranged in a chemical reaction. We can see this even better using models to represent the equation, as in Figure 1.

The elements hydrogen, oxygen, nitrogen, chlorine, bromine and iodine exist as diatomic molecules. Thus, they are written in equations as H_2, O_2, N_2, Cl_2, Br_2 and I_2. All other elements are shown as single atoms and represented by a monatomic symbol in equations; i.e. Mg for magnesium, Cu for copper, S for sulphur, etc. There is one other important point to remember in writing equations. *Never change a formula* to make an equation balance. For example, the formula of magnesium oxide is always MgO. Mg_2O and MgO_2 do not exist. Equations can only be balanced by putting a number *in front* of the whole formula, i.e. 2MgO or 3MgO.

The copper dome on Brighton Pavilion is green because the copper has reacted with oxygen in the air to form copper oxide and then this has reacted very slowly with water to form green copper hydroxide. Write word equations and then balanced formula equations for these two reactions

Extra information in equations

- **State symbols** show the state of a substance. The symbol (s) after a formula indicates the substance is a solid; (l) is used for liquid; (g) for gas and (aq) for an aqueous solution (i.e. a substance dissolved in water). For example,

zinc + sulphuric acid → zinc sulphate + hydrogen

$$Zn(s) + H_2SO_4(aq) \rightarrow ZnSO_4(aq) + H_2(g)$$

Reaction conditions. The conditions needed for a reaction can be shown above the arrow in the equation. For example,

hydrogen peroxide $\xrightarrow{MnO_2\ catalyst}$ water + oxygen

$$2H_2O_2 \xrightarrow{MnO_2\ catalyst} 2H_2O + O_2$$

calcium carbonate (limestone) $\xrightarrow{heat}$ calcium oxide (lime) + carbon dioxide

$$CaCO_3 \xrightarrow{heat} CaO + CO_2$$

QUESTIONS

1 Look at the photographs above. Answer the questions in the captions to the photos.

2 Write word equations, and then balanced formula equations, for the following reactions.
(a) aluminium with oxygen to give aluminium oxide (Al_2O_3)
(b) copper oxide with sulphuric acid to give copper sulphate ($CuSO_4$) and water
(c) nitrogen with hydrogen to give ammonia (NH_3)
(d) charcoal (carbon) burning in oxygen to give carbon dioxide
(e) iron with chlorine to give iron chloride ($FeCl_3$)

3 Sulphuric acid (H_2SO_4) can be manufactured from sulphur in three stages.
- Burn sulphur in air (oxygen) to form sulphur dioxide (SO_2).
- React sulphur dioxide with oxygen to form sulphur trioxide (SO_3).
- React sulphur trioxide with water to form sulphuric acid.

(a) How many atoms of the different elements are present in one molecule of (i) sulphuric acid, (ii) sulphur dioxide?
(b) Write word equations for the three stages in the manufacture of sulphuric acid.
(c) Write balanced formula equations for these three stages.

When natural gas burns, methane (CH_4) reacts with oxygen in the air to form carbon dioxide and water. Write a word equation and then a balanced formula equation for this reaction

13 FORMULAS AND EQUATIONS

The yellow substance in yellow 'no parking' lines is lead chromate, $PbCrO_4$. What does this formula tell us about lead chromate?

Using formulas

Formulas are useful in several ways.

1 They are shorter and quicker to write than the actual names of substances.
2 They enable us to compare the relative masses of different molecules. For example, using relative atomic masses for hydrogen, carbon and oxygen of 1, 12 and 16 respectively:
the relative molecular mass (relative formula mass) of water,
$H_2O = 1 + 1 + 16 = 18$
the relative formula mass of carbon dioxide, $CO_2 = 12 + 16 + 16 = 44$
3 They help us to calculate the mass of 1 mole of different substances. For example:
the mass of 1 mole of water, $H_2O = 18$ g
the mass of 1 mole of calcium carbonate, $CaCO_3$
$= 40.0 + 12.0 + 16.0 + 16.0 + 16.0 = 100$ g
4 They help us to calculate the masses of elements which combine to form a compound. For example, sulphuric acid, H_2SO_4, is manufactured from sulphur, S.
Thus, 1 mole of S (S = 32) gives 1 mole of H_2SO_4.
1 mole of S is 32 g, and
1 mole of H_2SO_4 is $(2 \times 1) + 32 + (4 \times 16) = 98$ g.
So 32 g of sulphur produce 98 g of sulphuric acid.

This photo shows one mole of common salt (sodium chloride NaCl), which has a mass of $23 + 35.5 = 58.5$ g

This sulphur is being stored after mining

Using equations

Look at the equation for the reaction of sodium with chlorine.

$$2Na + Cl_2 \rightarrow 2NaCl$$

This equation tells us that:

- sodium (Na) reacts with chlorine (Cl_2) to form sodium chloride (NaCl),
- 2 atoms of Na react with 1 molecule of Cl_2 to give 2 particles of NaCl,

- 2 moles of Na react with 1 mole of Cl_2 to give 2 moles of NaCl,
- using relative atomic masses (Na = 23 and Cl = 35.5); 2 × 23 g Na react with 2 × 35.5 g Cl to give 2 × 58.5 g NaCl.

This example shows that equations can tell us

1. the reactants and the products
2. the formulas of these substances
3. the numbers of particles (or moles) of the reactants and products
4. the masses of reactants and products

It is important in industry to know the amounts of reactants and products. Industrial chemists need to know how much product they can make from a given amount of starting material. For example, using an equation and relative atomic masses, we can calculate how much lime (calcium oxide) can be obtained from 1 kg of pure limestone (calcium carbonate).

This lime kiln converts limestone into lime

$$\text{calcium carbonate} \rightarrow \text{calcium oxide} + \text{carbon dioxide}$$

$$CaCO_3 \rightarrow CaO + CO_2$$

$$\therefore \text{1 mole } CaCO_3 \rightarrow \text{1 mole } CaO$$

$$(40 + 12 + 3 \times 16)\text{ g } CaCO_3 \rightarrow (40 + 16)\text{ g } CaO$$

$$100\text{ g } CaCO_3 \rightarrow 56\text{ g } CaO$$

$$\therefore 1000\text{ g } CaCO_3 \rightarrow 560\text{ g } CaO$$

So 1 kg of pure limestone produces 0.56 kg of lime.

Using equations, it is also possible to calculate the volumes of gases which react. Chemists have found that *1 mole of any gas occupies 24 dm^3 at room temperature (20°C) and atmospheric pressure.* (At *standard* temperature and pressure, or s.t.p., (0°C and 1 atmosphere), 1 mole of a gas occupies 22.4 dm^3.)

For example, ammonia (NH_3) is manufactured from nitrogen (N_2) and hydrogen (H_2).

$$\text{nitrogen} + \text{hydrogen} \rightarrow \text{ammonia}$$

$$N_2 + 3H_2 \rightarrow 2NH_3$$

$$\therefore \text{1 mole } N_2 + \text{3 moles } H_2 \rightarrow \text{2 moles } NH_3$$

i.e. $24\text{ dm}^3 \text{ nitrogen} + 3 \times 24\text{ dm}^3 \text{ hydrogen} \rightarrow 2 \times 24\text{ dm}^3 \text{ ammonia}$

So $1\text{ dm}^3 \text{ nitrogen} + 3\text{ dm}^3 \text{ hydrogen} \rightarrow 2\text{ dm}^3 \text{ of ammonia}$

QUESTIONS

1 60 g of a metal M (M = 60) combine with 24 g of oxygen (O = 16).
(a) How many moles of O react with one mole of M?
(b) What is the formula of the oxide of M?

2 The fertiliser 'nitram' (ammonium nitrate) has the formula NH_4NO_3.
(a) What are the masses of nitrogen, hydrogen and oxygen in 1 mole of nitram? (N = 14, H = 1, O = 16)
(b) What are the percentages of nitrogen, hydrogen and oxygen in nitram?

3 Nitram is made from ammonia and nitric acid.

$$\underset{\text{ammonia}}{NH_3} + \underset{\text{nitric acid}}{HNO_3} \rightarrow \underset{\text{ammonium nitrate}}{NH_4NO_3}$$

(a) What does this equation tell you?
(b) What mass of NH_3 reacts with HNO_3 to give 1 mole of nitram? (N = 14, H = 1, O = 16)
(c) What mass of NH_3 reacts with HNO_3 to give 1 kg of nitram?
(d) What volume of NH_3 at s.t.p. produces 1 mole of nitram?

SECTION C: THINGS TO DO

Activities

Why do people take soluble aspirins? Why do the tablets fizz when added to water?

1 Soluble aspirin

When a soluble aspirin tablet is added to water, the mass of the solution after the tablet has dissolved is less than the mass of the tablet plus the mass of the water before mixing.

1 Why do you think there is a loss in mass when the soluble aspirin is added to water?

2 Describe an experiment which you could carry out to test your ideas for the loss in mass. (Draw diagrams if it will help.) You should try to use materials and apparatus which are readily available in the lab.

3 What do you think happens to particles in the soluble aspirin when it is added to water?

4 Soluble aspirins dissolve faster in warm water than in cold water. Why is this?

2 Extracting tin from tinstone

The Production Manager at a tin smelter has to meet certain production targets. The smelter must produce 595 tonnes of tin each month. In order to do this, the manager must calculate how much purified tinstone (tin oxide) he must buy to produce 595 tonnes of tin. The basis for his calculation is the equation

$$\begin{array}{ccccccc} \text{tin oxide} & + & \text{coke (carbon)} & \rightarrow & \text{tin} & + & \text{carbon monoxide} \\ SnO_2 & + & 2C & \rightarrow & Sn & + & 2CO \end{array}$$

SnO_2 means that tin oxide contains 2 atoms of oxygen for every 1 atom of tin.

CO means that carbon monoxide contains 1 atom of carbon for every 1 atom of oxygen.

Using relative atomic masses, the manager can now calculate how much tin can be obtained from a certain amount of tin oxide (SnO_2). The relative formula masses of SnO_2 and CO can be calculated by adding together the correct number of relative atomic masses for the atoms of the elements in each compound.

Relative atomic mass of tin	= 119
Relative atomic mass of oxygen	= 16
Relative atomic mass of carbon	= 12
∴ the relative formula mass of SnO_2	$= 119 + (2 \times 16) = 151$

1 What is the relative formula mass of CO?

When we put relative masses below the formulas in the equation, we get:

$$\begin{array}{lccccccc} & SnO_2 & + & 2C & \rightarrow & Sn & + & 2CO \\ \text{Relative masses:} & 151 & + & ? & \rightarrow & 119 & + & ? \end{array}$$

This means that 151 tonnes of tin oxide (tinstone) can produce 119 tonnes of tin.

If the Production Manager had a monthly production target of 119 tonnes of tin, then 151 tonnes of tinstone are required.

2 What mass of tinstone is needed to meet the production target of 595 tonnes of tin per month?

3 What mass of coke (carbon) is needed each month?

4 The smelter normally operates all night and day. It produces about 1 tonne of tin per hour. Is the production target possible? Explain your answer.

5 In 1990, tin from the smelter was sold at £5000 per tonne. Suppose the smelter produces 600 tonnes of tin in December. What is the value of this tin?

6 What concerns will the Production Manager have regarding the tin smelter and environmental issues?

The disused tin mine at Wheal Coates, near St Agnes in Cornwall. Why did the mines have tall chimneys? Why is tinstone no longer mined in Cornwall?

3 Finding the formula of red copper oxide

Jason and Meera had just completed their experiment with black copper oxide. Their results showed that its formula was CuO. After the experiment, Jason read about a second oxide of copper called red copper oxide. So he and Meera decided to investigate its formula.

They took a weighed amount of red copper oxide and reduced it to copper. In order to get a more reliable result, they carried out the experiment five times, starting with different amounts of red copper oxide. Here are their results.

Expt Number	Mass of red copper oxide taken /g	Mass of copper in the oxide /g
1	1.43	1.27
2	2.10	1.87
3	2.86	2.54
4	3.55	3.15
5	4.29	3.81

Table 1 The results of experiments to investigate the formula of red copper oxide

Recording the results

1 Start a spreadsheet program on a computer and open up a new spreadsheet for your results.

2 Enter the experiment numbers and the masses of copper oxide and copper as in the diagram below.

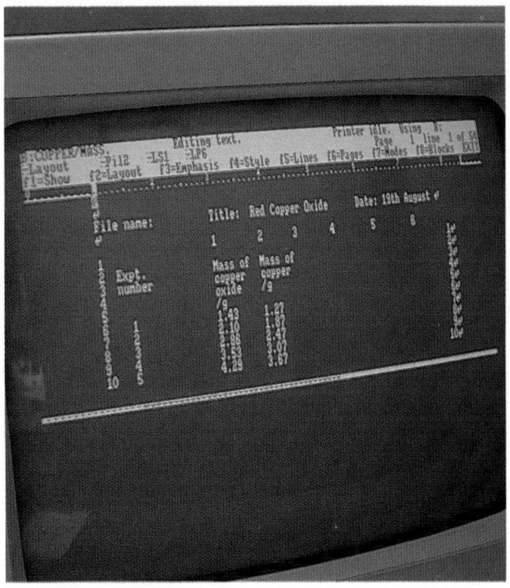

3 Enter a formula in column 4 to work out the mass of oxygen in the red copper oxide taken. (If you cannot do this, ask your teacher to help you type it in.)

4 Enter a formula in column 5 to find the number of moles of copper in the oxide (Cu = 63.5).

5 Enter a formula in column 6 to find the number of moles of oxygen in the oxide (O = 16).

6 From the spreadsheet, plot a line graph of moles of copper (y-axis) against moles of oxygen (x-axis). (If you cannot pick graphs directly from the spreadsheet, draw the graphs by hand.)

7 Look at your graph.
(a) What is the average value for $\frac{\text{moles of copper}}{\text{moles of oxygen}}$ in red copper oxide?

(b) How many moles of copper combine with one mole of copper in red copper oxide?
(c) What is the formula for red copper oxide?

8 If possible, print a copy of your spreadsheet and your graph.

Study Questions

1 The kinetic theory of matter states that all substances contain particles which are moving. Use this theory to explain the following:
(a) When a balloon is taken out of doors on a cold day, it gets smaller.
(b) Water boils at less than 80°C on the top of Mount Everest.
(c) Snow and ice sometimes disappear on a cold winter's day without leaving any trace of water.
(d) Solids, such as salt and sugar, can dissolve in water.
(e) Dogs can be trained to find hidden drugs such as cannabis.

2 (a) What name is used for the smallest particle of a compound which can freely exist?
(b) What name is used for the smallest particle of an element which has the properties of that element?
(c) Which particles in an atom are used in chemical bonding?
(d) Sometimes we talk about 'splitting the atom' to create a new element. What part of the atom is actually 'split'?
(e) If the formula for water is H_2O, how many atoms are there in one molecule of water?
(f) Write the formula for six molecules of water.

3 In the boxes below, ● and ○ represent different atoms.

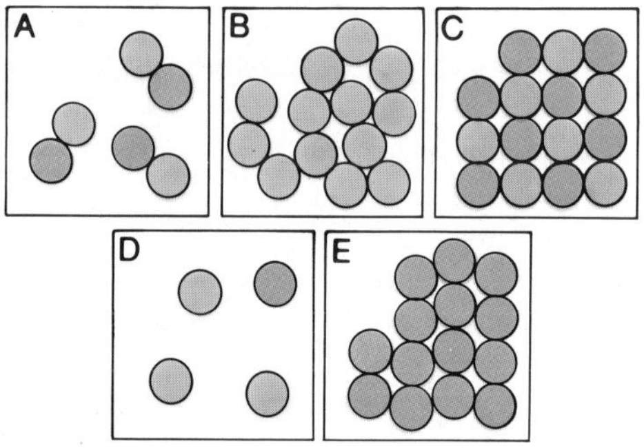

Which box contains
(a) a close packed metal,
(b) a solid compound,
(c) a liquid,
(d) a mixture,
(e) a diatomic molecule?

4 What atoms and molecules do the following contain? Air, acetic acid (CH_3COOH) in vinegar, sulphuric acid (H_2SO_4), limestone ($CaCO_3$).

5 Copper carbonate decomposes, when it is heated, into copper oxide and carbon dioxide according to the equation

$$CuCO_3 \rightarrow CuO + CO_2$$

An empty test tube was first weighed. Some copper carbonate was put in it, and it was weighed again. The diagrams below show the scale of the top-pan balance for these weighings.

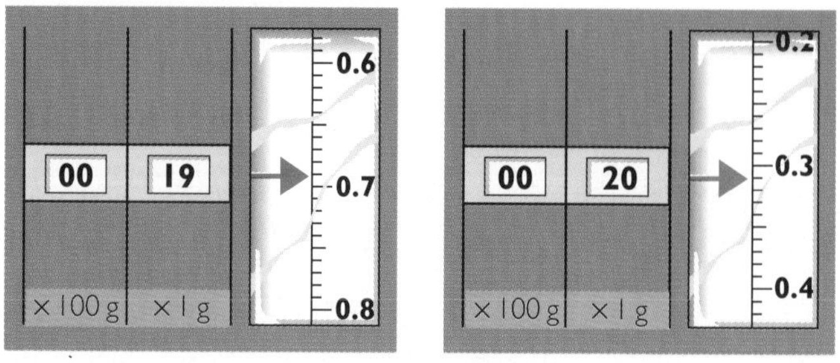

(a) Complete the results table below.

	Mass/g
Mass of empty test tube	——g
Mass of test tube + copper carbonate	——g
Mass of copper carbonate taken	——g

(b) Calculate the mass of 1 mole of
(i) copper carbonate,
(ii) copper oxide (Cu = 64, C = 12, O = 16).
(c) Calculate the expected mass of copper oxide remaining in the experiment in which the test tube was heated.
(d) The test tube was allowed to cool before it was reweighed. The mass remaining was found to be 0.45 g. Compare this result with the expected result obtained in part (c) and account for any difference.
(e) Why was the test tube allowed to cool before it was weighed?
(f) What would you do to ensure that all the copper carbonate had decomposed?

6 A solid compound contains 39% potassium, 1% hydrogen, 12% carbon and 48% oxygen, by weight.
(a) What is the ratio of the weights of K:H:C:O in the compound?
(b) What is the ratio of moles of K:H:C:O in the compound?
(c) What is the formula of the solid?

7 Tinstone, SnO_2, is an important mineral from which tin can be obtained.
(a) How much does 1 mole of tinstone weigh?
(b) How much tin does 1 mole of tinstone contain?
(c) What is the percentage of tin in tinstone?
(d) How much tin is there in 5000 kg of tinstone?
(e) Where are the world's major deposits of tinstone found? (You will need to refer to an encyclopaedia.)
(f) When tinstone is converted into tin, is the tinstone *oxidised*, *reduced*, *synthesised*, or *decomposed*?

8 (a) Rust is hydrated iron oxide. Its formula can be written as $Fe_xO_y.\ ZH_2O$. Describe an experiment you could do to find the values of x, y and z.
(b) A sample of rust contained 56.0 g of iron, 24.0 g of oxygen and 27.0 g of water. What is the formula of this rust?

SECTION D

Electricity and electrolysis

This aerial photo shows a hydroelectric power station most of which is built within the mountainside. Power stations like this generate electricity from a renewable resource, and can be hidden underground so as not to spoil the countryside

1 ELECTRICITY IN EVERYDAY LIFE

A city at night. How could we survive without electricity?

Look at all the electrical appliances in this kitchen. It is difficult to imagine life without electrical appliances

It is hard to imagine life without electricity. Every day we depend on electricity for cooking, for lighting and for heating. At the flick of a switch, we use electric fires, electric kettles and dozens of other electrical gadgets. All these electrical appliances use mains electricity. The electricity is generated in power stations from coal, oil or nuclear fuel. Heat from the fuel is used to boil water. The steam produced drives turbines and generates electricity. In this way, energy in the fuel is converted into electrical energy.

In addition to the many appliances which use mains electricity, there are others like torches, radios and calculators that use electrical energy from cells and batteries.

Electricity can also be used to manufacture some important chemicals. For example, salt (sodium chloride) cannot be decomposed into sodium and chlorine using heat, but it can be decomposed using electricity. Sodium and chlorine are manufactured by passing electricity through molten sodium chloride.

$$\text{sodium chloride} \xrightarrow{\textit{electricity}} \text{sodium} + \text{chlorine}$$

These uses of electricity show why it is so useful to our society.

- It can be used to transfer energy easily from one place to another.
- It can be converted into other forms of energy and used to warm a room, light a torch or cook a meal.

Electric charges

Comb your hair quickly with a plastic comb and then use the comb to pick up tiny bits of paper. Why does this happen?

Atoms are composed of three particles – **protons**, **neutrons** and **electrons**. The centre of the atom, called the **nucleus**, contains protons and neutrons. Protons are positive but neutrons have no charge. Electrons occupy the outer parts of the atom and move around the nucleus. The negative charge on one electron just balances the positive charge on one proton so atoms have equal numbers of protons and electrons. For example, hydrogen atoms have 1 proton and 1 electron. Carbon atoms have 6 protons, 6 electrons and 6 neutrons. Aluminium atoms have 13 protons, 13 electrons and 14 neutrons (Figure 1).

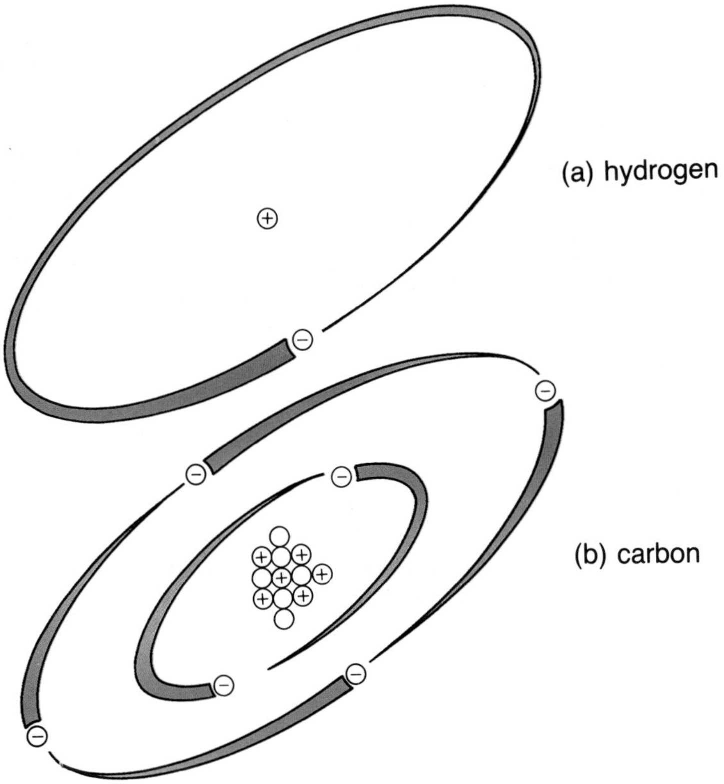

Figure 1 Protons, neutrons and electrons in a hydrogen atom and a carbon atom (⊕ = proton, ○ = neutron and ⊖ = electron)

When you comb your hair quickly with a plastic comb, the comb pulls electrons off the atoms in your hair. The comb will then have more electrons than protons so it has a negative charge overall. Your hair will have fewer electrons than protons and so it has a positive charge overall.

When you bring the charged comb close to tiny pieces of paper, the negative charge on the comb repels electrons from the area of the paper nearest to it (Figure 2(a)). This part of the paper therefore becomes positive and it is attracted to the comb because *unlike charges attract*. If the paper is light enough, it can be picked up (Figure 2(b)).

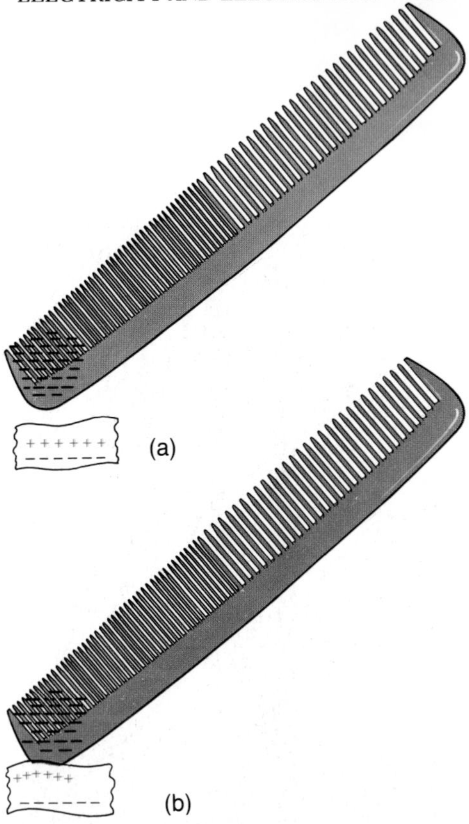

Figure 2

The plastic comb has just been rubbed on a woollen jumper. What is making the water bend?

QUESTIONS

1 Electricity is used more widely than gas for our energy supplies. Why is this?
2 Talk to an elderly person about what it was like in their home before there were so many electrical gadgets. Prepare a report of your conversation for the rest of your class.
3 Lithium atoms have 3 protons, 3 electrons and 4 neutrons. Draw a picture of a lithium atom similar to that for a carbon atom in Figure 1.
4 Look at the photo above. Why does the water bend?

2 ELECTRIC CURRENTS

Why are plastic and glass materials used to hold the cables to pylons in the Grid System?

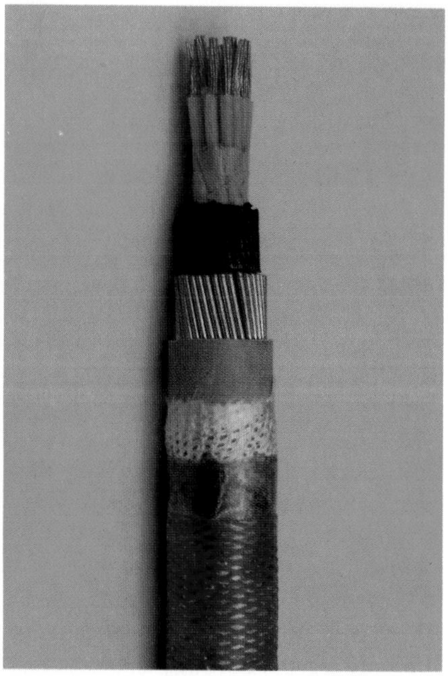

The cut-away end of this electrical cable shows the layers of different material which go to make it up. Bundles of copper wire are at the centre. Copper is used because it is such a good conductor. The layers of material around the copper provide insulation and protection

Materials, such as plastic combs, which hold their charge on rubbing and do not allow electrons to pass through them are called **insulators**. Plastics like polythene, cellulose acetate and PVC are used to insulate electrical wires and cables. Materials that allow electrons to pass through them are called **conductors** (Table 1).

Conductors			Insulators
Good	**Moderate**	**Poor**	
metals, e.g. copper aluminium iron	carbon silicon germanium	water	rubber air plastics, e.g. polythene PVC

Table 1 Some conductors and insulators

Metals, in particular, are good conductors. Thus, copper, iron and aluminium, which are easily made into wire, are used for fuses, wires and cables in electrical machinery. Copper is used more than any other metal in electrical wires and cables because electrons move through it easily – it is a very good conductor. When an electric current flows through the wire, electrons in the copper atoms are attracted towards the positive terminal of the battery. At the same time, extra electrons are repelled into the copper wire from the negative terminal (Figure 1). *An electric current is simply a flow of electrons*. Electrons flow through the metal rather like water flows through a pipe or traffic moves along a road.

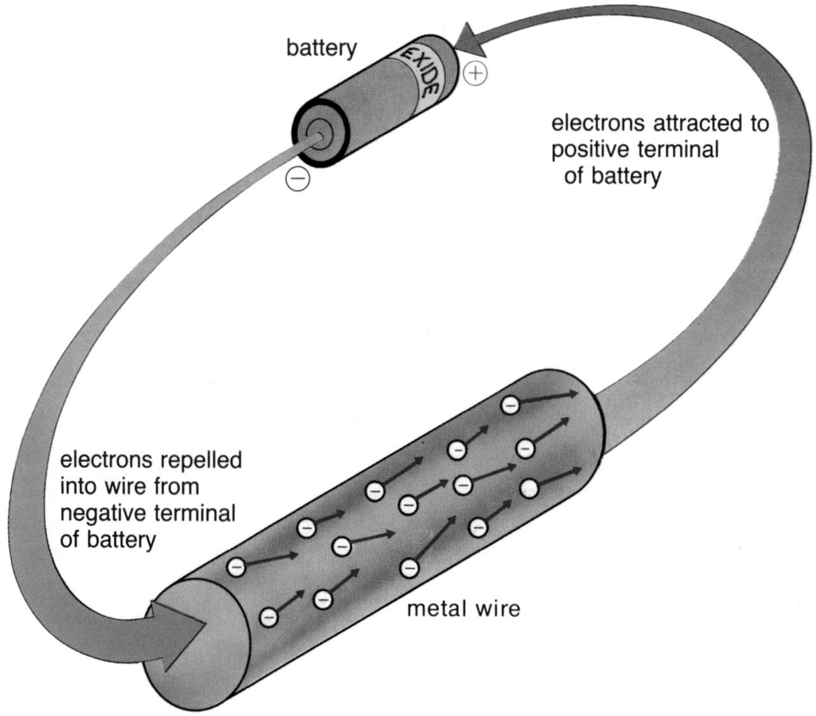

Figure 1

Measuring electricity

We could measure the amount of electricity (electric charge) that has flowed along a wire by counting the number of electrons which pass a certain point. The charge on one electron is, however, much too small to be used as a practical unit in measuring the quantity of electricity. The practical unit normally used is the **coulomb** (C), which is about six million, million, million (6×10^{18}) electrons.

If one coulomb of charge passes along a wire in one second, then the rate of charge flow (i.e. the electric current) is 1 coulomb per second or 1 **ampere** (A). If 3 coulombs pass along the wire in 2 seconds, then the current is $\frac{3}{2}$ coulombs per second, or $\frac{3}{2}$ A. If Q coulombs flow along a wire in t seconds, the electric current (I) is given by,

$$I = \frac{Q}{t}$$

This equation can be rearranged to give

$$Q = I \times t$$

i.e. charge in coulombs = current in amps × time in seconds

$$\therefore \begin{aligned} 1 \text{ A for } 1 \text{ sec} &= 1 \times 1 = 1 \text{ C} \\ 2 \text{ A for } 1 \text{ sec} &= 2 \times 1 = 2 \text{ C} \\ 2 \text{ A for } 2 \text{ sec} &= 2 \times 2 = 4 \text{ C} \end{aligned}$$

The equation charge = current × time can be compared to the flow of water along a pipe since:

amount of water passed = rate of flow of water (i.e. current) × time

QUESTIONS

1 What is (i) an electric current, (ii) a conductor, (iii) an insulator?

2 (a) Name 4 elements which conduct electricity when solid.
(b) Name 4 elements which conduct electricity when liquid.
(c) Name 4 elements which do not conduct electricity when solid.

3 Design an experiment to compare the conduction of electricity by thin copper wire and thick copper wire.
(a) Draw a diagram of the apparatus you would use.
(c) Say what you would do.
(c) Say what measurements you would make.
(d) Say how you would compare the two wires.

4 Look at the photo above. This shows a silicon chip. Silicon chips are built up using transistors. Transistors are made from semiconductors. Find out about semiconductors, transistors and silicon chips. What are they? How do you think they work? What are they used for?

5 The current in a small torch bulb is about 0.25 A. How much electricity flows if the torch is used for 15 minutes?

6 The current in a car headlamp is 4 A.
(a) What is the current, in coulombs per second?
(b) What is the current, in electrons per second? (Assume 1 coulomb = 6×10^{18} electrons.)
(c) How much electricity (electric charge) flows if the head lamp is used for 1 hour?

3 WHICH SUBSTANCES CONDUCT ELECTRICITY?

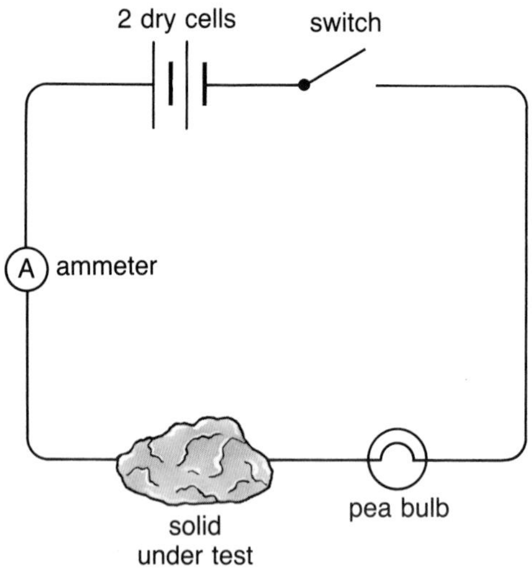

Figure 1

Which solids conduct?

You could use the apparatus in Figure 1 to test whether a solid conducts electricity. If the solid conducts, what happens when the switch is closed?

Experiments show that *the only common solids which conduct electricity well are metals and graphite.* When metals and graphite conduct electricity, electrons flow through the material, but there is no chemical reaction. No new substances form and there is no change in weight. No solid *compounds* conduct electricity.

Which liquids conduct?

Pure water does not conduct electricity. But water containing a little sulphuric acid *will* conduct electricity. Unlike metals, the water changes when it conducts. It is decomposed into hydrogen and oxygen. Electricity is a form of energy, like heat. We can use it to boil water, to cook food and to cause chemical reactions. Energy in the electric current causes the decomposition of the water. This can be summarised in a word equation as:

$$\text{water} \xrightarrow[\textit{energy}]{\textit{electrical}} \text{hydrogen} + \text{oxygen}$$

The decomposition of a substance such as water, by electricity, is called **electrolysis**. The compound which is decomposed is called an **electrolyte** and we say that it has been **electrolysed**.

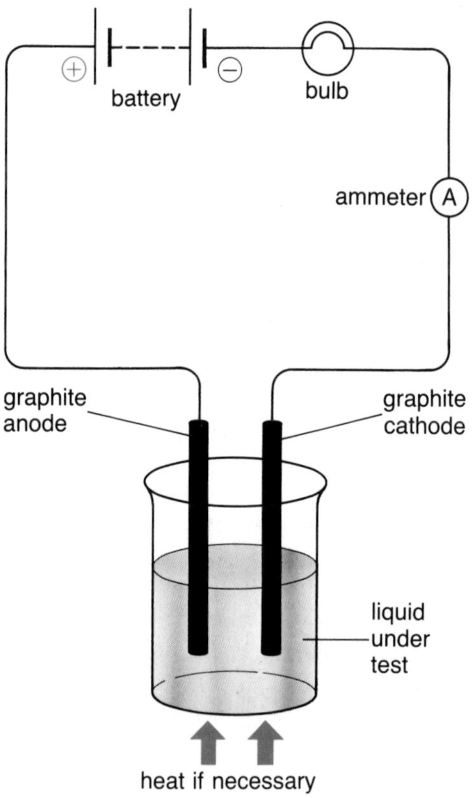

Figure 2

Pure liquids	Does the liquid conduct?	Aqueous solutions	Does the solution conduct?
Bromine	no	alcohol (C_2H_6O)	no
Mercury	yes	sugar ($C_{12}H_{22}O_{11}$)	no
Molten sulphur	no	sulphuric acid (H_2SO_4)	yes
Molten zinc	yes	acetic acid ($C_2H_4O_2$)	yes
Alcohol (C_2H_6O)	no	copper sulphate ($CuSO_4$)	yes
Tetrachloromethane (CCl_4)	no	sodium chloride ($NaCl$)	yes
Molten sodium chloride ($NaCl$)	yes	potassium iodide (KI)	yes
Molten lead bromide ($PbBr_2$)	yes		

Table 1 Testing to see which liquids and aqueous solutions conduct

Figure 2 shows how we can test the conductivity of liquids. The terminals through which the current enters and leaves the electrolyte are called **electrodes**. The electrode connected to the positive terminal of the battery is positive itself. This is called the **anode**. The electrode connected to the negative terminal of the battery is negative itself. This is called the **cathode**. Table 1 shows the results of tests on various liquids and aqueous solutions. Look at the results in Table 1.

1 Do the liquid metals conduct electricity?
2 Do the liquid non-metals conduct electricity?
3 Do the compounds containing only non-metals (non-metal compounds) conduct (i) when liquid, (ii) in aqueous solution?
4 Do the compounds containing both metals and non-metals (metal/non-metal compounds) conduct electricity (i) when liquid, (ii) in aqueous solution?

The answers to these questions and the results of the experiment are summarised in Table 2.

	Elements		**Compounds**	
Substance	**Metals and graphite**	**Non-metals except graphite**	**Metal/ non-metal**	**Non-metal**
Examples	Fe, Zn	Br_2, S	NaCl, $CuSO_4$	C_2H_6O, CCl_4
Solid	yes	no	no	no
Liquid	yes	no	yes	no
Aqueous solution	—	—	yes (and acids)	no (except acids)

Table 2 The conduction of electricity by elements and compounds

Notice the following points from these results:

- Metal/non-metal compounds conduct electricity when they are molten (liquid) and as aqueous solutions. These compounds are decomposed during electrolysis.
- Non-metal compounds do *not* conduct in the liquid state or in aqueous solution (except aqueous solutions of acids).

QUESTIONS

1 Explain the following words clearly. *electrolysis, electrolyte, electrode, anode, cathode*

2 *Calcium, carbon disulphide, copper sulphate solution, lead, carbon, water, methane (natural gas), phosphorus, dilute sulphuric acid*
From this list name (i) two metals, (ii) two non-metals, (iii) two electrolytes, (iv) two pure liquids at 20°C, (v) two elements which conduct, (vi) two compounds that are gases at 110°C, (vii) three compounds that are non-electrolytes.

3 Michael Faraday was one of the first scientists to investigate electrolysis. Find out about his life and work. Prepare a short talk about Faraday for the rest of your class. If possible, show drawings, diagrams and photos in your talk.

Michael Faraday – one of the first scientists to investigate electrolysis

Aluminium is manufactured by electrolysis from aluminium oxide. In this photo the workman is tapping molten aluminium from the container, or cell, where electrolysis occurs

4 INVESTIGATING ELECTROLYSIS

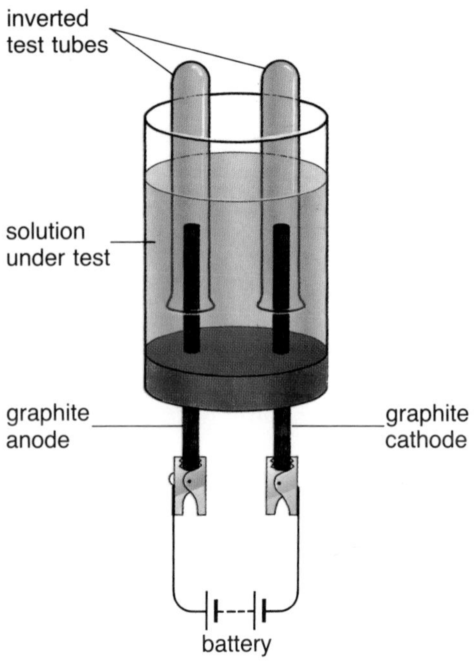

Figure 1

The products of electrolysis

When compounds are electrolysed, new substances are produced at the electrodes. For example, when electricity is passed through molten sodium chloride, pale green chlorine gas comes off at the anode and sodium forms at the cathode. When copper sulphate solution is electrolysed using the apparatus in Figure 1, a pink deposit of copper appears on the cathode. Bubbles of gas stream off the anode and collect in the inverted test tube. This gas relights a glowing splint. This shows that it is oxygen.

Table 1 shows the products formed at each electrode when various liquids and aqueous solutions are electrolysed. Remember that water in the aqueous solutions may be electrolysed.

Substance electrolysed	Product at anode	Product at cathode
Molten lead bromide	brown fumes of bromine	lead
Molten sodium chloride	pale green chlorine gas	sodium
Aqueous potassium iodide	iodine which colours the solution brown	hydrogen
Aqueous copper sulphate	oxygen	copper (deposited on the cathode)
Hydrochloric acid	chlorine	hydrogen
Aqueous zinc bromide	bromine which colours the solution brown	zinc (deposited on the cathode)
Aqueous copper chloride	chlorine	copper (deposited on the cathode)

Table 1 The products formed at the electrodes when some liquids and aqueous solutions are electrolysed

1 Which elements are produced at the anode?
2 Which elements are produced at the cathode?

The compounds are decomposed by electrical energy in the electric current. An element is produced at each electrode. This is very different from the conduction of electricity by metals, which are not decomposed during conduction. The first two electrolyses in the table can be summarised as:

$$\text{lead bromide} \xrightarrow[\textit{energy}]{\textit{electrical}} \text{lead} + \text{bromine}$$

$$\text{sodium chloride} \xrightarrow[\textit{energy}]{\textit{electrical}} \text{sodium} + \text{chlorine}$$

When metal/non-metal compounds and acids conduct electricity

- metals or hydrogen are formed at the cathode
- non-metals (except hydrogen) are formed at the anode

Explaining electrolysis in terms of ions

Sodium and chlorine are manufactured by electrolysis of molten sodium chloride. When molten sodium chloride is electrolysed, sodium is produced at the cathode and chlorine is produced at the anode. Sodium particles in the electrolyte have been attracted to the negative cathode, so they are probably positively charged. Scientists write these positive sodium particles in sodium chloride as Na^+. At the same time, chlorine is produced at the positive anode. So chlorine particles in the electrolyte must be negatively charged. These negatively charged chloride particles in sodium chloride are written as Cl^-. Charged particles, like Na^+ and Cl^-, which move to the electrodes during electrolysis are called **ions**. During electrolysis, Na^+ ions near the cathode combine with negative electrons on the cathode forming neutral sodium atoms (Figure 2).

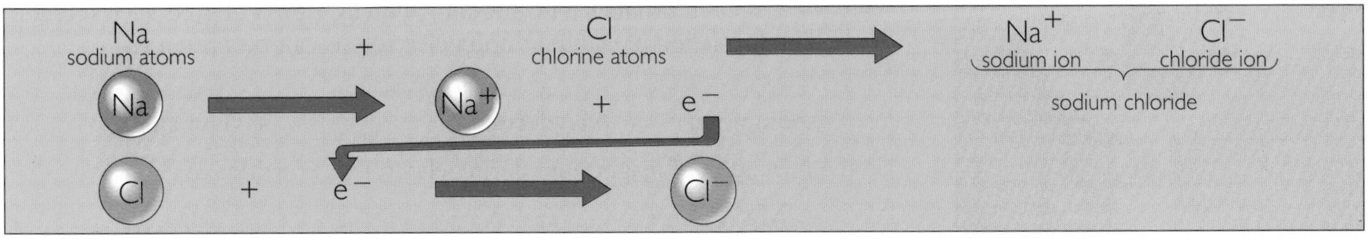

Figure 2
The electrolysis of other molten and aqueous substances can also be explained in terms of ions

Na^+	+	e^-	$\longrightarrow$	Na
sodium ion in sodium chloride electrolyte		electron on cathode from battery		sodium atom in metal

At the anode, Cl^- ions lose an electron to the positive anode and form neutral chlorine atoms (Figure 2).

Cl^-	$\longrightarrow$	e^-	+	Cl
chloride ion in electrolyte		electron given to anode		chlorine atom

The chlorine atoms then join up in pairs to form molecules of chlorine gas, Cl_2.

Cl	+	Cl	$\longrightarrow$	Cl_2
chlorine atom		chlorine atom		chlorine molecule

QUESTIONS

Write *true* or *false* next to each answer in questions **1** to **3**.

1 When pure lead bromide is electrolysed,
(a) it must be molten.
(b) lead forms at the anode.
(c) decomposition occurs.
(d) a brown gas forms at the anode.
(e) the process is exothermic.

2 The following substances are electrolytes:
(a) copper sulphate solution
(b) sugar solution
(c) dilute sulphuric acid
(d) copper
(e) molten wax
(f) liquid sulphur

3 When potassium chloride solution is electrolysed, the products include
(a) chlorine at the anode.
(b) hydrogen at the cathode.
(c) oxygen at the cathode.
(d) potassium at the cathode.
(e) chlorine at the cathode.
(f) potassium at the anode.

4 Plan an experiment to coat a steel paper clip with copper.

5 CHARGES ON IONS

The electrolysis equipment in which Faraday studied the charges on ions. Where are the electrodes? What are the vertical tubes for?

During electrolysis,

- metals and hydrogen are produced at the cathode,
- non-metals (except hydrogen) are produced at the anode.

The negative cathode attracts positive ions and the positive anode attracts negative charges. So we can deduce that:

- Metals and hydrogen have positive ions. These ions are called **cations** because they are attracted to the cathode.
- Non-metals (except hydrogen) have negative ions. These ions are called **anions** because they are attracted to the anode.

Electrolysis experiments show that it requires twice as much electricity (twice as many electrons) to produce a magnesium atom as it does to produce a sodium atom. The formation of sodium during electrolysis can be written as

Na^+	+	e^-	$\longrightarrow$	Na
sodium ion		one electron		sodium atom

So the formation of magnesium during electrolysis is written as

Mg^{2+}	+	$2e^-$	$\longrightarrow$	Mg
magnesium ion		two electrons		magnesium atom

In this way we can build up a list of ions with their charges, like those in Table 1. Notice that copper can form two ions, Cu^+ and Cu^{2+}. We show this in the names of its compounds by using the names copper(I), pronounced 'copper one', and copper(II), pronounced 'copper two'. Thus copper forms two oxides, two chlorides, two sulphates, etc. The correct names for its two oxides are copper(I) oxide which is red, and copper(II) oxide which is black. Most of the common copper compounds are copper(II) compounds. These include copper(II) oxide and blue copper(II) sulphate. Iron can also form two different ions, Fe^{2+} and Fe^{3+}, and we use the names iron(II) and iron(III) for their respective compounds.

Copper has two oxides and two chlorides

Most metal ions have a charge of 2+. All the *common* metal ions without a charge of 2+ are shown in the table. These are Ag^+, Na^+ and K^+ with a charge of 1+ (to remember this say 'AgNaK') and Cr^{3+}, Al^{3+} and Fe^{3+} with a charge of 3+ (to remember this say 'CrAlFe').

Notice also in the table that some negative ions are made from a group of atoms. For example, nitrate, NO_3^-, contains one nitrogen atom and three oxygen atoms. Groups of atoms like this are called **radicals**.

Positive ions (cations)		Negative ions (anions)	
Hydrogen	H^+	Chloride	Cl^-
Sodium	Na^+	Bromide	Br^-
Potassium	K^+	Iodide	I^-
Silver	Ag^+	Nitrate	NO_3^-
Copper(I)	Cu^+	Hydroxide	OH^-
Copper (II)	Cu^{2+}	Oxide	O^{2-}
Magnesium	Mg^{2+}	Carbonate	CO_3^{2-}
Calcium	Ca^{2+}	Sulphide	S^{2-}
Zinc	Zn^{2+}	Sulphate	SO_4^{2-}
Iron(II)	Fe^{2+}	Sulphite	SO_3^{2-}
Iron(III)	Fe^{3+}		
Aluminium	Al^{3+}		
Chromium	Cr^{3+}		

Table 1 Common ions and their charges

Farmers often put lime on the soil of their fields. Lime is an ionic compound. Why is it added to the soil?

Bonds between ions

Metal/non-metal compounds are made of ions. They are called **ionic compounds**. They include common salt (sodium chloride), lime (calcium oxide) and iron ore (iron(III) oxide).

In ionic compounds, the charges on the positive ions just balance the charges on the negative ions. So because the sodium ion is Na^+ and the chloride ion is Cl^-, we can predict that the formula of sodium chloride will be Na^+Cl^- or simply NaCl. By balancing the charges in this way, we can work out the formulas of other ionic compounds. For example, the formula of lime is $Ca^{2+}O^{2-}$(CaO) and that of iron(III) oxide is $(Fe^{3+})_2(O^{2-})_3$ or simply Fe_2O_3. In iron(III) oxide, the six positive charges on two Fe^{3+} ions have been balanced by six negative charges on three O^{2-} ions.

QUESTIONS

1 Suppose your best friend has been absent from school. Write down what you would say to him or her to explain why an aluminium ion is written as Al^{3+}, whereas a silver ion is written as Ag^+.

2 Look at Faraday's electrolysis equipment in the photograph on the opposite page.
(a) Where are the electrodes?
(b) What are the vertical tubes for?
(c) What do you think would happen if copper(II) chloride solution was electrolysed in the equipment?
(d) How has our equipment for electrolysis changed since Faraday's time?

3 Write the symbols for the ions in the following compounds and then work out their formulas.
calcium chloride, copper(II) sulphide, aluminium oxide, potassium hydroxide, iron(III) iodide, chromium bromide

4 Write down the names and formulas of eight common ionic compounds. (Think about rocks, metal ores, fertilisers, salts.)

6 ELECTROLYSIS IN INDUSTRY

Obtaining metals by electrolysis

Reactive metals like sodium, magnesium and aluminium cannot be obtained by reducing their oxides to the metal with carbon (coke). These metals can only be obtained by electrolytic decomposition of their molten (fused) compounds. We cannot use electrolysis of their aqueous compounds because hydrogen (from the water) is produced at the cathode and *not* the metal. For example, during the electrolysis of aqueous sodium chloride, hydrogen (not sodium) is produced at the cathode.

Metals low in the reactivity series, such as copper and silver, can be obtained by reduction of their compounds or by electrolysis of their aqueous compounds. When their aqueous compounds are electrolysed, metals form at the cathode.

Aluminium is obtained from bauxite, impure aluminium oxide. The picture shows bauxite being mined

Manufacturing aluminium by electrolysis

Aluminium is manufactured by the electrolysis of *molten* aluminium oxide. This is obtained from bauxite.

Pure aluminium oxide does not melt until 2045°C. This would make its electrolysis very costly. The aluminium oxide is therefore dissolved in molten cryolite (Na_3AlF_6) which melts below 1000°C. Figure 1 shows a diagram of the electrolytic method used. Aluminium ions in the electrolyte are attracted to the carbon cathode lining the tank. Here they accept electrons to form aluminium:

$$\textbf{Cathode } (-) \qquad Al^{3+} + 3e^- \rightarrow Al$$

Molten aluminium collects at the bottom of the cell and is tapped off.

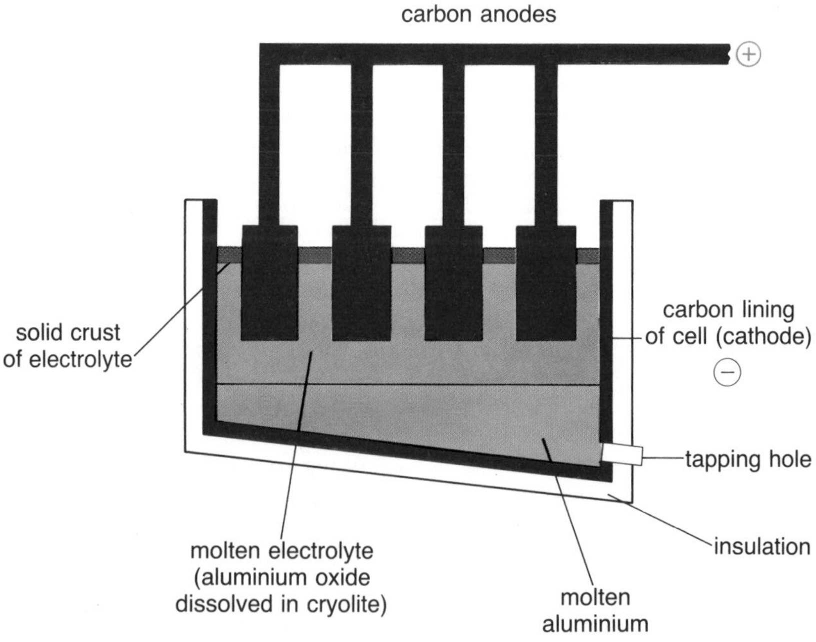

Figure 1
The electrolytic cell for aluminium manufacture

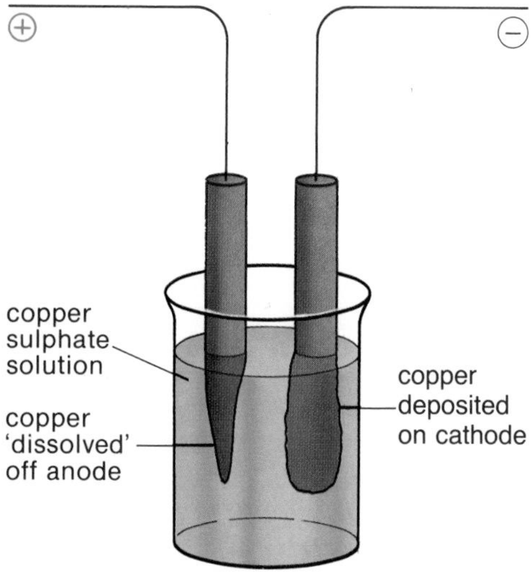

Figure 2
Purifying copper by electrolysis

Oxide ions (O^{2-}) are attracted to the carbon anodes. Here they give up their electrons forming oxygen atoms:

Anode (+) $O^{2-} \rightarrow O + 2e^-$

The oxygen atoms then combine in pairs to form oxygen gas (O_2).

Purifying copper by electrolysis

When copper sulphate solution is electrolysed with copper electrodes, copper is deposited on the cathode and the copper anode loses weight (Figure 2).

The aqueous copper sulphate contains copper ions (Cu^{2+}) and sulphate ions (SO_4^{2-}). During electrolysis, Cu^{2+} ions are attracted to the cathode where they gain electrons and deposit on the cathode:

Cathode (−) $Cu^{2+} + 2e^- \rightarrow Cu$

SO_4^{2-} ions are attracted to the anode, but they are not discharged. Instead, copper atoms, which make up the anode, give up two electrons each and go into solution as Cu^{2+} ions:

Anode (+) $Cu \rightarrow Cu^{2+} + 2e^-$

The overall result of this electrolysis is that the anode loses weight and the cathode gains weight – copper metal is transferred from the anode to the cathode.

This method is used industrially to purify crude copper. The impure copper is the anode of the cell. The cathode is a thin sheet of pure copper. The electrolyte is copper sulphate solution. The impure copper anode dissolves away and pure copper deposits on the cathode.

Impure copper anodes being transferred to an electrolysis tank for purification

QUESTIONS

1 Get into a group with 2 or 3 others. Suggest reasons for the following.
(a) Clay is the most abundant source of aluminium, but the metal is never extracted from clay.
(b) Aluminium extraction plants are usually sited near hydroelectric plants.

2 *The costs of manufacturing aluminium*
It takes about 16 units (kilowatt-hours) of electricity to produce 1 kg of aluminium.
(a) Suppose aluminium manufacturers pay 3p a unit for electricity. How much is the cost of electricity in producing 1 kg of aluminium?
(b) What other costs are involved in manufacturing aluminium besides the cost of electricity?

3 *The costs of marketing aluminium foil*
(a) Find the mass and the cost of a roll of aluminium foil in the shops.
(b) Calculate the cost of 1 kg of aluminium foil from the shops.
(c) Aluminium costs about £1 per kg from the factory. Why does the aluminium foil cost much more than the aluminium direct from the factory?

4 Magnesium is manufactured by electrolysis of molten magnesium chloride using a graphite anode and a steel cathode.
(a) Give the symbols and charges on the ions in the electrolyte.
(b) Draw a circuit diagram to show the directions in which the ions and electrons move during electrolysis.
(c) Write equations for the processes which occur at the anode and the cathode.
(d) Why do you think the anode is made of graphite and *not* steel which is cheaper?

7 ELECTROPLATING

This picture shows the bodywork of a car rising out of an electroplating bath. Notice the wire carrying the current from the bodywork

The method for purifying impure copper described in the last unit can be used industrially to coat (plate) articles with copper. The process is called **electroplating**. Several metals can be used for electroplating articles. The most commonly used metals, apart from copper, are chromium, silver and tin. If the article is to be plated with copper, the article is the cathode of the cell, the anode is a piece of copper and the electrolyte is copper sulphate solution. During the electroplating process, the copper anode dissolves away and copper deposits on the article at the cathode.

Although it is easy to deposit a metal during electrolysis, the conditions must be carefully controlled so that the metal sticks to the object to be plated. For example, chromium will not stick to iron (steel) or copper during electroplating. Thus, steel articles to be chromium plated (such as car bumpers and electric kettles) are first plated with nickel. This forms a firm deposit on the steel which can then be plated with chromium.

These three objects have been electroplated with either copper or chromium

The metal coating is deposited on the cathode and so the object to be plated must be used as the cathode during electrolysis. The electrolyte must contain a compound of the metal which forms the coating.

Figure 1 shows a nickel alloy fork being electroplated with silver. Notice that the fork is attached to a rotatable arm and that the electrolyte contains silver nitrate ($AgNO_3$).

During electrolysis, silver ions (Ag^+) in the electrolyte are attracted to the cathode (the fork) where they gain electrons and form a deposit of silver:

Cathode (−) $Ag^+ + e^- \rightarrow Ag$

The anode is a piece of silver. Nitrate ions (NO_3^-) in the electrolyte are attracted to the anode, but they are not discharged. Instead, silver atoms in the anode give up electrons to the anode and go into solution as Ag^+ ions.

Anode (+) $Ag \rightarrow Ag^+ + e^-$

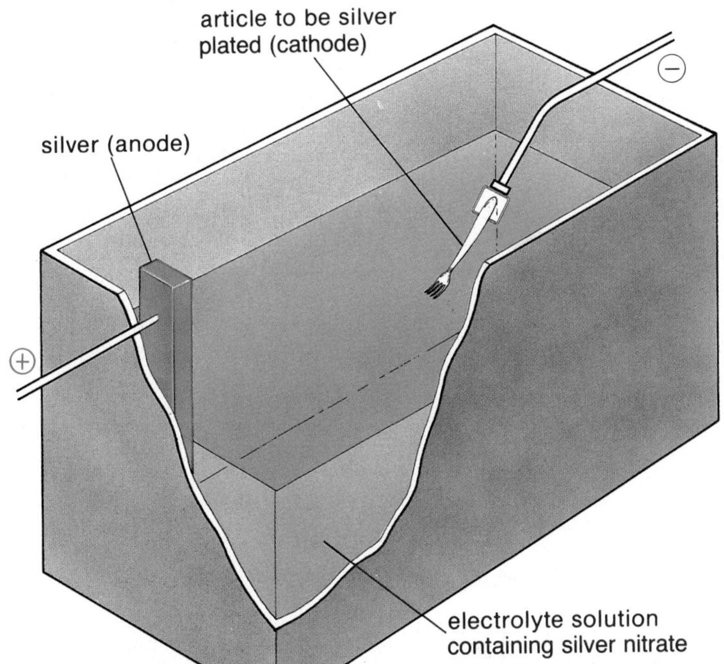

Figure 1
A fork being electroplated with silver

QUESTIONS

1 Explain why each of the following precautions must be taken during electroplating.
(i) The object to be plated must be clean and free of grease.
(ii) The object to be plated should be rotated.
(iii) The electric current must not be too large.
(iv) The temperature and the concentration of the electrolyte must be carefully controlled.

2 (a) The object to be electroplated is made the cathode in an electrolytic cell. Why is this?
(b) Suppose you want to nickel plate an article. What substance would you choose for (i) the anode, (ii) the electrolyte?

3 Steel bumpers on cars are electroplated with nickel and then with chromium.
(a) Why is the steel plated with a layer of nickel before chromium?
(b) Give two reasons for electroplating steel.

4 Some cutlery is stamped 'EPNS'. This stands for *e*lectro*p*lated *n*ickel *s*ilver. How is this plating done?

5 Articles of jewellery (e.g. bracelets) are often electroplated with gold, silver or copper. Why?

SECTION D: THINGS TO DO

Activities

1 Recycling aluminium

Aluminium is obtained from bauxite which is impure aluminium oxide. If we continue to use aluminium at our present rate, the bauxite will only last another 30 years. This means that we should think about saving our reserves of aluminium. There are two ways in which most of us waste aluminium.

- milk bottle tops
- aluminium foil

In this activity, you will try to estimate

- how much aluminium we use each year
- how much we could save by recycling aluminium.

How much aluminium do we use each year?

1 Estimate the number of bottle tops your family uses (on average) each week.

2 Estimate the area of aluminium foil (in square centimetres) that your family uses each week for cooking, wrapping food, etc.

3 Use your answer to question **1** to work out the mass of aluminium which your family uses each week in milk bottle tops. (Assume that one milk bottle top weighs 0.25 g.)

4 Use your answer to question **2** to work out the mass of aluminium your family uses each week in foil. (Assume 100 cm^2 of foil weighs 0.4 g.)

5 Use your answers in questions **3** and **4** to calculate the total mass of aluminium (milk bottle tops plus foil) which your family uses in one week.

6 How much aluminium does your family use in one year?

7 Estimate the total mass of aluminium which we use in the UK in one year. (Assume that the population of the UK is 56 million.)

Recycling aluminium cans

How much could we save by recycling aluminium?

8 Use your answer to question **7** to calculate the *total cost* of the aluminium which we use for milk bottle tops and foil each year in the UK. (Aluminium costs £800 per tonne from the factory. 1 tonne = 1000/kg.)

9 If possible, collect answers to question **8** from others in the class and calculate an average value.

10 Would we save the amount of money estimated in questions **8** and **9** if we recycled aluminium milk bottle tops and foil? Explain your answer.

Further questions to answer and discuss

11 About 35% of the aluminium used in the UK is recycled. What do you think are the difficulties in recycling more aluminium?

12 Which forms of aluminium are easiest to recycle? (Remember that aluminium has many other uses besides milk bottle tops and foil.)

13 What other common household materials could be recycled besides aluminium?

14 Why is it important to recycle resources, like aluminium?

2 World's largest hydroelectric project

Overseas news, 10.11.86.

Venezuela breaks grip of oil

Venezuela opened the world's largest hydroelectric complex yesterday at Guri, 650 kilometres south-east of Caracas.

The Guri dam project is designed to speed up Venezuela's industrial development and reduce its dependence on oil.

The dam holds back a gigantic reservoir covering an area about one fifth the size of Wales. The project cost an estimated £3500 million pounds and has taken 25 years to complete. In a ceremony at Guri, the President of Venezuela ordered the floodgates to be opened. This sent waters from the River Caroni rushing down the 160m-high spillway.

In his speech marking the event, the President said "We should feel well satisfied with what we have achieved in Guri. This will allow us to vary our economy and rescue it from dependence on oil." Power from Guri will supply 70% of Venezuela's energy needs and is expected to save 300 000 barrels of oil per day.

More importantly, the dam provides cheap electric power for the country's huge aluminium, iron and steel industries, centred at Ciudad Guayana 80 kilometres north-east of Guri.

What was once a collection of huts in scrubland on the edge of the River Orinoco is now a Venezuelan boomtown.

But, Guri is not just the saving of petroleum and assured energy at low cost. It also represents an opportunity for the growth and modernisation of many industries.

1 Read the newspaper report above and write down three phrases which show that Venezuela is breaking the 'grip of oil'.

2 Why do you think Venezuela wants to 'break the grip of oil'?

3 Describe briefly how hydroelectricity is generated.

4 The aluminium works at Ciudad Guayana are the eighth largest in the world. Why is the Guri project important to the aluminium works?

5 Design a poster for use throughout Venezuela to show the benefits of the Guri project.

Study Questions

1 When aluminium is exposed to the air, it becomes coated very quickly with a thin layer of oxide about 10^{-6} cm thick. This layer does not flake off, nor does it increase in thickness on standing. In order to protect the aluminium even more than its natural oxide layer does, it is possible to thicken the layer to 10^{-3} cm by a process called anodising. The aluminium is first degreased and then anodised by making it the anode during the electrolysis of sulphuric acid. The oxygen released at the anode combines with the aluminium and increases the thickness of the oxide layer.
(a) Why is anodising useful?
(b) Why does aluminium not corrode away like iron, even though aluminium is coated very quickly with a layer of oxide?
(c) Write a word equation for the reaction which takes place when aluminium is exposed to the air.
(d) Carbon tetrachloride (tetrachloromethane) can be used to degrease aluminium before anodising.
Name one other liquid which would be a suitable degreasing agent.
(e) Why is water not used to degrease aluminium?
(f) Why is it necessary to degrease the aluminium?
(g) How many times thicker is the oxide coating after anodising?
(h) Why is anodised aluminium especially useful as a building material?

2 This question concerns the properties of substances *A* to *E*.
Choose from *A* to *E*, the correct substance for
(a) the wiring for an electric light,
(b) the insulator for overhead electric cables,
(c) a quick-acting garden fertiliser,
(d) petrol,
(e) sugar,
(f) calcium.

3 Water containing a little dilute sulphuric acid was decomposed by electrolysis.
(a) What products are formed when water is decomposed? Write a word equation to summarise the results.
(b) What is meant by the word 'decompose'?
(c) Where does the energy needed to decompose the water come from?
(d) What substance is produced (i) at the anode, (ii) at the cathode?
(e) How would you test to see if the products you suggest are formed?
(f) What would happen if you changed round the connections from the battery to the electrodes?
(g) In some countries, hydrogen is obtained from solutions containing water by electrolysis. What do you think is the most expensive aspect of these electrolyses?
(h) Which countries are most likely to use electrolytic processes to obtain hydrogen?

4

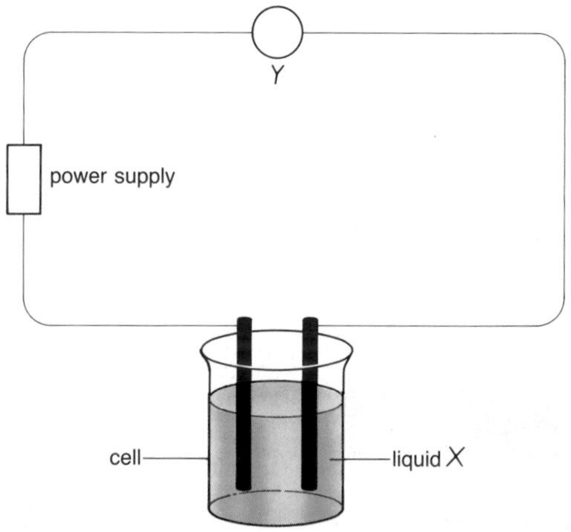

Liquid *X* is to be electrolysed.
(a) What type of power supply is suitable for this?
(b) What apparatus could be connected at *Y* to show that *X* conducts?
(c) Give the name of the charged particles which conduct the electric current in (i) the connecting wires, (ii) the electrolyte.
(d) A steel dish is to be coated with copper using electrolysis. What would you use in the cell as the (i) anode, (ii) cathode, (iii) electrolyte?

Substance	Does the pure substance conduct electricity at room temperature?	Is the substance soluble in water?	PH of aqueous solution	Does the aqueous solution conduct?
A	yes	reacts with water	12	yes
B	yes	no	—	—
C	no	yes	7	no
D	no	yes	7	yes
E	no	no	—	—

SECTION E

From metals to plastics

How are metals and plastics used in the construction of Concorde? Why were they chosen for these uses?

1 EVERYDAY MATERIALS

Why are metals like copper used for electrical wires and cables?
Most **metals**
- *are hard*
- *are strong*
- *have a high density*
- *conduct electricity*
- *can be hammered or pulled into different shapes*

Look at the things around you. Most of them can be classified as one of the following five materials.

Metals including alloys

Glass

Pottery (ceramics) including bricks and concrete

Plastics such as polythene, perspex and PVC

Fibres including cotton, wool, Nylon and wood

Why are cups and saucers made of ceramics?
Pottery *is*
- *fairly cheap*
- *hard*
- *strong*
- *inert (unaffected by hot, soapy water)*
- *brittle*

Why are lots of everyday articles now made from plastics?
Plastics
- *are fairly cheap*
- *have a low density*
- *can be transparent*
- *are easily moulded and coloured*
- *are flexible*

Why is glass used for returnable bottles?
Glass *is*
- *transparent*
- *hard*
- *brittle*
- *inert (and unaffected by the weather)*
- *fairly cheap*
- *easy to clean*

In this section, we shall be concentrating on the properties and uses of metals and plastics.

During the last 20 or 30 years, scientists have developed a whole range of new materials. Most of these new materials are based on the five types listed above.

Usually the new materials have special uses or superior properties to other materials. This is illustrated very well in the photos below.

Until the 1950s, tennis rackets were made of wood. This was the only lightweight material which could be bent into an oval shape for the racket head and which was strong enough to hold the racket strings taut (see right).

During the 1960s, a number of very lightweight metal alloys were produced. These were stronger than wood, so the strings could be tighter. This enabled players to return the ball faster (see above).

During the 1980s, even better rackets were produced. These were made from plastics reinforced with carbon (graphite) fibres. The frames of these rackets have the strength of carbon fibres and the flexibility of plastic. Fibre-reinforced plastics are examples of **composite materials**. Composites are made of two different materials. One of the first manufactured composite materials was reinforced concrete. This has the hardness and weight-bearing strength of concrete and the flexibility and strength of steel (see right).

QUESTIONS

1 Look at the photo below.

(a) What are the different cups, mugs and glasses made of?
(b) Why have different materials been used to make these drinking vessels?
(c) When would each of the different cups, mugs and glasses be used?

2 Make a summary of the properties of fibres like the summaries for metals, glass, plastics and pottery on page 100.

3 Most solids can be classed as metal, glass, ceramic, plastic or fibre.
(a) In which of these 5 classes would you put (i) brass, (ii) marble, (iii) perspex, (iv) rubber, (v) sugar, (vi) steel, (vii) cotton, (viii) wood.
(b) Which of these materials was most difficult for you to classify? Why was it difficult to classify?

4 The words below are used to describe materials.
brittle, flexible, lightweight, soft, dense, hard, opaque, soluble, dull, inert, reactive, stiff, elastic, insoluble, shiny, transparent
(a) What do the following mean? *elastic, flexible, inert, opaque*
(b) The words in the list can be put into pairs of opposites. Write down each of the eight pairs.
(c) For each word in the list, write down a material which could be described by that word.

5 (a) What properties made wood suitable for tennis rackets?
(b) Why do you think lightweight alloys were better than wood?
(c) What is a composite material?
(d) Plastic reinforced with graphite fibre is a composite material. Why is it so good for the frames of rackets?
(e) Why is pure plastic not used for the frames?
(f) Why is pure graphite not used for the frames?
(g) Give one other example of a composite material.

2 INTRODUCING METALS

Compare the colour of the new coins and the old coins in this photo. Why do coins slowly go darker?

Metals are among the most important and useful materials. Just look around and notice the uses of different metals – cutlery, cars, ornaments, handles and locks for doors, pipes and radiators, girders and bridges.

The reactivity of metals

Shiny aluminium window frames and ladders *quickly become covered* with a thin layer of aluminium oxide.

$$\text{aluminium} + \text{oxygen} \rightarrow \text{aluminium oxide}$$

In the same way, cars and nails made of iron are *slowly* covered with a layer of rust which is iron oxide. Things containing copper, like coins and brass, also react with oxygen in the air to form a thin layer of black copper oxide, but this process is *very slow*.

$$\text{copper} + \text{oxygen} \rightarrow \text{copper oxide}$$

Gold reacts even more slowly than copper. Gold jewellery stays shiny for years. The gold *never reacts* with oxygen in the air. These reactions with oxygen show that aluminium is more reactive than iron, iron is more reactive than copper and copper is more reactive than gold. So, we can arrange these metals in an order of reactivity called a **reactivity series** (Figure 1).

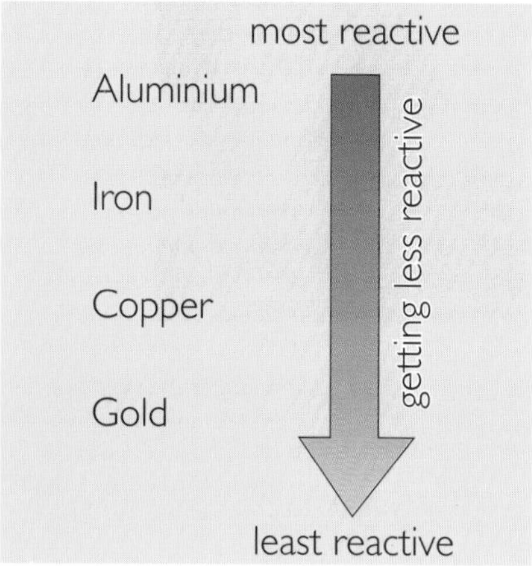

Figure 1

Metals and acids

Various foods, including vinegar, citrus fruits and rhubarb, contain acids. These acids will attack kitchen utensils and cutlery made of certain metals.

Table 1 shows what happens when different metals are added to dilute hydrochloric acid at room temperature (21°C). This acid is more reactive than the acids in foods, but it shows how different metals are attacked.

Which of these foodstuffs contain acids?

One of these spoons has been left in a glass of lemon juice for several days. Citric acid in the juice has attacked metals in the spoon

Notice in Table 1 that aluminium does not react at first. This is because its surface is protected by a layer of aluminium oxide. The oxide reacts slowly with the acid. This exposes aluminium, which reacts more vigorously.

The metals used most commonly for pans and cutlery are aluminium, copper, nickel and iron (steel). Copper is the only one which does not react with the acids in food. But copper is so expensive that aluminium and steel are used in most of the saucepans sold today. The thin oxide coating on aluminium helps to protect it from the weak acids in many foods. However, oxalic acid in rhubarb and acetic acid in vinegar do react with aluminium and 'clean' the saucepan during cooking. Iron and nickel, which are used in cutlery, are also attacked by acids in food. Ordinary kitchen cutlery becomes badly marked if it is left in lemon juice (which contains citric acid) or in vinegar.

Foods which contain acids are best stored in unreactive containers made of glass or plastic. Tin cans are also used to store acidic foods like pineapples and grapefruit. Tin cans are made of steel coated on both sides with tin and then lacquered on the inside. The lacquer forms an unreactive layer between the tin and its contents.

Copper is so unreactive that a copper cooking pot will last for many years

Metal used	**Reaction with dilute hydrochloric acid**	**Highest temperature recorded**
Aluminium	No reaction at first, but vigorous after a time. Hydrogen is produced rapidly	85°C
Copper	No reaction. No bubbles of hydrogen	21°C
Iron	Slow reaction, bubbles produced slowly from iron	35°C
Lead	Very slow reaction, a few bubbles appear on the lead	23°C
Magnesium	Very vigorous reaction, hydrogen is produced rapidly	95°C
Zinc	Moderate reaction, bubbles of hydrogen are produced steadily	55°C

Table 1 Reactions of metals with dilute hydrochloric acid

QUESTIONS

1 Look at Table 1.
(a) Which metal reacts (i) most vigorously, (ii) least vigorously?
(b) Write the metals in order of decreasing reactivity with dilute hydrochloric acid.

2 Which acids are present in (i) lemon juice, (ii) vinegar, (iii) rhubarb, (iv) sour milk?

3 (a) Why is copper a better metal for saucepans than aluminium?
(b) Why is copper not used for most pans today?
(c) Why is aluminium less reactive to acidic foodstuffs than expected?
(d) How are tin cans protected from acids in their contents?

4 Write word equations for the reactions of
(a) aluminium with dilute hydrochloric acid,
(b) magnesium with dilute sulphuric acid,
(c) zinc with dilute nitric acid.

5 Make a list of six important uses for metals. Say why metals are used in these six ways.

3 THE REACTIVITY SERIES

In the last unit we studied the reactions of metals with dilute hydrochloric acid. All the metals react except copper. The products of the reaction are hydrogen and the chloride of the metal.

$$\text{metal} + \text{hydrochloric acid} \rightarrow \text{metal chloride} + \text{hydrogen}$$

Using M as the symbol for the metal, and assuming M forms M^{2+} ions, we can write a general equation as

$$M(s) + 2HCl(aq) \rightarrow MCl_2(aq) + H_2(g)$$

A similar reaction occurs between metals and dilute sulphuric acid. This time the products are hydrogen and a metal sulphate. We can summarise these reactions as

	metal + acid → metal compound + hydrogen
(above Cu)	

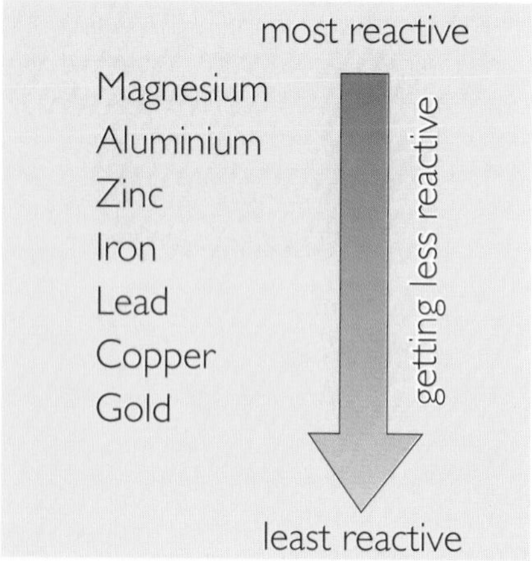

Figure 1

From these reactions of metals with dilute acids and the reactions of metals with air (oxygen) in Unit 1, we can draw up an overall reactivity series. This is shown in Figure 1.

Reactions of metals with water

We can check the order of metals in our reactivity series by investigating the reaction of metals with water. Table 1 summarises the reactions of six metals with water.

Metal	Reaction with water
Calcium	Sinks in the water. Steady stream of hydrogen produced. The solution becomes alkaline and cloudy due to the formation of calcium hydroxide
Copper	No reaction
Iron	Rusts very slowly, provided oxygen is present. Brown hydrated iron oxide is produced
Magnesium	Tiny bubbles of hydrogen appear on the surface of the magnesium after a few minutes. The solution slowly becomes alkaline
Potassium	A violent reaction occurs. A globule of molten potassium skates over the water surface, hissing and burning with a lilac flame. Hydrogen and potassium hydroxide form
Sodium	A vigorous reaction occurs. A globule of molten sodium skates about the water surface. Hydrogen and sodium hydroxide form

Table 1 The reactions of six metals with water

Potassium reacts vigorously with water, releasing hydrogen and producing a lilac flame

Look at the results in Table 1.

- Is the order of reactivity with water similar to that with acids?
- Use the results in Table 1 and Figure 1 to draw up a reactivity series including the extra metals in Table 1.

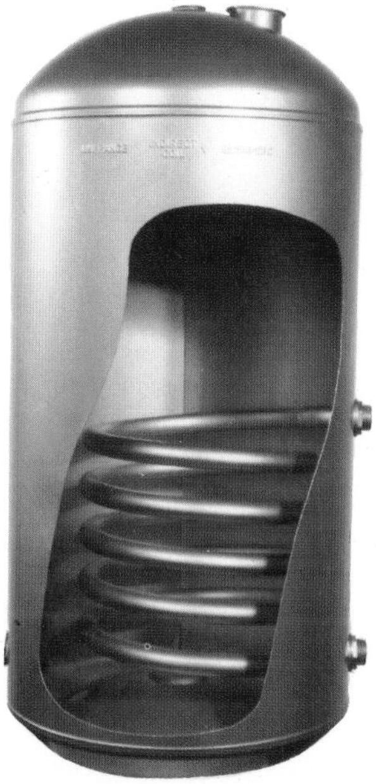

Why are hot water tanks made of copper and cold water tanks made of plastic or steel?

Summarising the reactions of metals

Notice that the order of reactivity of metals is the same with air (oxygen), with water and with acids. This is not really surprising because metal atoms react to form metal ions in each case.

$$M \rightarrow M^{2+} + 2e^-$$

The higher the metal in the reactivity series, the more easily it forms its ions.
Reaction with air (oxygen). Metals lose electrons to form metal ions. The electrons are taken by oxygen molecules (O_2) forming oxide ions (O^{2-}).

$$\text{metal} + \text{oxygen} \rightarrow \text{metal oxide}$$
$$2M + O_2 \rightarrow 2M^{2+}O^{2-}$$

Reaction with water. Here again, the metals lose electrons to form ions. The electrons are taken by water molecules which form oxide ions (O^{2-}) and hydrogen (H_2).

$$\text{metal} + \text{water} \rightarrow \text{metal oxide} + \text{hydrogen}$$
$$M + H_2O \rightarrow M^{2+}O^{2-} + H_2$$

What is the disadvantage of using iron (steel) for the boiler in a steam train?

The oxides of metals high in the reactivity series (e.g. Na_2O and CaO) react further with water to form solutions of their hydroxides. The reactivity series is a very useful summary of the reactions of metals. Metals at the top of the series want to lose electrons and form ions. Metals at the bottom of the series are just the opposite. Ions of these metals want to gain electrons and form atoms.

QUESTIONS

1 Look at the photographs of the hot water tank and the steam engine. Answer the questions in the captions to the photos.
2 Plan an experiment that you could do to find the position of lithium in the reactivity series. Say what you would do and how you would ensure that your test is fair.
3 (a) *X* is a metal which reacts with dilute hydrochloric acid but not with water. What metal could *X* be?
(b) Metal *P* will remove oxygen from the oxide of metal *Q*, but not from the oxide of metal *R*. Write *P*, *Q* and *R* in their order in the reactivity series (most reactive first).
4 Write word equations and balanced symbolic equations for the reaction of
(a) copper with oxygen,
(b) aluminium with oxygen,
(c) calcium with water,
(d) lead with dilute nitric acid,
(e) magnesium with dilute sulphuric acid.
5 Find out about the manufacture of chromium and titanium. How does the reactivity series help to explain the methods used?

4 THE STRUCTURE OF METALS

Zinc crystals on the surface of galvanised iron

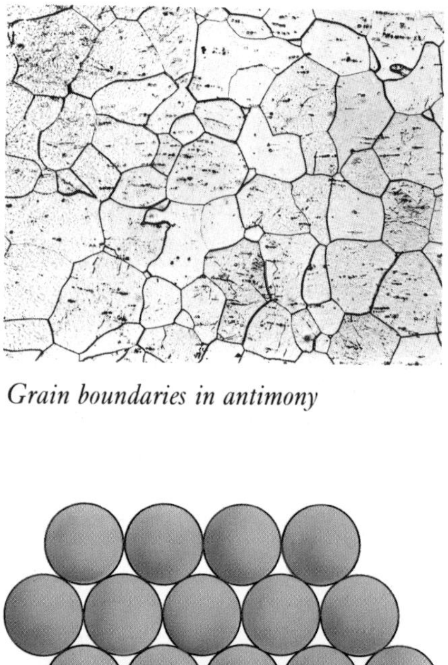

Grain boundaries in antimony

Look carefully at the surface of some galvanised iron (iron coated with zinc). Dustbins, farm gates and cold water tanks are often made of galvanised iron. You will see irregularly-shaped areas separated from each other by clear boundaries. The irregular areas are called **grains** and the boundaries between grains are **grain boundaries**. The grains in tin and zinc are usually easy to see, but the grains in most metals are too small to see with the naked eye. The oxide coating on many metals also makes it difficult to see the grains. But, if the metal surface is clean and smooth, the grains can be seen under a microscope.

X-ray analysis shows that the atoms in metal grains are packed in a regular order, but the grains themselves are irregularly-shaped crystals of the metal pushed tightly together.

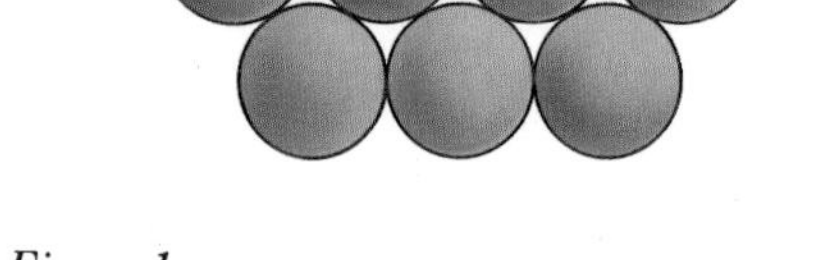

Figure 1

Metals usually have a high density. This suggests that the particles are packed close together. The high melting points and boiling points of most metals also indicate that the atoms are held closely together by strong forces of attraction. In fact, X-ray studies show that the atoms of most metals are packed as close together as possible. This arrangement is called **close packing**. Figure 1 shows a few atoms in one layer of a metal. Notice that each atom in the middle of the metal touches 6 other atoms in the same layer. When a second layer is placed on top of the first layer, atoms in the second layer sink into the dips between atoms in the first layer (Figure 2). This means that any one atom in the first layer can touch 6 atoms in its own layer, 3 atoms in the layer above it and 3 atoms in the layer below it. The total number of 'nearest neighbours' to one atom in a close-packed structure is therefore 12. The number of nearest neighbours to an atom or ion in a crystal is called its **coordination number**. Thus, we can say that the coordination number for atoms in a close-packed metal structure is 12.

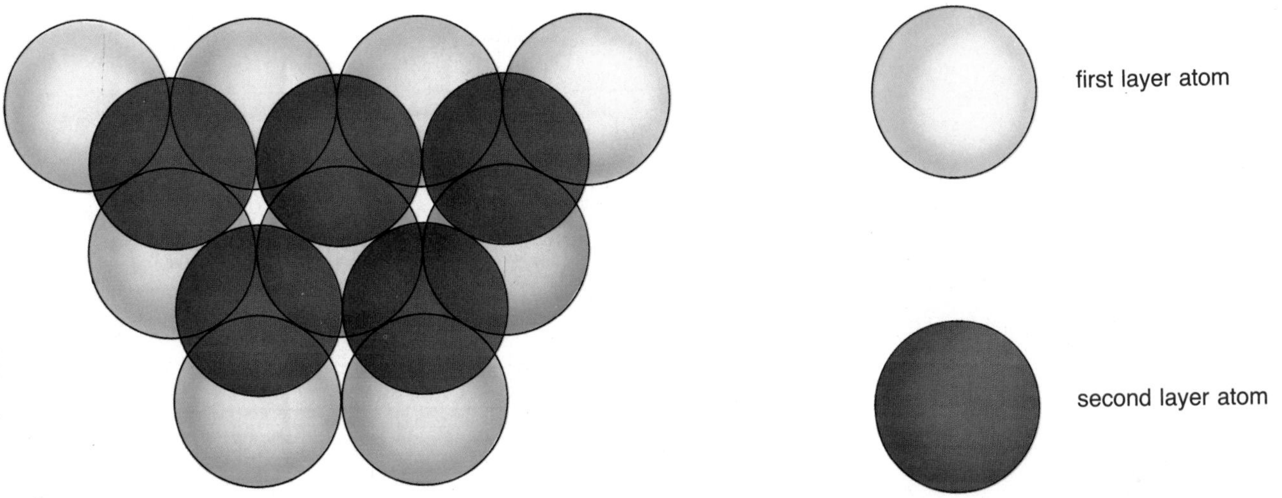

Figure 2

QUESTIONS

1 A good model of the close packing in a single layer of metal atoms can be made by blowing a slow stream of gas through a fine jet into a dish containing dilute soap solution (Figure 3). The small bubbles look like atoms in the layer of metal.
The large areas of regular packing in the 'bubble raft' represent a single crystal grain. The bubble raft in the diagram also shows the four main 'flaws' in metal structures.

- **Grain boundaries** are the narrow areas of disorder between one crystal grain and another.
- **Dislocations** occur when a row of atoms is displaced or it 'peters out' and the regular packing ceases.
- **Vacant sites** occur where an atom is missing from the metal structure.
- **'Foreign' atoms**. When atoms of another element form part of the metal structure, they break up the orderly arrangement. The same thing happens in the bubble raft where there are 'foreign' bubbles which are larger or smaller than the rest.

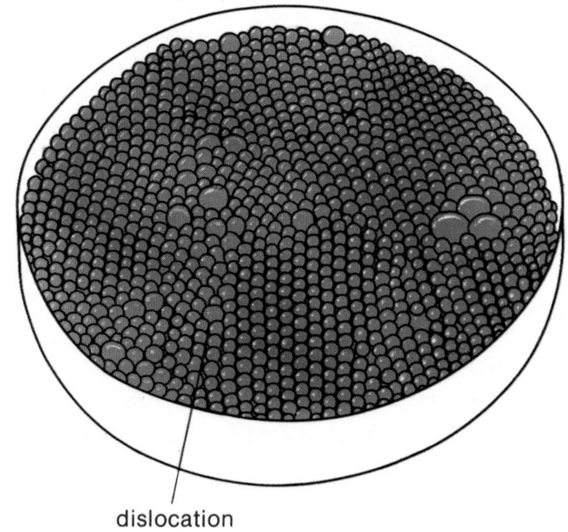

(a) Trace the rim of the bubble raft in the diagram above and mark the position of the dislocation.
(b) Also on your tracing, mark the position of a grain boundary, a vacant site and a foreign atom.

2 Make two tracings of Figure 2. Use these tracings to show two possible arrangements for a third layer of close-packed atoms.

3 (a) Explain why atoms in a close-packed metal structure have a coordination number of 12.
(b) Look closely at the structure of diamond in unit 3, section H. What is the coordination number of carbon atoms in diamond?
(c) Look closely at the structure of sodium chloride in unit 7, section H. What are the coordination numbers for Na^+ ions and Cl^- ions in the structure?

5 PROPERTIES OF METALS

A blacksmith depends on the malleability of metals to hammer and bend them into useful shapes

The properties of a metal depend on the size of its crystal grains and also on the way its atoms are arranged in the crystal.

In general, metals
- have high densities
- have high melting points
- have high boiling points
- are good conductors of heat
- are good conductors of electricity
- are malleable (can be hammered into different shapes)
- are ductile (can be pulled into wires)

- **Density**. The close packing of atoms in most metal crystals helps to explain their high densities.
- **Melting and boiling points**. High melting points and high boiling points suggest that there are strong forces holding the atoms together in metal crystals. Chemists think that the outermost electrons of each atom can move about freely in the metal. So the metal consists of positive ions surrounded by a 'sea' of moving electrons (Figure 1). The negative 'sea' of electrons attracts *all* the positive ions and this cements everything together. The strong forces of attraction between the moving electrons and the positive ions result in high melting points and high boiling points.
- **Conductivity**. When a metal is connected in a circuit, freely moving electrons in the metal move towards the positive terminal. At the same time, electrons can be fed into the other end of the metal from the negative terminal (Figure 2). This flow of electrons through the metal forms the electric current.

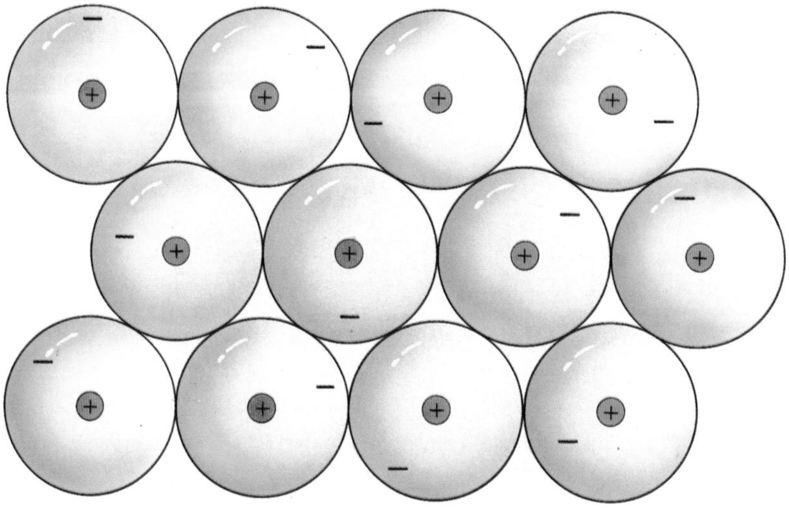

Figure 1
The outermost electrons of each atom move around freely in the metal structure

- **Malleable and ductile.** The bonds between atoms in a metal are strong but they are *not* rigid. When a force is applied to a metal crystal, the layers of atoms can 'slide' over each other. This is known as **slip**. After slipping, the atoms settle into position again and the close-packed structure is restored. Figure 3 shows the positions of atoms before and after slip. It shows how the atoms move when a metal is hammered into a different shape.

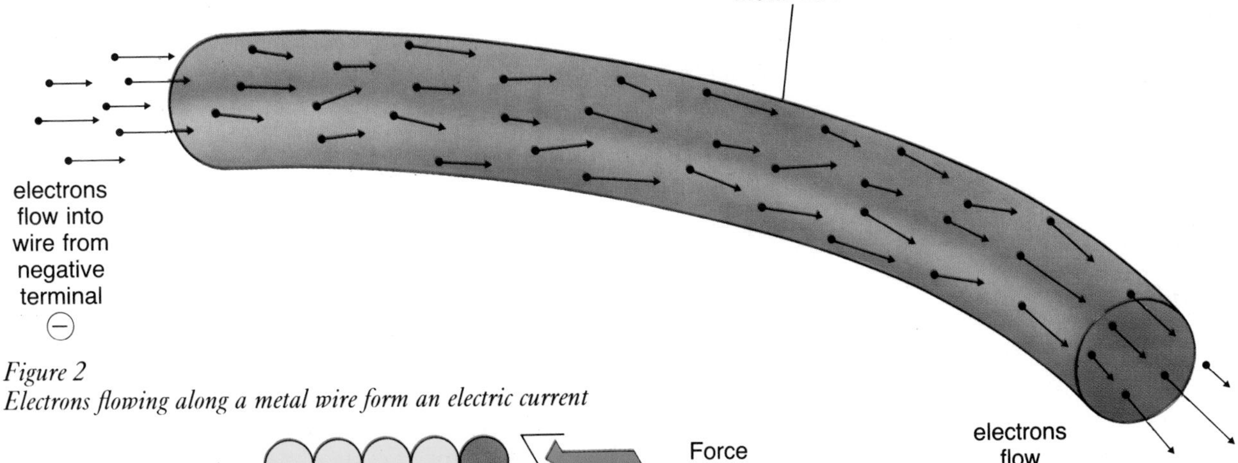

Figure 2
Electrons flowing along a metal wire form an electric current

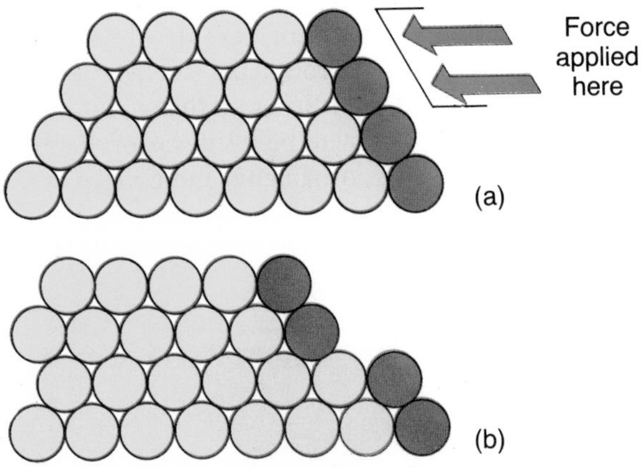

Figure 3
The positions of atoms in a metal crystal (a) before, and (b) after 'slip' has taken place

QUESTIONS

1 (a) Draw up a table showing the density, the melting point and the boiling point of lead, magnesium, tungsten and zinc.
(b) Use the data in your table to explain one important use for each metal.
2 (a) Explain what 'malleable', 'ductile' and 'slip' mean in your own words.
(b) Now find out what your dictionary says for the meanings of 'malleable', 'ductile' and 'slip'.
3 Discuss the following questions in groups of 3 or 4.
(a) Metals are malleable, polythene is plastic, rubber is elastic. What exactly do we mean by the words 'malleable', 'plastic' and 'elastic'? How would you describe clay?
(b) When blacksmiths are working, they often dip red-hot steel objects into cold water. Why do they do this? (Hint: Metals get harder as their crystal grains get smaller.)
(c) Sodium has a much lower density and melting point than most metals. How would you explain this?
4 Answer true or false to parts (a) to (f).
Some reasons for classifying magnesium as a metal are
(a) it burns to form an oxide.
(b) it reacts with non-metals.
(c) it reacts only with non-metals.
(d) it is magnetic.
(e) it conducts electricity.
(f) it has a high density.
5 Draw a diagram similar to Figure 3 to show how 'slip' occurs when a metal is pulled into a wire.
6 (a) Why is it helpful to picture the structure of metals as close-packed atoms?
(b) Sometimes it is helpful to picture metals as positive ions surrounded by a 'sea' of moving electrons. When is this model of metal structure useful?

6 ALLOYS

Metallurgists have found ways of making metals harder and stronger. They do this by preventing slip in the metal crystals. If slip cannot occur, then the metal is less malleable and less ductile. Metals can be made stronger by reducing the size of crystal grains and by alloying.

Reducing grain size

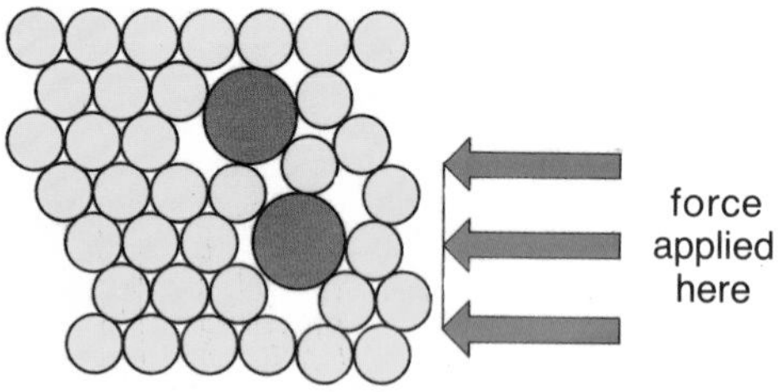

Figure 1
Slip cannot occur in the alloy because the atoms of different sizes cannot slide over one another

Slip does not occur across the grain boundaries in a metal. So metals with small grains are stronger, harder and less malleable than metals with larger grains. Grains can be made smaller by allowing the molten metal to solidify rapidly, or by heating a metal and then cooling it quickly (**quenching**).

Alloying

Metals can be made stronger by adding small amounts of another element. Brass is made by mixing copper and zinc. This alloy is much stronger than pure copper or pure zinc. The different-sized atoms break up the regular packing of metal atoms. This prevents slip and makes the metal tougher and less malleable (Figure 1). In steel, the lattice of iron atoms is distorted by adding smaller, carbon atoms. These form crystals of iron carbide which are very hard. The regions of iron carbide in the softer iron make steel very strong.

As soon as our ancestors had built furnaces that would melt metals, they began to make alloys. The first alloy to be used was probably bronze, a mixture of copper and tin. Bronze swords, ornaments and coins were being made as early as 1500 BC.

Only 0.2% carbon in iron produces steel strong enough for supports and girders in bridges like this one – the Forth Road Bridge

Bronze articles were made as early as 1500 BC

Nowadays, we depend on alloys. Most of the metallic materials we use are alloys and not pure metals. Almost all the metal parts in a car are alloys. If the bodywork and components were pure iron, they would rust very rapidly compared to present-day steels. They would also be very malleable and would buckle under strain.

Using alloys

It is possible to make alloys with very specific properties. Some are made for hardness. Some are resistant to wear and corrosion. Other alloys have special magnetic or electrical properties. Usually alloys are made by mixing the correct amounts of the constituents in the molten state. For example, solder is made by melting lead, mixing in tin and then casting the alloy into sticks.

The most important alloys are those based on steel. 'Manganese steel', containing 1% carbon and 13% manganese, can be made very tough by heating it to 1000°C and then 'quenching' in water. It is used for parts of rock-breaking machinery and railway lines. Stainless steel is another important alloy. Cutlery contains about 20% chromium and 10% nickel. Chromium stops the steel rusting; nickel makes it harder and less brittle.

This photo shows steel being quenched to make it tough. This kind of steel is used for railway lines, which have to be very strong

Brass is an alloy of copper and zinc. It is tough and will not chip or break, so it is ideal for making weights

During the last 30 years, aluminium alloys have been used more and more. These include *duralumin* which contains 4% copper. Aluminium alloys are light, strong and corrosion resistant, but they may cost six times as much as steel. They are used for aircraft bodywork, window frames and lightweight tubing. The best-known alloys of copper are brass and bronze. Coins are also copper alloys.

QUESTIONS

1 Make a table showing the elements present in the following alloys, and the main uses of each alloy.
brass, bronze, stainless steel, solder, duralumin

2 Look at Figure 2.
(a) Which process produces the strongest alloy – chill casting (i.e. rapid cooling), or sand casting (i.e. slow cooling), or the liquid alloy? Explain your answer.
(b) What percentage of aluminium produces the strongest alloy?

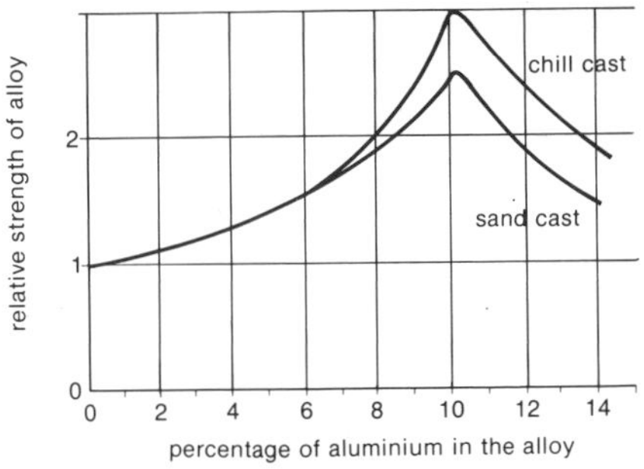

Figure 2
The effect of aluminium on the strength of copper alloys

(c) How many times stronger is the alloy than pure copper?
(d) What percentage of aluminium produces a sand cast alloy twice as strong as pure copper?
(e) Why do you think the strength of the alloy increases at first and then decreases as more aluminium is added to the alloy?

3 Describe an experiment that you could carry out to compare the strengths of two wires.

7 INTRODUCING PLASTICS

Fifty years ago, very few people had heard about plastics. Nowadays, plastics are very common. They are used for lots of everyday articles which were once made from metals, glass, ceramics and fibres.

Plastics are materials which can be moulded into different shapes when they are first made. On cooling, they set to form solids.

Plastics have changed the materials we now use for many different articles

There are only 20 or so important plastics, but these are manufactured and sold under thousands of different trade names. This makes it difficult to identify some plastics.

The properties and uses of plastics

The most important plastics are polythene, PVC, perspex, Nylon and Terylene. The properties and uses of three of these plastics are shown in Figure 1. As you can see, the uses of these three plastics are related to their properties.

About one third of all plastics are used in packaging. The building industry uses the largest amount of plastics. Plastics are used in buildings for cavity wall insulation, roof insulation, water pipes, water tanks, guttering, window frames and floor tiles.

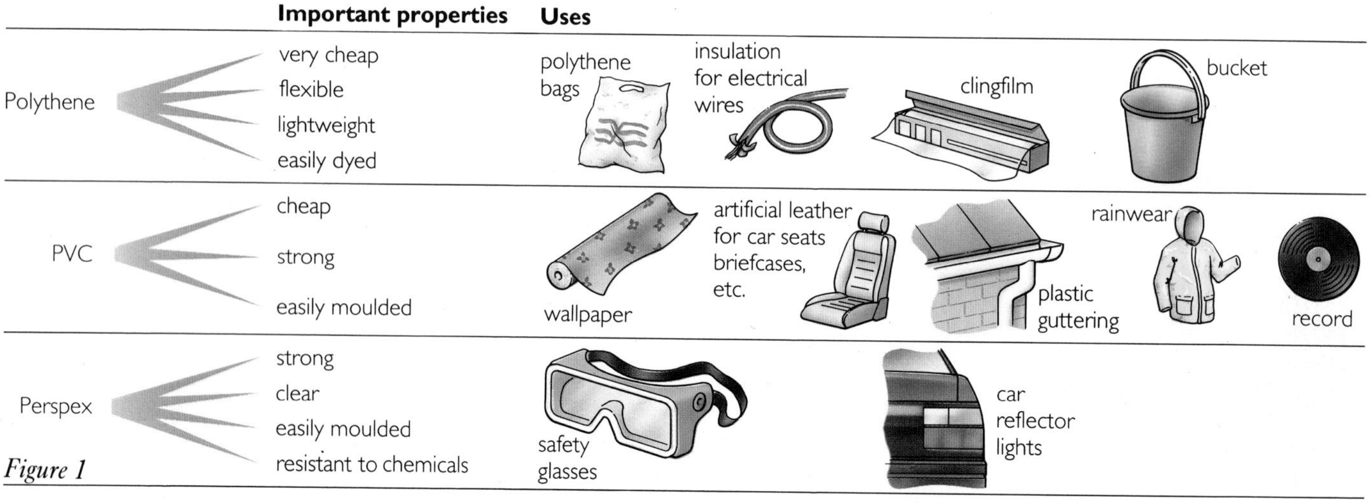

Figure 1

What are plastics?

Plastics are composed of large molecules called **polymers**. These polymer molecules may be long thin chains or 3-D structures with cross-links. Plastics are made by joining together small molecules called **monomers**.

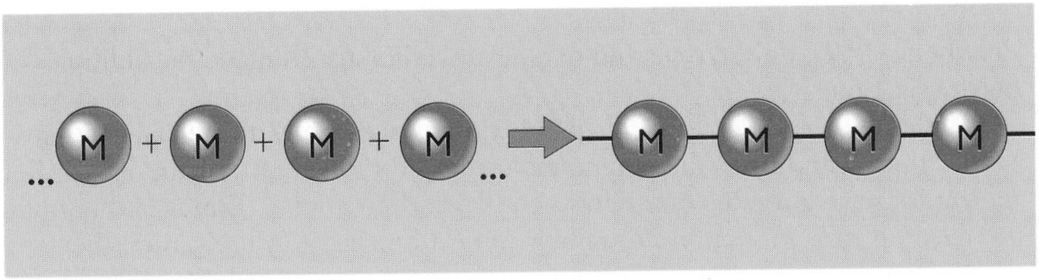

Some plastics like Nylon and Terylene are made of two different monomers.

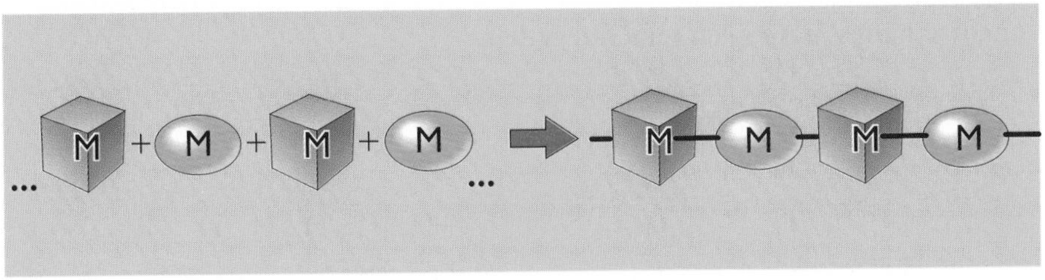

The process of joining together small molecules, called monomers, to form a large polymer is called **polymerisation**.

QUESTIONS

1 Look at the photo of the windsurfer below. The board and mast are made almost entirely from polymers.

(a) The board of the surfer is made from polyurethane foam covered with a polythene 'skin'. Why is polyurethane foam used for the inside? Why is a polythene 'skin' used?
(b) The mast is made of polyester (Terylene) reinforced with glass fibre. Why are these two materials used?
(c) The sail is made from Terylene. What are the advantages of using Terylene?

2 Here is a list of some uses for plastics in our homes.
roof insulation – *expanded polystyrene*
ceiling tiles – *expanded polystyrene*
cavity wall insulation – *polyurethane foam or expanded polystyrene*
outdoor wall cladding – *rigid PVC*
window frames – *rigid PVC*
corrugated roofing – *rigid PVC*
water tanks – *polythene or polypropene*
drainpipes and guttering – *rigid PVC*
carpet – *nylon*
kitchen worktops – *melamine*
crockery – *melamine*
damp-proof course – *polythene*
floor tiles – *PVC*
baths and washbasins – *acrylic (perspex)*
(a) Try to find out whether the articles in *your* home are made of the plastics listed.
(b) Make a list of the uses of other plastics in your home.

3 Discuss the following questions in groups of 3 or 4. Look at the uses of the plastics in question 2.
(a) For each of these uses say what materials might be used if plastics did not exist.
(b) For each of the uses try to decide why we now use plastics. What are the advantages of using plastics?
(c) For each of the uses try to decide why that particular plastic is chosen.

8 MAKING PLASTICS

Polythene sheeting emerging from an extrusion machine after manufacture

Plastics are made by polymerisation. There are two types of polymerisation:

- **additional polymerisation,**
- **condensation polymerisation.**

Addition polymerisation

Polythene, PVC, polystyrene and perspex (acrylic) are all made by addition polymerisation.

Polythene is made by heating ethene at high pressure with special catalysts. Under these conditions the ethene molecules add together (polymerise). Ethene molecules have a double bond between two carbon atoms. (Bonds and bonding are covered fully in section H, units 5 and 6.) During polymerisation, these double bonds 'open up' and the carbon atoms on separate ethene molecules join together to form a molecule of polythene. Polythene is short for poly(ethene). The name polythene is used because 'poly' means many. So polythene means 'many ethenes' have been joined together.

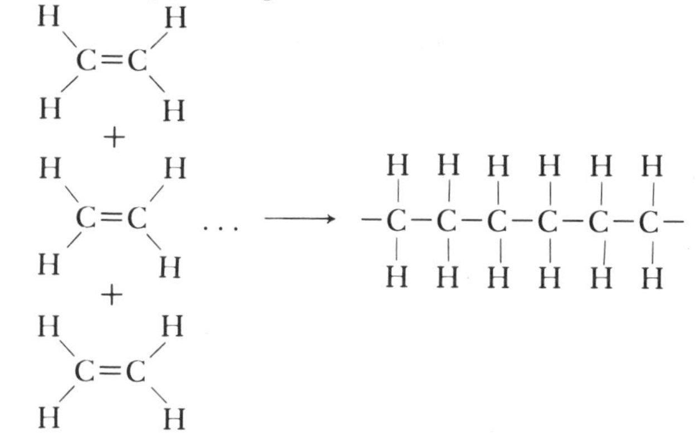

ethene molecules (monomers) — part of the polyethene molecule (polymer)

When ethene forms polythene, very long chains are produced. These long molecules contain anywhere between 500 and 50 000 carbon atoms.

Three other plastics formed by addition polymerisation are listed in Table 1. The table also shows the structure of the monomers which are used to make each polymer.

Recycled polythene is used in industries where the toxicity of plastic is not a critical factor

Plastic (polymer)	Monomer	
PVC	vinyl chloride (chloroethene)	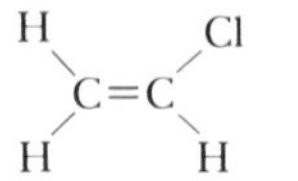
Polystyrene	Styrene	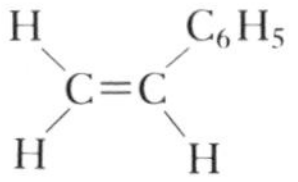
Perspex (acrylic)	methylmethacrylate	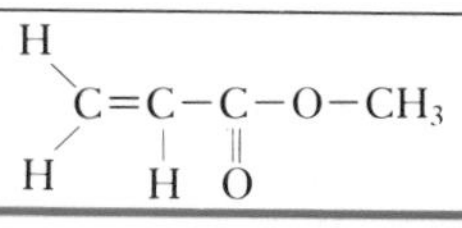

Table 1 Some plastics and the monomers used to produce them

Condensation polymerisation

Nylon and Terylene are made by condensation polymerisation. These polymers are made from two different monomers, not just one like addition polymers. The two different monomers have reactive atoms at the ends of their molecules. These react with each other joining the two monomers together. For example, Nylon is made from

diaminohexane

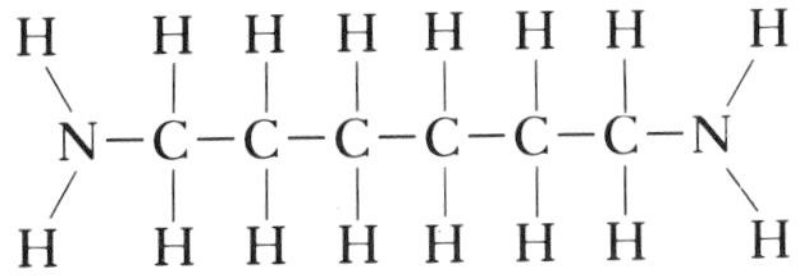

and

hexanedioyl-dichloride

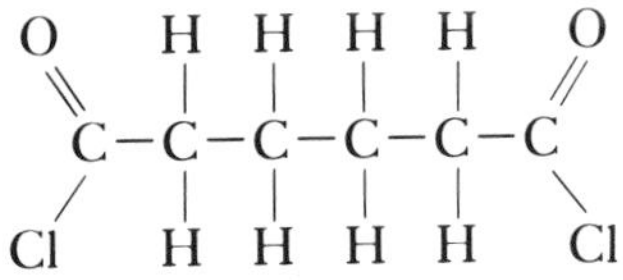

These complex molecules can be shown much simpler as

When one molecule of diaminohexane reacts with one molecule of hexanedioyldichloride, they join together. Hydrogen chloride is also produced.

But this reaction can occur again and again at the ends of the product, forming a very long polymer.

Figure 1 shows a method which you could use to make Nylon.

Most plastics consist of molecules with very long, thin chains. Sometimes, however, the long polymer molecules can form bonds with each other at points along the chain. This produces cross-linked three-dimensional structures which are harder and more rigid. Bakelite, melamine, resins and epoxyglues are examples of these plastics.

Once a plastic has been produced, it must then be turned into a useful article. This is usually done by

- moulding the warm, soft plastic under pressure, or
- **extruding**, or 'pushing out', the warm, soft plastic into different shapes by forcing it through nozzles or between rollers.

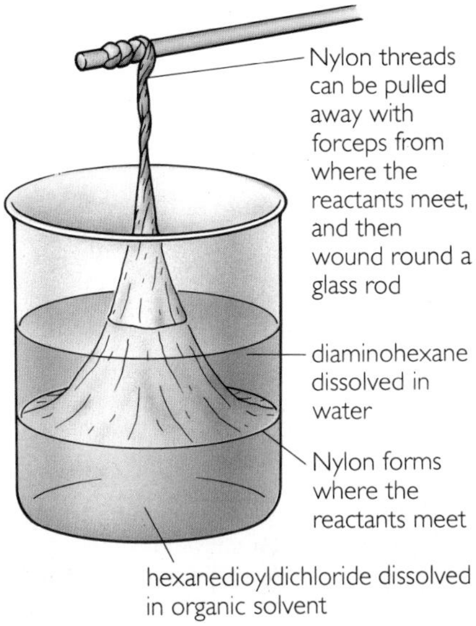

Figure 1
Making Nylon

QUESTIONS

1(a) Use molecular models to make three ethene molecules.
(b) Using these three ethene molecules, break the double bonds and produce a short portion of polythene chain. (The equation on the opposite page will help you with this.)
2 Look at the structures of the monomers in Table 1.
(a) Which part of their structures do they have in common?
(b) What do you think happens when these monomers polymerise to form polymers?
(c) Draw a short section of the polymer chain for PVC. Show six carbon atoms in the chain.
3 Find out what you can about the processes used to make everyday articles from plastics. Some processes to look out for are:
vacuum forming, extruding, moulding, foaming, thermoforming.

9 THE STRUCTURE OF PLASTICS

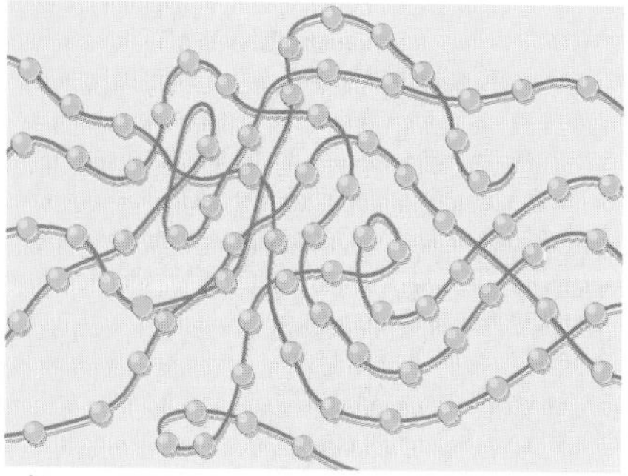

Figure 1
Tangled chains of long, thin molecules in a thermosoftening plastic

When polythene and Nylon are warmed, they become *soft* and can be moulded. As they cool, they set hard again. Because of this, they are called **thermosoftening plastics**. PVC, polystyrene and acrylic are also thermosoftening plastics.

Thermosoftening plastics contain long, thin molecules which form tangled chains (Figure 1).

In these plastics there are strong forces between the atoms along each chain, but weak forces between one chain and another. This means that the chains can move over each other easily on stretching, flexing or warming.

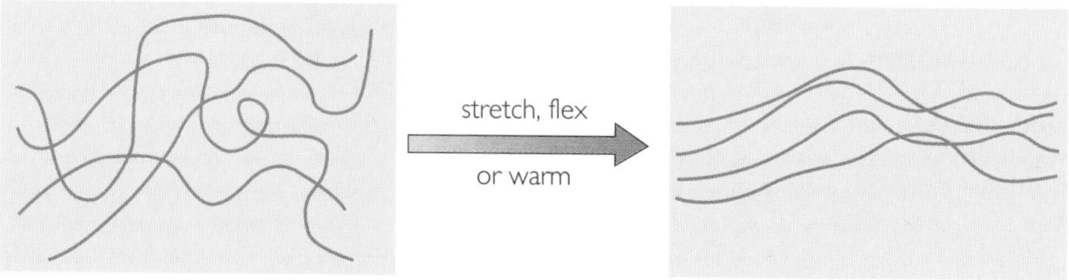

This plastic piping is made from a thermosoftening plastic. It has been warmed and then moulded into the required shape

So, thermosoftening plastics

- stretch easily,
- are flexible,
- soften on warming,
- melt at low temperatures,
- can be shaped by warming and then moulded or extruded.

When Bakelite, melamine and resin are heated, they do *not* become soft. Once these plastics have been produced and moulded, they set hard and cannot be re-melted. Because of this, they are called **thermosetting plastics**. Urea-formaldehyde resin, polyester resin and epoxy glues are also thermosetting plastics.

Thermosetting plastics contain a network of large cross-linked molecules (Figure 2).

In these cross-linked plastics, there are strong forces between the atoms along each chain *and* also strong cross-link forces between one chain and another. This means that the chains cannot move over each other on stretching or warming.

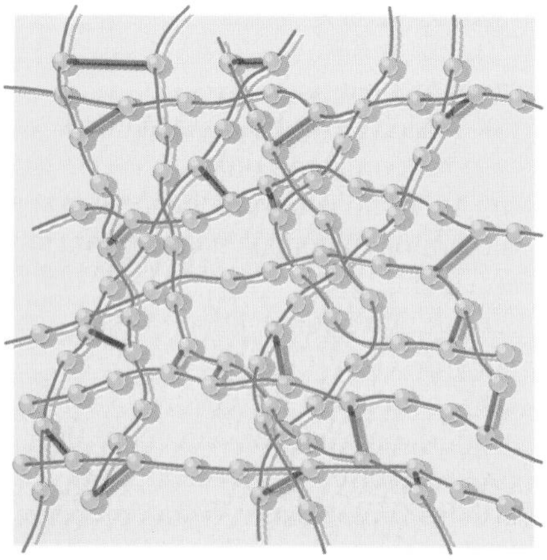

Figure 2
Cross-links in a thermosetting plastic

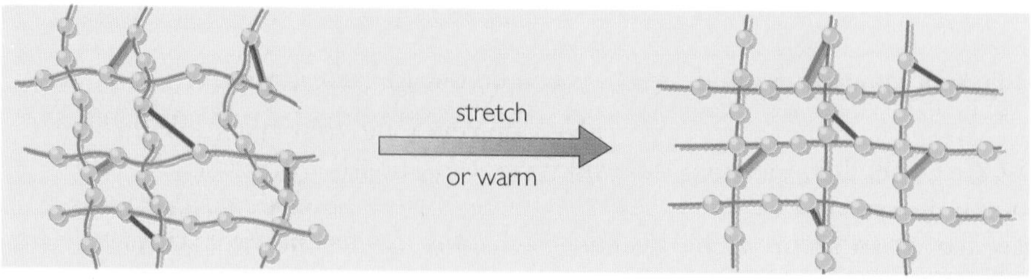

So, thermosetting plastics

- are hard and rigid,
- do not flex,
- burn or char on heating,
- do not soften or melt on heating.

Some thermosetting polymers form cross-links and harden at room temperature when a catalyst is added. One example of this is the polyester resin used in repair kits for car bodywork (see right). The polyester molecules form cross-links and harden when mixed with the catalyst (hardener).

Bakelite is a thermosetting plastic. It was one of the first plastics to be developed and used for everyday objects like this sugar jar. It is not used widely these days because it is relatively fragile and expensive to produce, compared with modern plastics

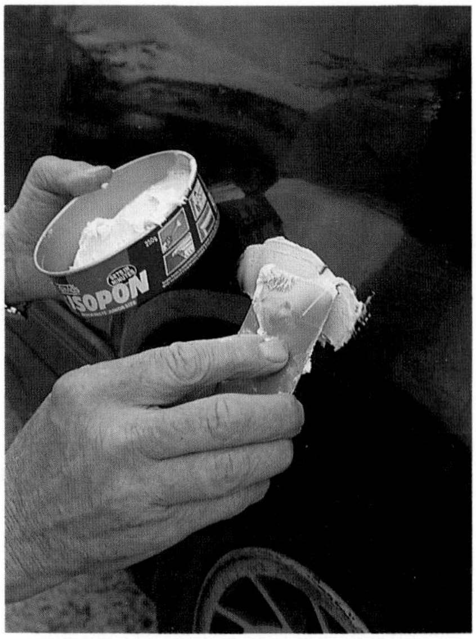

QUESTIONS

1 (a) Thermosoftening plastics can be compared to chocolate when it is warmed. Why is this?
(b) Thermosetting plastics can be compared to an egg when it is warmed. Why is this?
(c) What happens to the polythene molecules in clingfilm when it is stretched around some cheese?

2 The properties of six important plastics are shown in the table below. Use the data in the table to suggest which plastic could be used to manufacture
(a) containers for lemon juice,
(b) laboratory safety screens,
(c) containers for soft cheese,
(d) the handles for screwdrivers,
(e) the inside surface of frying-pans,
(f) tea mugs.
In each case explain why you chose the particular plastic.

Plastic	Relative strength	Flexibility	Max temperature for use /°C	Resistance to dilute acids	Resistance to oils	Clarity	Cost /£ per kg
High density polythene	4	fairly stiff	150	excellent	good	poor	0.8
Low density polythene	1	very floppy	70	good	good	poor	0.7
Perspex	9	stiff	90	good	good	excellent	1.6
Polycarbonate	9	stiff	140	excellent	poor	good	3.0
PTFE	30	fairly flexible	250	excellent	excellent	poor	20.0
Urea-formaldehyde resin	9	very stiff	75	poor	good	poor	0.9

10 PLASTIC WASTE

What should we do to prevent plastic rubbish from littering our streets and countryside?

In some areas, plastic waste is a serious problem. Plastic bottles, polythene bags and other plastic articles litter our streets and the countryside. Farmers have even complained of cows choking and dying after swallowing plastic bags.

Unfortunately, the mess from plastic waste is made worse because most plastics are **non-biodegradable**. This means that they cannot be broken down by bacteria and the weather, like paper, wood and steel. So plastic rubbish lies around for years and years. In the last few years, biodegradable plastics have been developed. These are now used for bags and other wrappings.

Getting rid of plastic waste

There are 3 ways in which we deal with plastic waste at present.

- **Landfill**

Most plastic waste is dumped with other waste into rubbish tips.

A landfill site for refuse

- **Incineration**

Plastics produce a lot of energy when they burn. So plastics are burnt in incinerators with other combustible waste. The heat produced can be used to heat homes and factories or to generate electricity.

Unfortunately, poisonous fumes are produced when some plastics are burnt. These fumes must be removed before the waste gases are released into the air.

- **Recycling**

About 2 million tonnes of plastic waste is produced in the UK each year. At present, only a small amount of this is recycled. Thermosoftening plastics are fairly easy to recycle. They can be melted or softened and then remoulded. This is shown on the left-hand side of Figure 1.

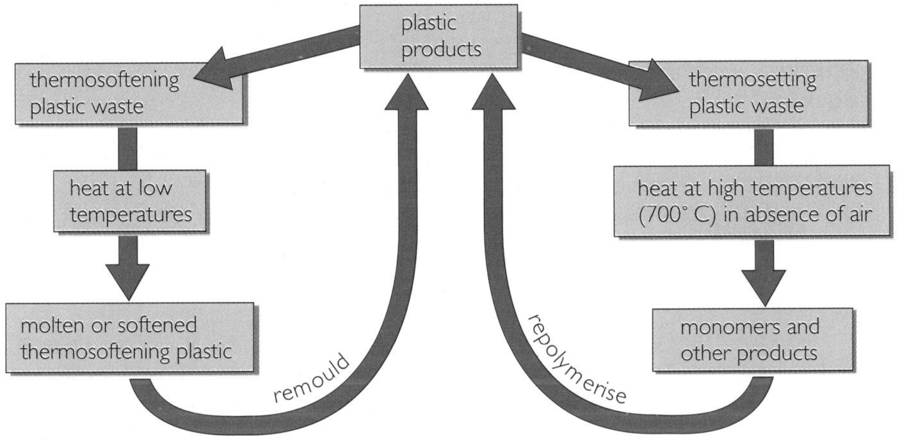

Figure 1
Recycling plastics

Thermosetting plastics are more difficult to recycle. They cannot be melted without decomposing. They are usually recycled by heating to about 700 °C in the absence of air. This process is called **pyrolysis**. Under these conditions, the polymers cannot burn. Instead, they decompose to form their original monomers and other simple molecules. The original molecules can then be re-polymerised. This is shown on the right-hand side of Figure 1.

At present, it is uneconomical to recycle most plastic waste. This is largely because of the cost of collecting and separating it, and the cost of energy in the recycling process. It is still cheaper to make new plastics from crude oil.

However, as oil becomes more scarce and more expensive, recycling will become more economical. This is why research into pyrolysis and other recycling processes is being carried out.

Recycling polythene

QUESTIONS

1 A small packaging company produces 50 kg of plastic waste each week. This could be used in an incinerator to provide hot water and heating for the factory. The incinerator would cost £1000.

Material	Heat produced by burning 1 kg
Polythene	40 000 kJ
Polystyrene	40 000 kJ
PVC	20 000 kJ
Heating oil	40 000 kJ

(a) Use the table to calculate how many kilograms of heating oil the plastic waste could save each week. Explain your calculation.
(b) What would this save in the cost of heating oil per week? (Assume heating oil costs £2 per kg.)
(c) How long would it take the company to save enough money to pay off the cost of the incinerator?

2 If plastics are not burnt completely, poisonous substances are produced. Some plastics burn to produce acidic gases containing hydrogen chloride.

Design an incinerator in which plastics could be burnt completely. The incinerator must not expel any unburnt plastic or acidic gases.

When foam-filled furniture burns, the fumes produced are very poisonous

3 Get into a small group with 2 or 3 others.
(a) Make a list of the main problems of recycling plastic waste.
(b) Make another list showing the advantages of recycling plastic waste.

SECTION E: THINGS TO DO

Activities

The Eiffel Tower is made of steel. Why is steel used? What other important structural materials do we have for large buildings and bridges?

1 The importance of iron and steel

Iron is the most important metal for industry and society. There are four reasons for this.

- **It is strong**. Although pure iron is weak, it can be alloyed with carbon to make steel, which is very strong. Steel has a high tensile strength. This means that it does not crack, break, stretch or bend easily. Large forces can deform steel, but it returns to its original shape when the force is removed.
- **It is easy to work**. Iron is very ductile and very malleable when it is hot. This makes it very useful for manufacturing different articles.
- **It forms alloys**. Iron can form a large range of alloys with a wide variety of different properties.
- **It is cheap**. Iron is the cheapest of all metals.

Because of these four factors, iron is used in greater quantity than any other metal.

1 List the factors which make iron more important than any other metal.

2 What do the following mean?
(i) tensile strength
(ii) ductile
(iii) malleable

3 The prices of metals are given in the 'commodities' section of some daily papers and also on teletext. Look up the prices for iron (steel), aluminium, copper and zinc. How does the price of iron (steel) compare with the other metals?

4 Suggest three reasons why iron is the cheapest of all metals.

5 List five important uses of iron (steel).

2 Plascup – a new material for throw-away cups

When was the last time you used a disposable (throw-away) cup? Did you burn your fingers? Did you enjoy the drink? Suppose you are the manager of a firm which makes throw-away cups for hot or cold drinks. A new material called *Plascup* has been produced with the following properties.

- It is easily coloured with dyes.
- It is a poor conductor of heat.
- It is easily cut with a knife.
- It starts to soften at 90°C.
- It is biodegradable.

1 Which property or properties make *Plascup* a good material for throw-away cups? Give a reason for your answer.

2 Which property or properties make *Plascup* unsuitable for throw-away cups? Give a reason for your answer.

3 Suggest one other property of *Plascup* that you might want to know before deciding to use it for throw-away cups.

What are the advantages of using throw-away food containers? What are the disadvantages?

3 Mining and producing copper

Look carefully at the information below.

Copper

What is the main mineral ore? Copper pyrites (chalcopyrites) a mixture of CuS and FeS forming $CuFeS_2$.
How is it found? Underground, mixed with other copper ores such as chalcocite (Cu_2S). About 0.5% of the ore is copper.
Where is it mined? One third of all copper ores mined in the West comes from Chile and the USA. The other major mining countries are Canada, Zambia and Zaire.
How is it mined? By open pit mining or underground mining.
What is the world production of copper? See Table 1. About one seventh of the copper produced is recycled from scrap. This is called **secondary copper**. Copper produced from its ores is called **primary copper**.

Country	Production in tonnes
USA	1 500 000
Japan	935 000
Chile	879 000
Zambia	522 000
Canada	505 000
Belgium	400 000
West Germany	379 000
UK (secondary copper only)	140 000

Table 1 The major producers of copper shown in order. The figures are from 1984.

Mining copper ore in Arizona. The different colours are due to malachite (green copper(II) carbonate), azurite (blue hydrated copper (II) carbonate), and cuprite (red copper (I) oxide)

1 What are the chemical names for
(a) CuS and FeS,
(b) Cu_2S?

2 What percentage of the ore is copper?

3 What mass of the ore is needed to produce a 1 kg piece of copper?

4 Only a small percentage of the ore is copper. Suggest three problems that this might cause.

5 The copper ore mined in Chile in 1984 was sufficient to produce nearly 1 300 000 tonnes of copper.
(a) How much copper did Chile actually produce in 1984?
(b) How do you account for the difference between the actual copper produced and the possible production of 1 300 000 tonnes?

6 What is the difference between primary copper and secondary copper?

7 Look at the countries where copper is mined. Which country is surprisingly absent from the major producers of the metal?

8 Make a bar chart showing the total copper production in 1984 for the seven most important producers.

Study Questions

1 Consider the following five alloys, *A*, *B*, *C*, *D* and *E*.

Alloy	Elements in the alloy	Properties
A	lead and tin	melts at 203°C
B	bismuth, cadmium, lead and tin	melts at 70°C
C	carbon, iron and tungsten	unaffected at high temperatures
D	copper and zinc	golden colour, does not tarnish
E	aluminium and lithium	low density, high strength

Which alloy would you use for
(a) joining electrical wires?
(b) making jewellery?
(c) 'plugging' an automatic fire sprinkler?
(d) making aircraft bodywork?
(e) making a drill for bricks and stone?

2 Look at the table below. It shows metals and alloys which are used in different parts of a car. For each of the parts listed, say why that particular metal or alloy is used.

Car part/component	Metal/alloy used
Bodywork and petrol tank	mild steel (iron with 0.15% carbon)
Pistons	aluminium alloy
Door handles, surrounds to headlamps	chromium-plated steel
Contact breakers	platinum
Lamp filaments	tungsten
Radiator	copper

3 (a) From the reactivity series choose a metal that
(i) occurs uncombined in the Earth's crust,
(ii) reacts vigorously with cold water,
(iii) is used in flares.
(b) Tin is slightly more reactive than lead. Predict what you would *see* when tin foil is placed in dilute hydrochloric acid.
(c) A gas may be made by passing steam over heated zinc using the apparatus shown below.

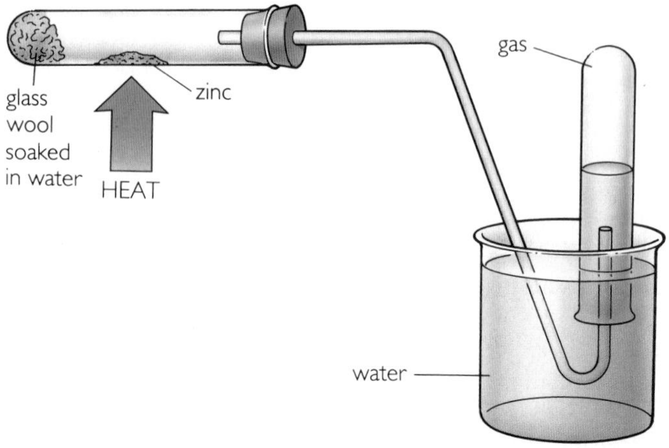

(i) Name the gas collected.
(ii) Write a word equation and a symbolic equation for the reaction.
(iii) Why should you wear safety goggles when carrying out this experiment?
(iv) Which substance has acted as a reducing agent in this reaction? Explain your choice.
(d) Chromium (Cr) may be prepared by heating chromium(III) oxide with aluminium powder.
(i) What does this tell you about the relative positions of chromium and aluminium in the reactivity series?
(ii) Write a word equation and a symbolic equation for the reaction.

4 (a) Draw a section of a polythene molecule.
(b) What bonds are present in polythene?
(c) What holds the molecules together in a sheet of polythene?
(d) Polythene is
(i) light,
(ii) easily moulded,
(iii) a good electrical insulator,
(iv) resistant to water, acids and alkalis.
Explain why polythene shows these properties.
(e) State 2 uses for polythene and explain how these uses relate to its properties.

5 (a) What is a plastic?
(b) What is the difference between thermosoftening plastics and thermosetting plastics?
(c) Give one example of a thermosoftening plastic and one example of a thermosetting plastic. Mention one use for each plastic.
(d) How do the structures of thermosoftening plastics and thermosetting plastics differ?

SECTION F

Patterns and properties

The periodic table is like a jigsaw. In a jigsaw the pieces must be arranged correctly to give a pattern or a picture. In the periodic table, elements are arranged in order of relative atomic mass to give a pattern in properties

1 PATTERNS OF ELEMENTS

Why are graphite strips used around the nose cones of aircraft and rockets like this?

One useful way of classifying elements is as metals and non-metals (see section A, unit 4). Unfortunately, it is not easy to classify some elements in this way. Take, for example, graphite and silicon. These two elements have high melting points and high boiling points (like metals) but they have low densities (like non-metals). They conduct electricity better than non-metals but not as well as metals. Elements with some properties like metals and other properties like non-metals are called **metalloids**.

Because of this limitation in classifying elements neatly as metals and non-metals, scientists looked for patterns in the properties of smaller groups of elements.

Mendeléev's periodic table

During the nineteenth century, several scientists tried to find some pattern in the properties of elements.

The most successful of these attempts was made by the Russian chemist Dmitri Mendeléev in 1869.

Mendeléev wrote down all the known elements in order of their relative atomic masses. He then arranged the elements in horizontal rows so that elements with similar properties appeared in the same vertical column. Because of the periodic repetition of elements with similar properties, Mendeléev called his arrangement a **periodic table**.

Figure 1 shows part of Mendeléev's periodic table. Notice that elements with similar properties, such as sodium and potassium, fall in the same vertical column. Which other pairs or trios of similar elements appear in the same column of Mendeléev's table?

Dmitri Mendeléev was the first chemist to successfully arrange the elements into a pattern linking their properties and relative atomic masses

- The vertical columns of similar elements are called **groups**.
- The horizontal rows are called **periods**.

Mendeléev had some brilliant ideas in connection with his periodic table.

- He left gaps in his table so that similar elements were in the same vertical group. Four of these gaps are shown as asterisks in Figure 1.
- He suggested that elements would be discovered to fill the gaps. The four gaps in Figure 1 are now filled by the elements scandium, gallium, germanium and technetium.
- He predicted the properties of the missing elements from the properties of elements above and below them in his table.

Within 15 years of Mendeléev's predictions, three of the missing elements had been discovered. Their properties were very similar to his predictions.

The success of Mendeléev's predictions showed that his ideas were probably correct. His periodic table was quickly accepted as an important summary of the properties of elements.

	GROUP							
	I	II	III	IV	V	VI	VII	VIII
Period 1	H							
Period 2	Li	Be	B	C	N	O	F	
Period 3	Na	Mg	Al	Si	P	S	Cl	
Period 4	K Cu	Ca Zn	* *	Ti *	V As	Cr Se	Mn Br	Fe Co Ni
Period 5	Rb Ag	Sr Cd	Y In	Zr Sn	Nb Sb	Mo Te	* I	Ru Rh Pd

Figure 1

QUESTIONS

1(a) Why did scientists start to use the term 'metalloid'?
(b) Name three metalloids.

2 Look at the table. This shows the predictions which Mendeléev made in 1871 for the element in period 4 below silicon. This element was discovered in 1886.
(a) What is the name for this element?
(b) Use a data book to check Mendeléev's predictions.
(c) How accurate were Mendeléev's predictions?

3 (a) Use a data book to look up the properties of aluminium (Al) and yttrium (Y).
(b) Make predictions for the element between Al and Y in Mendeléev's periodic table.
(c) Use a modern periodic table to identify the element between Al and Y.
(d) Check your predictions for this element against its actual properties.

Mendeléev's predictions for the element below silicon

1 grey metal

2 density 5.5 g cm^{-3}

3 relative atomic mass = average of relative atomic mass of Si and Sn

$$= \frac{28.1 + 118.7}{2} = 73.4$$

4 melting point higher than that of tin – perhaps about 800°C

5 formula of oxide will be XO_2.
Density of XO_2 = 4.7 g cm^{-3}

6 the oxide XO may also exist

2 MODERN PERIODIC TABLES

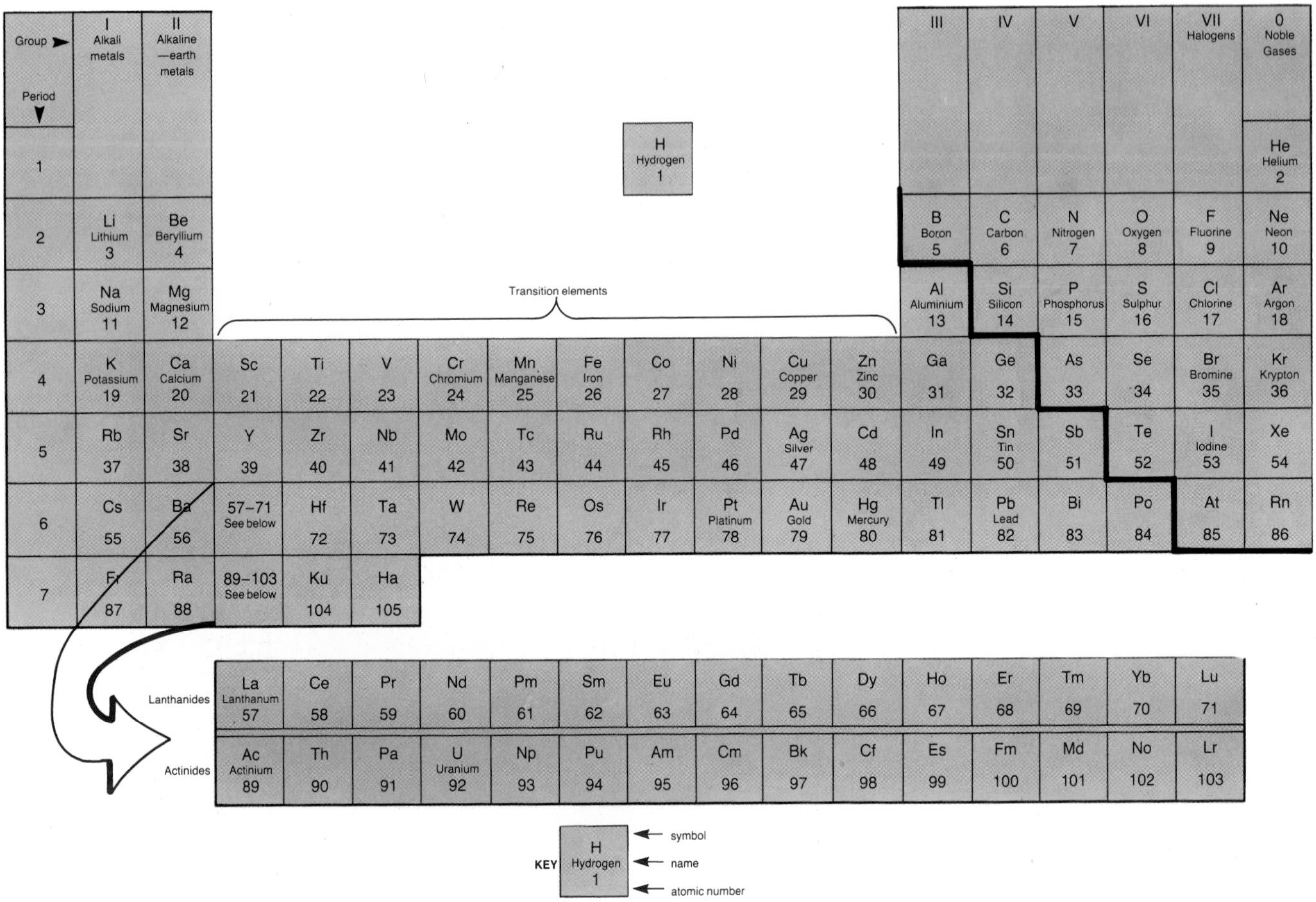

Figure 1
The modern periodic table

All modern periodic tables are based on the one produced by Mendeléev in 1869. A modern periodic table is shown in Figure 1. The elements are numbered along each period, starting with period 1, then period 2, etc. The number given to each element is called its **atomic number**. Thus hydrogen has an atomic number of 1, helium 2, lithium 3, etc. (You will learn more about atomic numbers in section K.) Notice the following points about the modern periodic table.

1 The most obvious difference between modern periodic tables and Mendeléev's is the position of the **transition elements**. These have been taken out of the simple groups and placed between group II and group III. Period 4 is the first to contain a series of transition elements. These include chromium, iron, nickel, copper and zinc.

2 Some groups have names as well as numbers. These are summarised in Table 1.

3 Metals are clearly separated from non-metals. The 20 or so non-metals are packed into the top right-hand corner above the red steps in Figure 1. Some elements close to the steps are metalloids. These elements have some properties like metals and some properties like non-metals.

Group number	Group name
I	alkali metals
II	alkaline earth metals
VII	halogens
O	noble (inert) gases

4 Apart from the noble gases, the most reactive elements are near the left-hand and right-hand sides of the periodic table. The least reactive elements are in the centre. Sodium and potassium, two very reactive metals, are at the left-hand side. The next most reactive metals, like calcium and magnesium, are in Group II, whereas less reactive metals (like iron and copper) are in the centre of the table. Carbon and silicon, unreactive non-metals, are in the centre of the periodic table. Sulphur and oxygen, which are nearer the right-hand edge, are more reactive. Fluorine and chlorine, the most reactive non-metals, are very close to the right-hand edge.

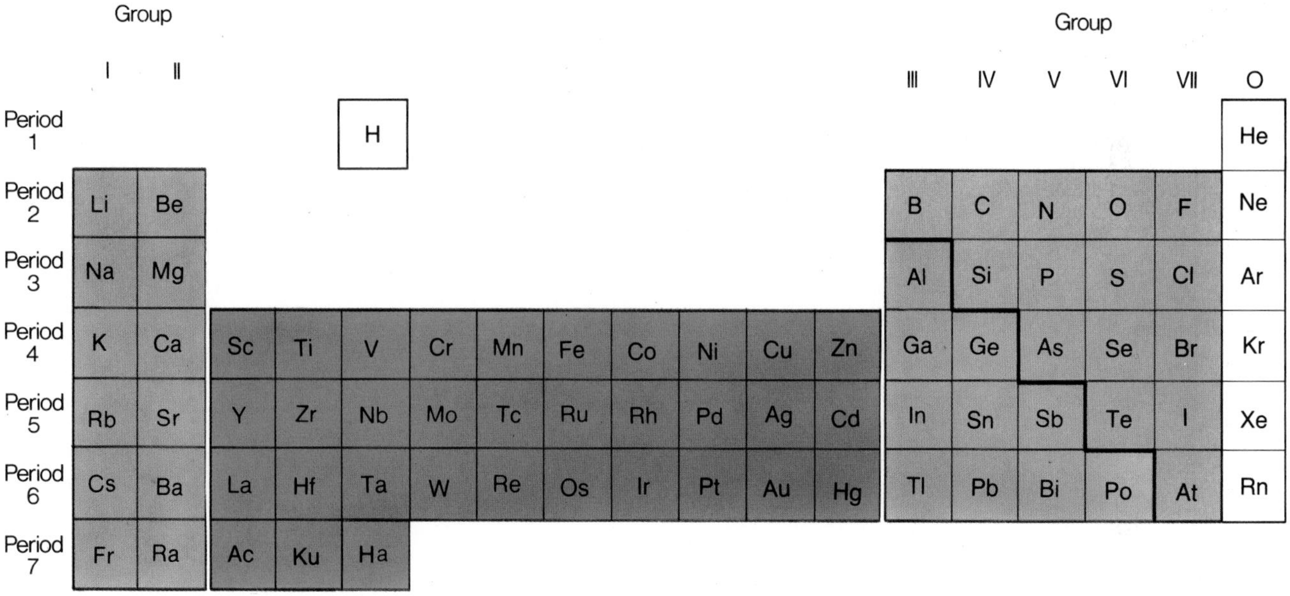

Figure 2
Blocks of similar elements in the periodic table

QUESTIONS

1 In modern periodic tables, you can pick out 5 blocks of elements with similar properties. These blocks are shown in different colours in Figure 2.
(a) The five blocks of similar elements are called non-metals, noble gases, poor metals, reactive metals and transition metals. Which name belongs to which coloured block in Figure 2?
(b) Which groups make up the reactive metals?
(c) In the transition metals, the elements resemble each other across the series as well as down the groups. Pick out two sets of three elements to illustrate these similarities.
(d) Some poor metals have properties like non-metals. Give two examples of this.
(e) The noble gases are very unreactive. The first noble gas compound was not made until 1962. Today, several compounds of them are known.
(i) The noble gases were once called 'inert gases'. Why was this?
(ii) Why do you think their name was changed to 'noble gases'?
(iii) Why are there no noble gases in Mendeléev's periodic table?
2 Draw an outline of the periodic table similar to Figure 2. On your outline, indicate where you would find (i) metals, (ii) non-metals, (iii) metalloids, (iv) elements with atomic numbers 11 to 18 inclusive, (v) the alkaline-earth metals, (vi) the most reactive metal, (vii) the most reactive non-metal, (viii) elements that might be used as disinfectants, (ix) gases, (x) magnetic elements, (xi) elements used in jewellery, (xii) elements with the highest densities.
3 Suppose you are Mendeléev. The year is 1869. You are just about to announce the discovery of your periodic table by writing a letter to the President of the Russian Academy of Sciences. Write down what you would say.

THE ALKALI METALS

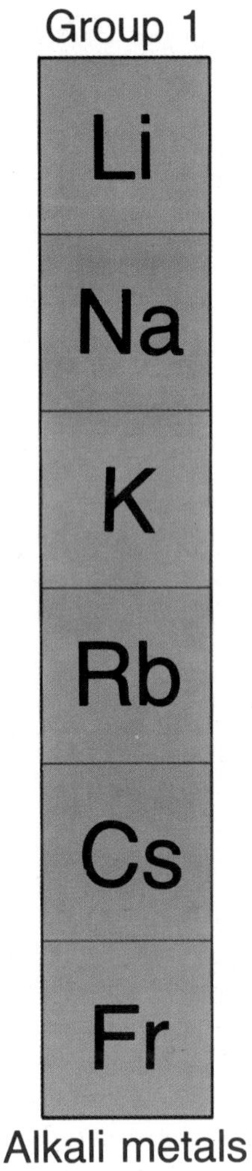

Figure 1

The elements in Group I are called the alkali metals because they react with water to form alkaline solutions. Lithium, sodium and potassium are the best known alkali metals. The other elements in Group I are rubidium, caesium and francium (Figure 1).

The alkali metals are so reactive that they must be stored under oil to protect them from oxygen and water vapour in the air. Some properties of lithium, sodium and potassium are summarised in Table 1.

Property	**Character**
Appearance	Shiny but quickly form a dull layer of oxide
Strength	Soft metals – easily cut with a knife
M.p. and b.pt.	Low compared with other metals
Density	Less than 1.0 g cm^{-3} – float on water
Reaction with air	Burn vigorously forming white oxides, e.g. sodium + oxygen → sodium oxide $4Na + O_2 \rightarrow 2Na_2O$
Reaction with cold water	Lithium reacts steadily, sodium vigorously, potassium violently. The products are hydrogen and an alkaline solution of the metal hydroxide, e.g. sodium + water → sodium hydroxide + hydrogen $2Na + 2H_2O \rightarrow 2NaOH + H_2$
Valency (combining power)	All have a valency of one Oxides are Li_2O, Na_2O, K_2O Ions are Li^+, Na^+, K^+

Table 1 Similarities of lithium, sodium and potassium

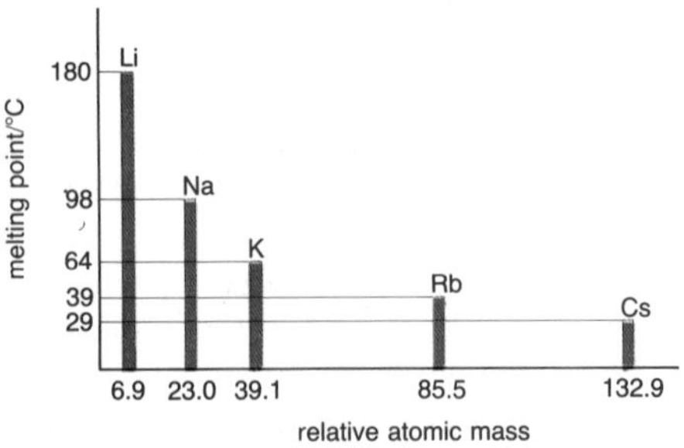

Figure 2

Notice how the elements react more vigorously with water from lithium to potassium. This is due to an increase in metallic character down the group. The ions of all alkali metals have a charge of +1. So all their compounds will have similar formulas.

Notice also in Table 1 that alkali metals have some unusual properties for metals

- Their melting points and boiling points are unusually low
- Their densities are so low that they float on water
- They are soft enough to be cut with a knife

Some of the physical properties of lithium, sodium and potassium are given in Table 2.

Element	Relative atomic mass	Melting point /°C	Boiling point /°C	Density /g cm^{-3}
Lithium	6.9	180	1330	0.53
Sodium	23.0	98	892	0.97
Potassium	39.1	64	760	0.86

Table 2 The physical properties of lithium, sodium and potassium

The large cylinder carries liquid sodium which is used as a coolant for nuclear reactors

These properties illustrate another important feature of the periodic table – *although the elements within a group are similar, there is a steady change in property from one element to the next.* The graph in Figure 2 shows the steady change in the melting points of the alkali metals very neatly.

Once we know the general properties of the elements in a group, and how these properties vary from one element to the next, we can predict the properties of other elements in the group.

Rock salt being processed above ground

The uses of sodium compounds

Most sodium compounds are made from sodium chloride. In hot countries, impure sodium chloride is left if sea water is allowed to evaporate. Sodium chloride also occurs in salt beds beneath the Earth's surface. The sodium chloride can be obtained from these salt beds by *solution mining.* This involves piping hot water down to the salt bed to dissolve the salt. Concentrated salt solution is then pumped to the surface. Impure salt is used for de-icing roads. Pure salt is used in cooking. Sodium and chlorine can be produced by electrolysing molten salt (section D, unit 4). Liquid sodium is used as a coolant in fast nuclear reactors and sodium vapour provides the yellow glow in street lamps.

QUESTIONS

1 Write word equations and then balanced equations with symbols for the reaction of potassium with (i) oxygen, (ii) water, (iii) chlorine.

2 Rubidium (Rb) is in Group I below potassium. Use the information in Tables 1 and 2 to predict (i) its melting point, (ii) its boiling point, (iii) its density relative to that of water, (iv) its relative atomic mass, (v) the symbol of its ion, (vi) the formula of its oxide and chloride, (vii) its reaction with water, (viii) how it burns in air.

3 Find out about (i) the manufacture of sodium, (ii) the use of liquid sodium as a coolant, (iii) the use of sodium vapour in street lamps, (iv) the manufacture of other sodium compounds from sodium chloride.

4 Draw a poster or diagram illustrating or summarising the manufacture and uses of sodium and its compounds.

5 A metal M is in Group III of the periodic table. Its chloride has a formula mass of 176.5.

(a) Write the formula of the chloride of the metal.

(b) Calculate the atomic mass of the metal. (Chlorine has relative atomic mass = 35.5.)

4 THE TRANSITION METALS

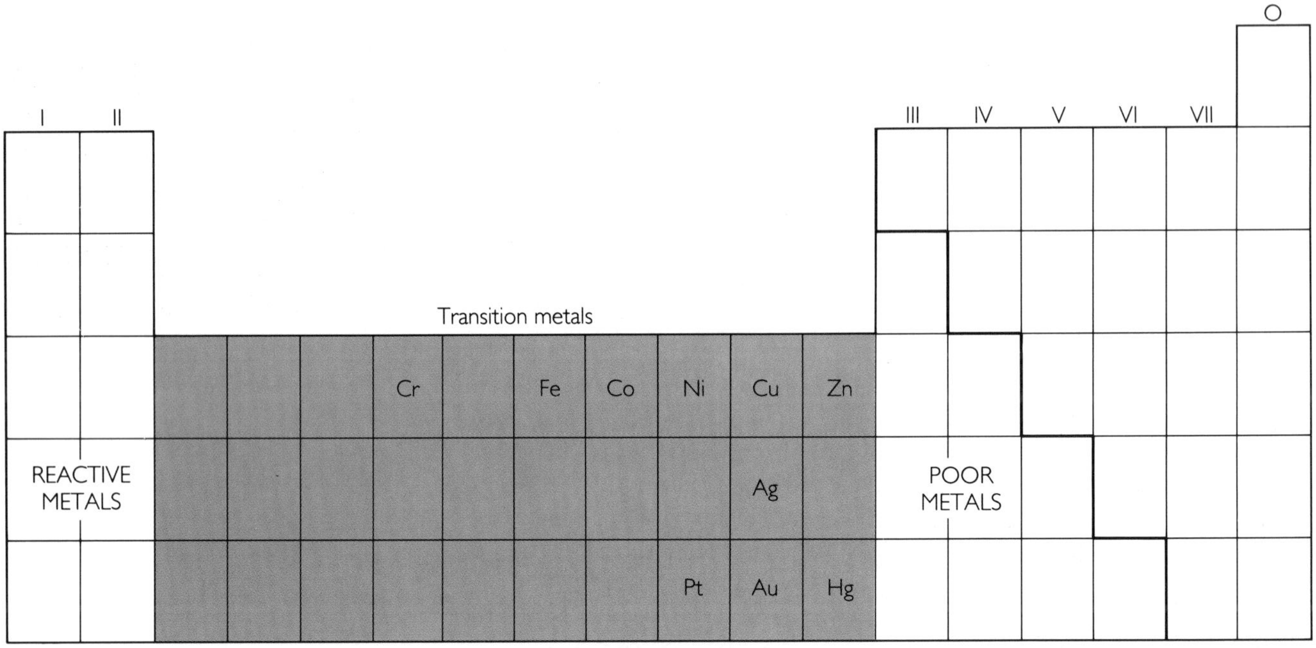

Figure 1

Figure 1 shows some of the common transition metals in the periodic table. The transition metals separate the reactive metals in Groups I and II from the triangle of poor metals.

> The transition metals have very similar properties. Unlike other parts of the periodic table, there are similarities in the transition metals across the periods as well as down the groups.

The most important transition metals are iron and copper. Iron is the most widely used metal. About 700 million tonnes of it are manufactured every year throughout the world. It is used as steel in girders and supports for bridges and buildings, in vehicles, in engines and in tools.

After iron and aluminium, copper is the third most widely used metal. About 9 million tonnes are manufactured each year. Copper is a good conductor of heat and electricity. It is also malleable (it can be made into different shapes) and ductile (it can be drawn into wires). Because of these properties, copper is used in electrical wires and cables and in hot water pipes and radiators. The uses of copper are increased by *alloying* it with other metals. Alloying copper with zinc produces brass which is harder and stronger than pure copper. Alloying copper with tin produces bronze. This is stronger and easier to cast into moulds than pure copper. There is more about alloys in section E.

This sculpture has been made from copper. Copper is a popular material with sculptors because it is easy to shape, and its colour gives an attractive finish

General properties of transition metals

Some properties of iron and copper are shown in Table 1. Compare these with the properties of alkali metals in Tables 1 and 2 in the last unit.

Element	Melting point /°C	Boiling point /°C	Density /g cm^{-3}	Reaction with water	Formulas of oxides	Symbols of ions	Salts
Iron	1540	3000	7.9	Does not react with pure water. Reacts slowly with steam	FeO Fe_2O_3	Fe^{2+} Fe^{3+}	Fe^{2+} salts are green Fe^{3+} salts are yellow or brown
Copper	1080	2600	8.9	No reaction with water or steam	Cu_2O CuO	Cu^+ Cu^{2+}	Cu^{2+} salts are blue or green

Table 1 Properties of iron and copper

The information in Table 1 illustrates some typical properties of transition metals.

1 High melting points and boiling points – much higher than alkali metals.

2 High densities – much higher than alkali metals.

3 Hard strong metals (high tensile strength), unlike the soft alkali metals.

Iron is used as steel in girders and supports such as those in this oil rig under construction

4 Fairly unreactive with water, unlike alkali metals. None of the transition metals react with cold water, but a few of them react slowly with steam, like iron.

5 More than one valency. Most of the transition metals can have more than one valency: Iron forms Fe^{2+} ions (valency 2) and Fe^{3+} ions (valency 3); Copper forms Cu^+ ions (valency 1) and Cu^{2+} ions (valency 2). Alkali metals only form ions of 1 + charge (valency 1).

6 Coloured compounds. Transition metals usually have coloured compounds with coloured solutions. In contrast, alkali metals have white salts with colourless solutions.

7 Catalytic properties. Transition metals and their compounds can act as **catalysts** (see section J). Iron or iron(III) oxide (Fe_2O_3) is used as a catalyst in the *Haber process* to manufacture ammonia (NH_3).

QUESTIONS

1 Make a table contrasting the properties of transition metals with those of alkali metals.

2 Write equations with symbols for the following word equations.
(a) iron + oxygen → iron(III) oxide
(b) iron + chlorine → iron(III) chloride
(c) iron + hydrogen chloride → iron(II) chloride + hydrogen

3 What is (i) an alloy, (ii) brass, (iii) bronze?

4 Use this text book and a data book to collect the following information for iron, and for copper.
(a) density
(b) malleability
(c) strength (tensile strength)
(d) conductivity
(e) reaction with air and water
(f) cost
Use this data to explain the important uses of copper and iron.

5 THE HALOGENS

VI	VII	0
		He
O	F	Ne
S	Cl	Ar
	Br	Kr
	I	Xe
	At	Rn

The elements in Group VII of the periodic table are called **halogens** (Figure 1). They are a group of reactive non-metals. The common elements in the group are fluorine (F), chlorine (Cl), bromine (Br) and iodine (I). The final element, astatine (At), does not occur naturally. It is an unstable radioactive element which was first synthesised in 1940.

The halogens are so reactive that they never occur free in nature. They occur combined with metals in salts such as sodium chloride (NaCl), calcium fluoride (CaF_2) and magnesium bromide ($MgBr_2$). This is the reason for their name, 'halogens', which means 'salt-formers'.

Sources of the halogens

The halogens occur in compounds with metals as negative ions – fluoride (F^-), chloride (Cl^-), bromide (Br^-) and iodide (I^-). The most widespread compound containing fluorine is fluorspar or fluorite (CaF_2). The main source of chlorine is sodium chloride (NaCl) in sea-water and in rock salt. Each kilogram of sea-water contains about 30 g of sodium chloride. Sea-water also contains small amounts of bromides and traces of iodides. Certain seaweeds also contain iodine but the main source of iodine is sodium iodate ($NaIO_3$). This is found in Chile mixed with larger quantities of sodium nitrate ($NaNO_3$).

Laminaria is a type of seaweed which contains iodine. You can often find laminaria at low tide on rocky beaches around Britain. Some types of seaweed are edible. They are an important source of the small amounts of iodine we need for a healthy diet

Blue John is a stone found in the hills of Derbyshire. It is mainly fluorite with streaks of blue running through it. These blue streaks are caused by traces of oil in the rocks

Patterns in physical properties

Table 1 lists some physical properties of chlorine, bromine and iodine.

How do the following properties of the halogens change as their relative atomic mass increases?

1 state at room temperature
2 colour of vapour
3 melting points (see Figure 2)
4 boiling points (see Figure 2)

Element	Relative atomic mass	Colour and state at room temperature	Colour of vapour	Structure	M.p. /°C	B.p. /°C
Chlorine	35.5	pale green gas	pale green	Cl_2 molecules	−101	−35
Bromine	79.9	red brown liquid	orange	Br_2 molecules	−7	58
Iodine	126.9	dark grey solid	purple	I_2 molecules	114	183

Table 1 Physical properties of chlorine, bromine and iodine

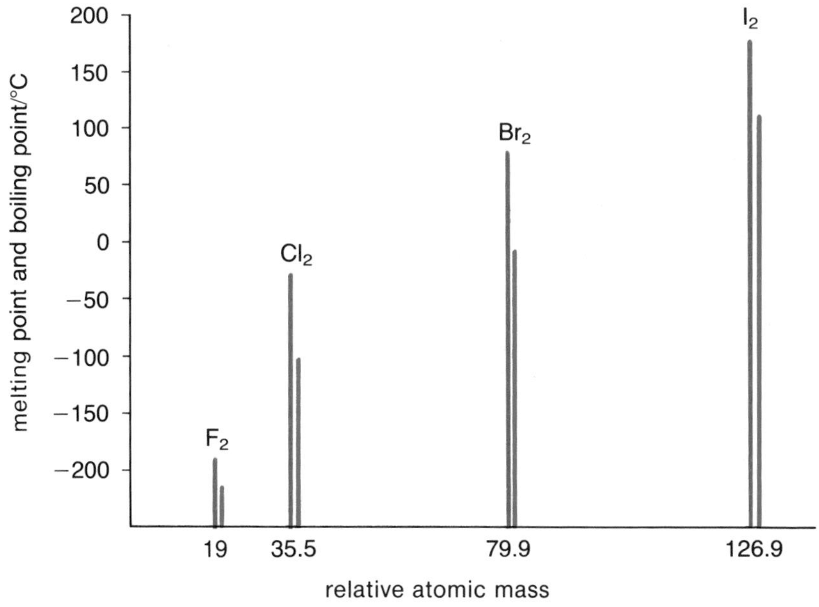

Figure 2
Melting points and boiling points of the halogens

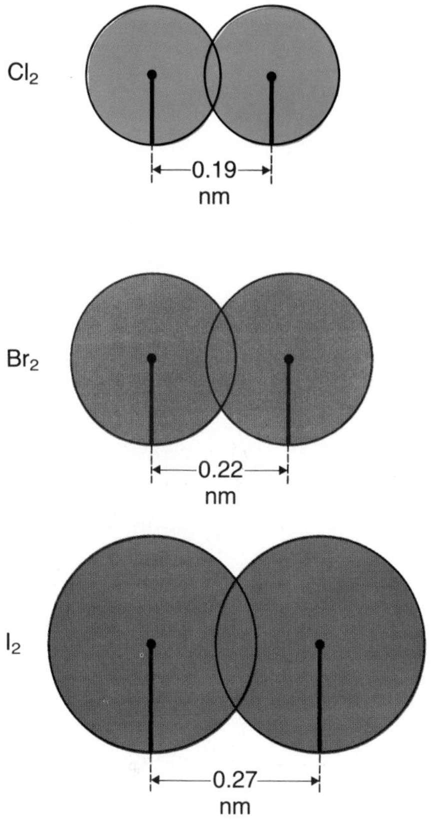

Figure 3

The halogens form diatomic molecules, F_2, Cl_2, Br_2 and I_2. The relative sizes of these molecules (Figure 3) explain the changes in volatility (i.e. melting points and boiling points) with increasing relative atomic mass.

As the molecules get larger and heavier, it is harder to break up the orderly packing within the solid to form a liquid. It is also more difficult to separate larger molecules and form the gas. This is because there are stronger intermolecular forces between the larger molecules than the smaller ones. Thus, more energy is needed to melt and to boil the larger molecules. Hence the melting points and boiling points increase with relative molecular mass. F_2 and Cl_2, the smallest halogen molecules, are gases at room temperature, Br_2 is a liquid and I_2 is a solid.

QUESTIONS

1 What are the most important sources of (i) chlorine, (ii) bromine, (iii) iodine?
2 List the characteristic physical properties of the halogens.
3 Use the information in this unit to predict the following properties of fluorine and astatine.
colour, state at room temperature, melting point, boiling point, relative atomic mass, structure
4 The following terms have been used in this unit. Discuss the meaning of each term with 2 or 3 others in your class.
unstable radioactive element, synthesised, diatomic molecules, volatility, intermolecular forces

6 REACTIONS OF CHLORINE AND THE HALOGENS

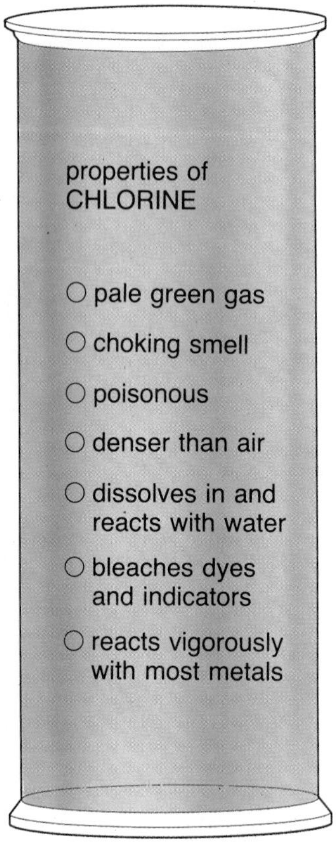

Figure 1

Chlorine

Chlorine is the most important element in Group VII. Chlorine and its compounds have many more uses than the other halogens and we shall look at these in the next unit. The important properties and reactions of chlorine are summarised in Figure 1. *Notice that chlorine has a choking smell and that it is poisonous (toxic).* Always treat it with care.

Test for chlorine

The best test for chlorine is to show that it will bleach moist litmus paper. If blue litmus paper is used, the paper first turns red (because chlorine reacts with water to form an acidic solution), and then goes white.

Test for chloride ions

When silver nitrate solution, containing silver ions (Ag^+ (aq)) is added to a solution containing chloride ions, a white precipitate of silver chloride is produced.

$$Ag^+(aq) + Cl^-(aq) \rightarrow AgCl(s)$$

The white silver chloride slowly turns purple/grey in sunlight. This reaction is used as a test for chloride ions.

Reactions with water

When chlorine is bubbled into water, it reacts to form a mixture of hydrochloric acid and hypochlorous acid.

chlorine	+	water	→	hydrochloric acid	+	hypochlorous acid
Cl_2	+	H_2O	→	HCl	+	$HClO$

The solution produced is very acidic. It is also a strong bleach which quickly turns litmus paper or universal indicator paper white.

Bromine reacts less easily, but in a similar way to chlorine. The solution produced is orange and acidic. It bleaches less rapidly than chlorine water.

Iodine dissolves in water only very slightly. The solution is a pale yellow colour. It is only slightly acidic and it bleaches very slowly.

Household bleaches, like Domestos, usually contain chlorine itself and chlorine compounds. These compounds are so reactive that they remove coloured stains by reacting with them to make colourless substances.

Household bleaches, like Domestos, usually contain chlorine itself and chlorine compounds. Because these substances are so reactive, they must be handled with great care. Bleach bottles always have hazard warnings printed on the label, and many have writing in Braille too, to warn blind or partially sighted people to be careful

Reactions with iron

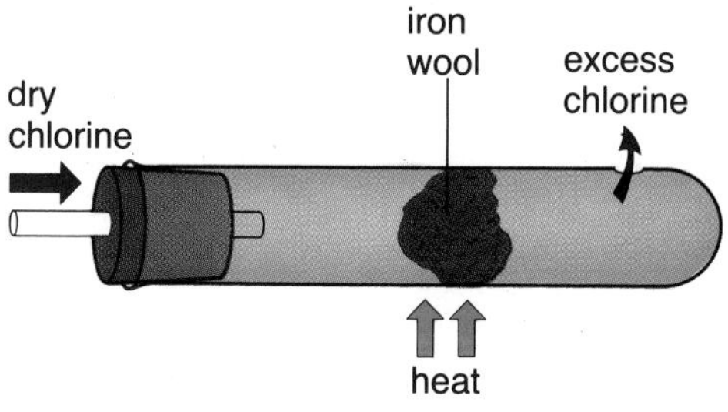

Figure 2

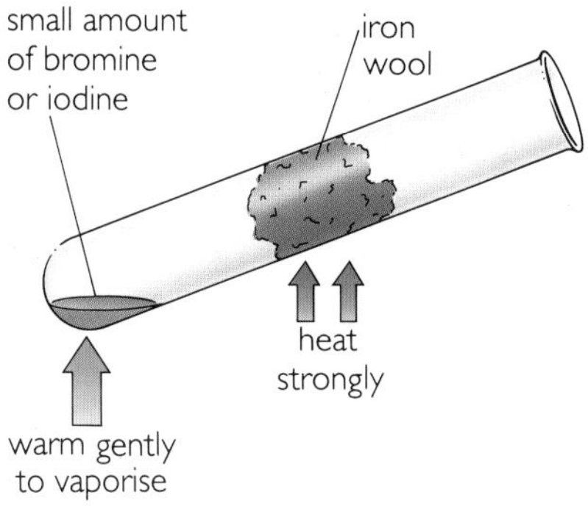

Figure 3

The reaction between chlorine and iron can be studied using the apparatus in Figure 2. The iron wool is heated to start the reaction and then the Bunsen is removed. The iron glows red hot as it reacts with the chlorine. The product is brown iron(III) chloride which collects in the tube.

iron	+	chlorine	→	iron(III) chloride
2Fe	+	$3Cl_2$	→	$2FeCl_3$

Figure 3 shows how we can investigate the reactions of iron with bromine or iodine. When iron is heated in bromine vapour, the reaction is less vigorous than that with chlorine. The iron glows but it must be heated all the time. The product is iron(III) bromide. When the experiment is repeated with iodine, the iron reacts very slowly to form iron(II) iodide.

Industrial bleaches containing chlorine and chlorine compounds are used to bleach paper and cloth. If chlorine-based products are allowed to leak into the environment, they can cause serious pollution

Comparing reactivities

The reactions of halogens with water, with iron and as bleaching agents show their relative reactivities very clearly. Chlorine is more reactive than bromine, and bromine is more reactive than iodine. Notice that *the halogens get less reactive as their relative atomic mass increases*. This is opposite to the trend in Group I. The alkali metals get *more* reactive with increasing relative atomic mass.

QUESTIONS

1 Describe experiments which show the relative reactivity of bromine, chlorine and iodine. Write word equations and balanced equations for the reactions you describe.

2 Write the five halogen elements in order of reactivity (the most reactive first).

3 Arrange the following pairs of elements in order of decreasing vigour of reaction under the same conditions. *lithium and iodine, potassium and chlorine, potassium and fluorine, sodium and chlorine, sodium and iodine*

4 The following poem by Vernon Newton is called 'Mistress Fluorine'.

Fervid Fluorine, though just nine,
Knows her aim in life: combine!
In fact, of things that like to mingle,
None's less likely to stay single.

(a) Use a dictionary to find the meaning of 'fervid' (line 1) and 'mingle' (line 3).
(b) Why is fluorine described as 'fervid'?
(c) Why is fluorine described as 'just nine'? (line 1)
(d) What does the poem say about the properties of fluorine?
(e) Try to write a short poem about any other element.

7 THE USES OF CHLORINE

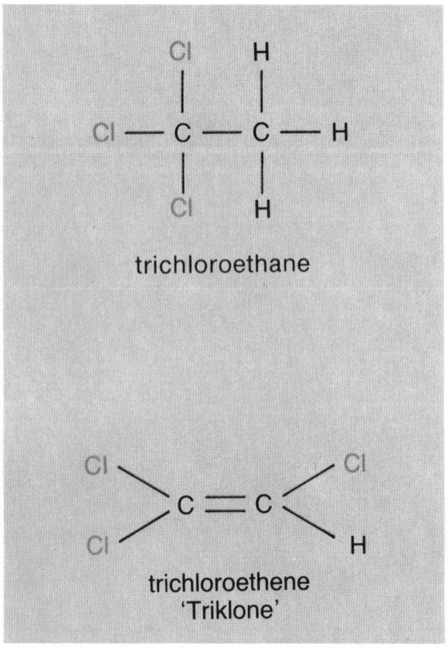

Figure 1
Solvents containing chlorine

Chlorine and its compounds have more uses than the other halogens. They are important in industry, agriculture, medicine and our homes. Two important uses of the element are in water sterilisation and in bleaches. Chlorine is added in small quantities to our water supplies. The chlorine kills bacteria in the water, but it does not affect humans or animals. Larger quantities of chlorine are added to swimming baths. The use of chlorine in domestic bleaches was discussed in unit 6 of this section.

Uses of chlorine compounds

- **Hydrochloric acid.** Large quantitites of chlorine are used to make hydrochloric acid. Chlorine is first burnt in hydrogen to form hydrogen chloride.

  $$\text{hydrogen} + \text{chlorine} \rightarrow \text{hydrogen chloride}$$
  $$H_2 + Cl_2 \rightarrow 2HCl$$

  The hydrogen chloride is then dissolved in water to produce hydrochloric acid. This is the cheapest industrial acid. It is used to clean the rust from steel sheets before galvanizing and to produce ammonium chloride.
- **Solvents**. Many of the solvents used to remove oil and grease from clothing and machinery are compounds containing chlorine (Figure 1). Tetrachloroethene is used in dry-cleaning clothes and trichloroethene ('Triklone') is used to degrease machinery. The 'thinner' (solvent) for Tippex is trichloroethane.

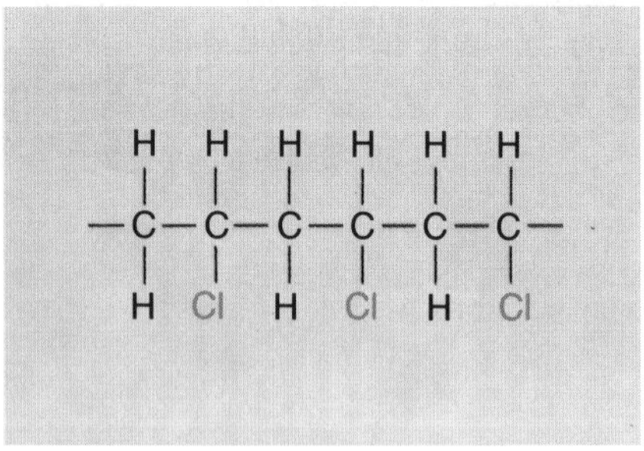

Figure 2
A section of PVC polymer

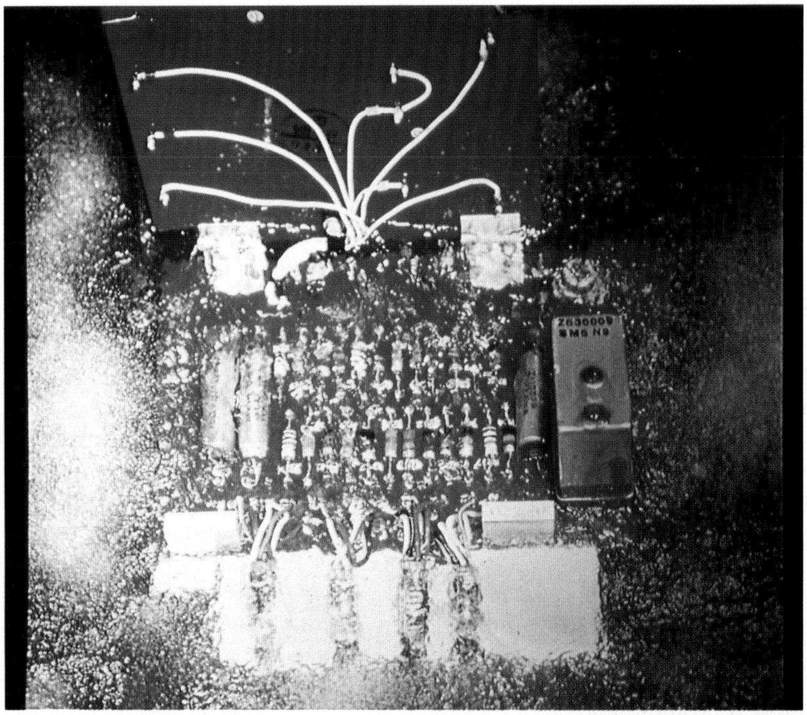

Circuit boards being de-greased and cleaned by dipping in Arklone, a chlorine containing solvent

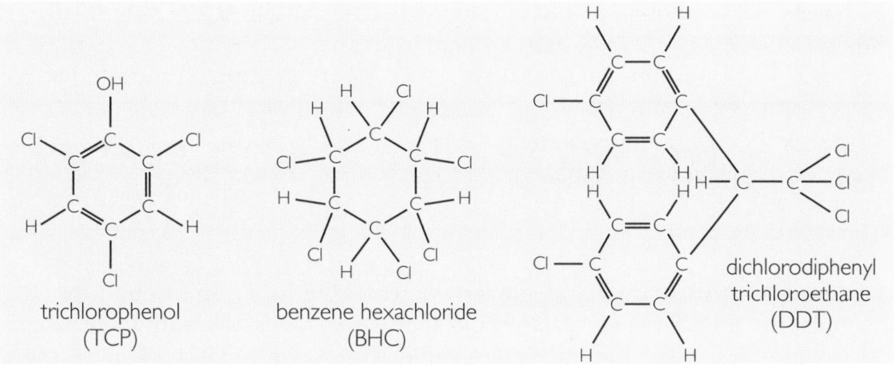

Figure 3
Disinfectants and insecticides containing chlorine

All these products are made from PVC

- **Plastics**. One of the most important plastics is PVC (polyvinyl chloride (Figure 2)). PVC plastics are used for raincoats, coverings for tables and shelves ('Fablon'), floor tiles ('Vinyl' tiles), upholstery, records and electrical insulation (section E, units 7 to 9).
- **Disinfectants**. Two of the most widely used general disinfectants are TCP (tricholorophenol – Figure 3) and Dettol, which is a similar chlorine-containing compound. These are used in many homes for treating cuts, gargling and cleaning sinks and toilets.
- **Insecticides**. During the 1940s, DDT (dichlorodiphenyl trichloroethane – Figure 3) was produced and used as an insecticide. It killed lice, flies and mosquitoes which were attacking crops and causing typhoid, dysentery and malaria. During the 1960s, however, it became clear that DDT was getting into food chains and killing birds and animals. The use of DDT was therefore banned in some countries including the UK. Its use as a pesticide has been replaced by various other chlorine compounds such as BHC (Figure 3).

DDT being used to de-louse captured soldiers during the Korean War

QUESTIONS

1 (a) How is hydrochloric acid manufactured?
(b) What are its uses?
2 Read the following poem from *The Penguin Book of Limericks*.

A mosquito was heard to complain
That a chemist had poisoned his brain;
The cause of his sorrow
Was 4,4-dichloro-
Diphenyltrichloroethane.

(a) Why do you think 4,4-dichlorodiphenyltrichloroethane is called DDT?
(b) What elements does DDT contain?
(c) What is the molecular formula of DDT?
(d) What use of DDT does the poem illustrate?
(e) During the 1960s, the bodies of birds of prey were found to contain unusually large amounts of DDT. Why was this?
(f) Why has our use of DDT changed since the 1940s?
(g) Design a poster to show the dangers of using large amounts of DDT.
3 Use molecular models to build the structures of the substances in Figures 1, 2 and 3. How do your molecular models compare with the structures drawn in the Figures?

8 PATTERNS IN THE PERIODIC TABLE

So far, we have studied a group of reactive metals (Group I) and a group of reactive non-metals (Group VII).

> In each group, the elements have similar properties with a gradual trend in properties from one element to the next.

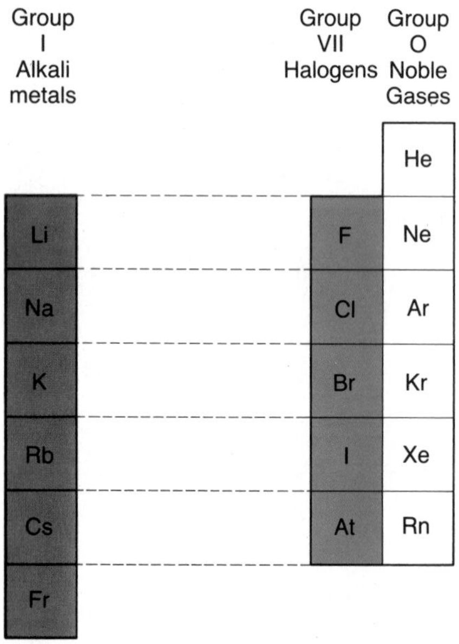

Figure 1

The groups are like families of similar elements (Figure 1). In Group I for example, the metals are all reactive, but they get *more* reactive with increase in relative atomic mass. Lithium reacts steadily with water, but potassium reacts violently. Imagine how francium will react with water.

In Group VII, fluorine is the most reactive. Here, the elements get *less* reactive with increase in relative atomic mass.

Electron structures

Chemical reactions involve changes in the electrons belonging to atoms. This led to the idea that noble gases must have very stable electron structures because they are so unreactive. They do not want to lose or gain electrons.

Scientists think that electrons occupy the outer parts of atoms in layers or **shells**. The first shell is filled and stable when it contains two electrons, like helium. The second shell is filled when it contains eight electrons. So neon with 10 electrons has 2 electrons in the first shell and eight electrons in the second shell. We say that its electron structure is 2, 8. The third shell is also stable when it contains 8 electrons. So argon is stable because its first, second and third shells are all filled with 2, 8 and 8 electrons respectively. We write the electron structure of argon as 2, 8, 8.

Figure 2 shows the first 20 elements in the periodic table. The electron structure of each element is written below its symbol. When the first shell is full, at helium, electrons go into the second shell. So the electron structure of lithium is 2, 1; beryllium is 2, 2; boron is 2, 3; etc. When the second shell is full at neon, electrons start to fill the third shell and so on. From these electron structures, we can explain why elements in the same group are so similar.

About 80 years ago, all airships were filled with hydrogen, because it is so light. However, hydrogen is highly flammable, so disasters like the one in the photo happened frequently. Now helium is used in airships. It is slightly heavier than hydrogen, but it is an unreactive Group 0 gas, so it cannot burn or explode

Group I: the alkali metals

Each alkali metal follows a noble gas in the periodic table. So all alkali metals have one electron in their outer shell (Figure 3).
By losing this one electron, their atoms form positive ions (Li^+, Na^+, K^+) with a stable electron structure like a noble gas. For example, the electron structure of Na^+ ions is 2, 8. This is like neon. *So alkali metals have similar properties because they have similar electron structures.*

- They are very reactive because they are so keen to lose the single electron in the outer shell.
- They all form ions with a charge of 1+ and so the formulas of their compounds are similar.

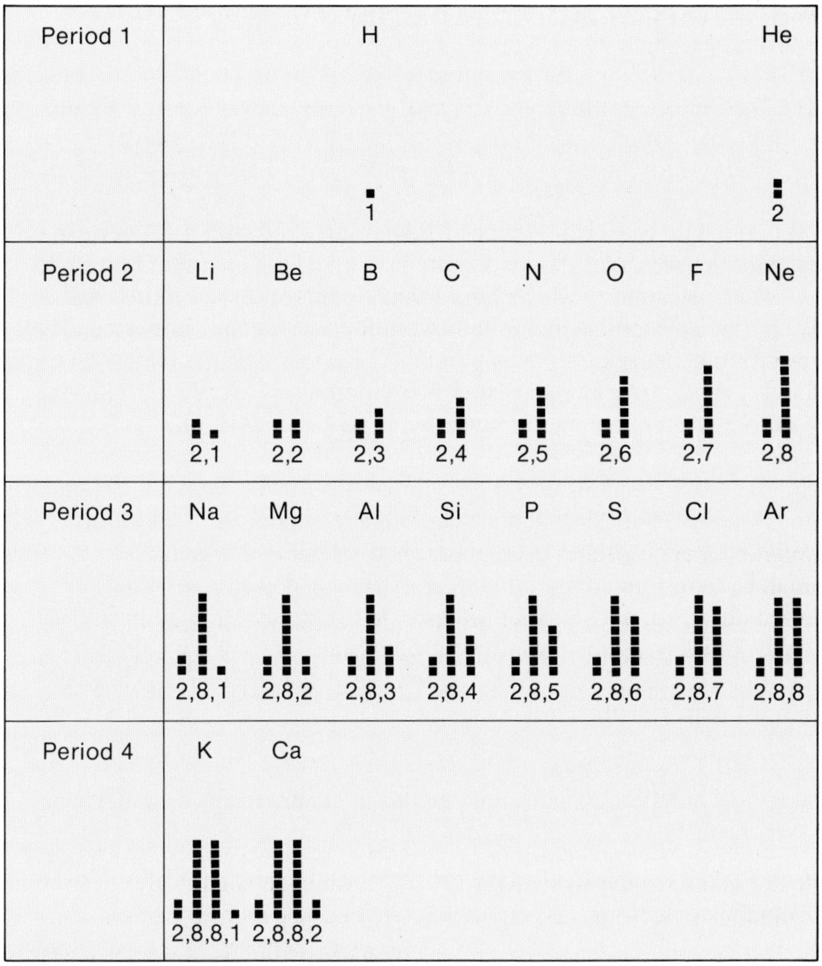

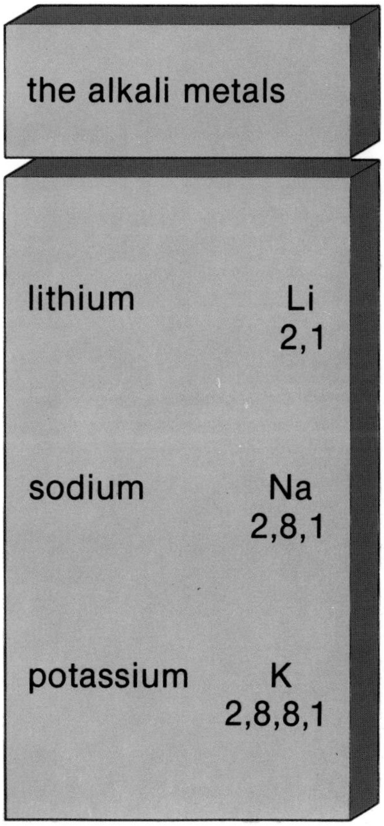

Figure 3 ↑

Figure 2 ←

Group VII: the halogens

Halogens come just before a noble gas in the periodic table. So halogen atoms have seven electrons in their outer shells. *Their similar electron structures lead to similar chemical properties.*

By gaining one electron, they form negative ions (F^-, Cl^-, Br^-) with stable electron structures like the next noble gas.

- They are reactive because they are so keen to gain one electron.
- They form ions with a charge of 1^-, so the formulas of their compounds are similar.

QUESTIONS

1 (a) Write down the electron structures of magnesium and calcium.
(b) How many electrons are there in the outer shell of an atom of an element in Group II?
(c) What charge will stable ions of Group II elements have?
(d) How many electrons are there in the outer shell of oxygen atoms?
(e) What is the charge on stable ions of oxygen?
(f) Why does magnesium react with oxygen to form an ionic compound, $Mg^{2+}0^{2-}$?

2 How many protons, neutrons and electrons will the following have? (i) N, (ii) N^{3-}, (iii) Ca, (iv) Ca^{2+}.

3 Make a summary of the key points from Section F (Patterns and Properties). You are allowed only one sheet of A4 paper.

SECTION F: THINGS TO DO

Activities

1 The discovery of anti-knock compounds

Between 1910 and 1920, motor car engines became more complicated. This resulted in 'knocking' or 'pinking'. The mixture of petrol vapour and air in a car cylinder should *not* explode until the piston is at the top of the cylinder. Knocking occurs when the mixture explodes too early. This causes the engine to work in fits and starts.

In 1916, an engineer called Thomas Midgley got a job with an American oil company. Midgley was asked to investigate how knocking could be prevented. He tried various substances which were soluble in petrol to see if they reduced knocking. Iodine and iodine compounds were found to reduce knocking, but they were rather expensive.

So Midgley tried other halogens and their compounds (Table 1). Chlorine, bromine and their compounds did not work as well as iodine and its compounds.

Next, Midgley tried elements and their compounds from Group VI. Selenium compounds reduced knocking better than iodine compounds. Tellurium compounds reduced knocking even better than selenium compounds. Unfortunately, tellurium compounds were extremely poisonous. Midgley studied the patterns in anti-knock properties very carefully. He deduced that the best anti-knock substance would be a compound of lead. This prediction turned out to be correct. Since then, tetraethyllead(IV) ($Pb(C_2H_5)_4$) has been added to most petrol to prevent knocking.

Group IV	Group V	Group VI	Group VII
C	N	O	F
Si	P	S	Cl
Ge	As	Se	Br
Sn	Sb	Te	I
Pb	Bi	Po	At

Table 1 Groups IV, V, VI, and VII in the periodic table

1 Iodine and its compounds have not been used commercially as anti-knock agents. Why is this?

2 Tellurium and its compounds have not been used commercially as anti-knock agents. Why is this?

3 Why was it pointless to test compounds of elements in Groups I, II and III (e.g. sodium chloride) as anti-knock agents?

4 How do anti-knock properties change (i) down a group, (ii) across a period?

5 Why did Midgley predict that the best anti-knock substance would be a compound of lead?

6 Why do you think Midgley never tested compounds of polonium and astatine?

7 Why is unleaded petrol being used more and more nowadays?

2 Ecology or economics – a mining enquiry

The facts

Geologists working for the mining company Zinc UK have discovered large deposits of zinc ore in an attractive coastal area of Southshire (Figure 1). The ore contains up to 24% zinc which is very rich indeed. There is about 50 million tonnes of ore, 40 metres below the surface, in an area of land owned by an absentee landlord. This could be mined at a rate of about 2 million tonnes per year. The project would create about 200 new jobs in Southshire.

At present there are otters and deer in Southshire. Should wildlife be disturbed in our quest to use the Earth's resources?

The issues

(i) *Environmental.* The area is regarded as one of outstanding natural beauty. It supports many wild plants and animals including deer and otter. It is a tourist attraction for visitors to the small seaside resorts along the coast. Most jobs in the area are related to dairy farming and tourism. Ten miles from the proposed mine there is a new housing estate at Whitford with room for expansion. Whitford has a few shops, a coffee bar, a pub and a primary school. The seaside resorts are small because of the poor road links and because there is no railway. The mining company have, however, promised to build a major road into the area if they are given permission to mine. The road would go directly to the motorway 25 miles away. Mining noise would be kept to a minimum but blasting would occur every other day at 15:00 hours.

(ii) *Economic and social.* Unemployment in Southshire is slightly above average. Most of the work on local farms and in the seaside resorts is seasonal. In the last few years, the seaside resorts have had fewer visitors.

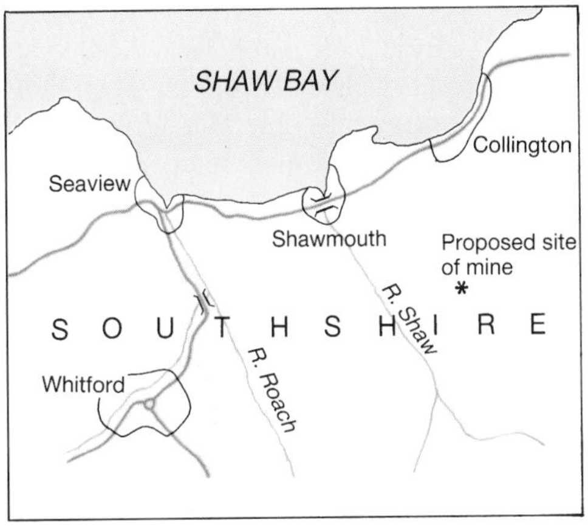

The people

1 A local Councillor from Shawmouth
2 A teenager from Seaview
3 A young mother from Whitford
4 A farm worker
5 The Secretary of the local environmental group
6 The absentee landlord
7 A representative of the mining company
8 A landlady in Collington

The enquiry

The MP for Southshire has decided to hold a public enquiry to hear the views of the eight people listed above. The enquiry will help to decide whether Zinc UK should be allowed to mine the ore. Suppose that you are one of the people named above. List the advantages and disadvantages that the mining operation might bring to you. Then write out the statement that you would make at the public enquiry.

Although mining sometimes damages and scars the countryside, areas can often be restored when finished. This attractive watersports area in the Thames Valley was once a gravel pit. After extracting the gravel for buildings and roads, the pit has been filled with water for sailing and fishing

Study Questions

1 (a) How was Mendeléev's periodic table similar to that of Newlands?
(b) How did Mendeléev's periodic table differ from that of Newlands?
(c) Why was Mendeléev's periodic table so successful and so readily accepted by other scientists?
(d) How do modern periodic tables differ from the one suggested by Mendeléev?

2 This question concerns the following families of elements in the periodic table.
A Group I – the alkali metals
B Group IV – containing carbon and lead
C Group VII – the halogens
D Group 0 – the noble gases
E the transition elements
Select from *A* to *E*, the family containing an element which
(i) melts at 25 K and boils at 27 K.
(ii) reacts with water forming a solution of pH 12.
(iii) forms a pink chloride, nitrate and sulphate.
(iv) is a liquid at room temperature and forms a white soluble sodium salt.
(v) forms no compounds.
(vi) is a solid, a poor conductor and boils above 3000°C.
(vii) is a good conductor of electricity and floats on water.
(viii) is a metalloid and melts on heating with a Bunsen.
(ix) has two crystalline forms at room temperature.
(x) forms an oxide which catalyses the decomposition of hydrogen peroxide very effectively.

3 This question concerns the following outline of the periodic table:
For each of the elements described below, choose the *one* blank area (*A* to *E*) in the table above where the element is

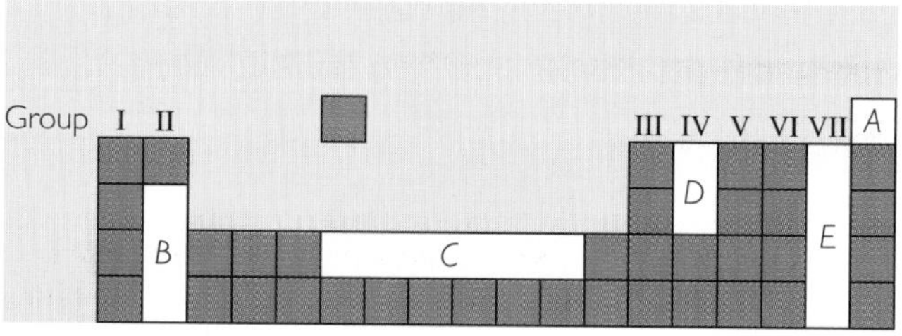

most likely to be found.
(i) A pale green element which reacts with most metals forming crystalline compounds.
(ii) An element which forms an oxide which dissolves in water to give a solution of pH 9.
(iii) An element which exists as a liquid over a temperature range of less than 10°C.
(iv) An element which reacts with dilute sulphuric acid producing hydrogen but does not react with water.
(v) An element which forms two different solid chlorides, one of which is green and the other yellow.
(vi) An element which is present in sand.
(vii) A colourless, odourless element which is gaseous at room temperature.
(viii) An element with a very high melting point which produces a gaseous acidic oxide.
(ix) A silvery-grey element which reacts vigorously with water producing hydrogen.
(x) An element which reacts with water forming an acidic solution.

4 (a) One important requirement for an anaesthetic is that its boiling point should be between about 40°C and 60°C.
(i) What is the disadvantage of a boiling point below 40°C for an anaesthetic?
(ii) What is the disadvantage of a boiling point above 60°C?
(b) The table below shows the anaesthetic effect, toxicity (i.e. harmful effect on the body) and flammability of methane and its four chlorine-containing derivatives. How do (i) the anaesthetic effect, (ii) the toxicity, (iii) the flammability, depend on the number of chlorine atoms in the molecule?
(c) Suggest how the chlorine atoms influence the toxicity.

Compound	Anaesthetic effect	Toxicity	Flammability
Methane, CH_4	none	none	flammable
Chloromethane, CH_3Cl	weak	little	flammable
Dichloromethane, CH_2Cl_2	moderate	little	non-flammable
Trichloromethane, $CHCl_3$	strong	toxic in large quantities	non-flammable
Tetrachloromethane, CCl_4	strong	very toxic	non-flammable

5 The element rubidium (Rb) has similar properties to sodium and potassium and is in the same group in the periodic table.
(a) Write down the formula of (i) rubidium oxide, (ii) rubidium hydroxide, (iii) rubidium nitrate.
(b) Describe what happens and write equations for the reactions which occur when (i) rubidium is added to water, (ii) rubidium hydroxide is added to hydrochloric acid.
(c) Is rubidium hydroxide an alkali? Explain your answer.

SECTION G

Acids, bases and salts

Oranges and other citrus fruits contain citric acid. This gives them a sharp taste

1 ACIDIC AND ALKALINE SOLUTIONS

What protective clothing is this technician wearing to protect herself from the acid she is pipetting?

The sharp taste of lemon juice and vinegar is caused by acids. Some acids can be identified by their sharp taste, but it would be too dangerous to rely on taste for all acids. Think of the dangers of tasting sulphuric acid or nitric acid.

A more sensible test for acids is to use indicators like litmus or Universal Indicator. Indicators change colour depending on how acidic or how alkaline a solution is.

- Acidic solutions make litmus turn red and give an orange or red colour with Universal Indicator (Figure 1).
- Alkaline solutions make litmus turn blue and give a green, blue or violet colour with Universal Indicator.

It would be very clumsy to use the colour of an indicator to describe how acidic something is. So scientists use a scale of numbers called the **pH scale.** On this scale,

- Acidic solutions have a pH below 7.
- Alkaline solutions have a pH above 7.
- Neutral solutions have a pH of 7.

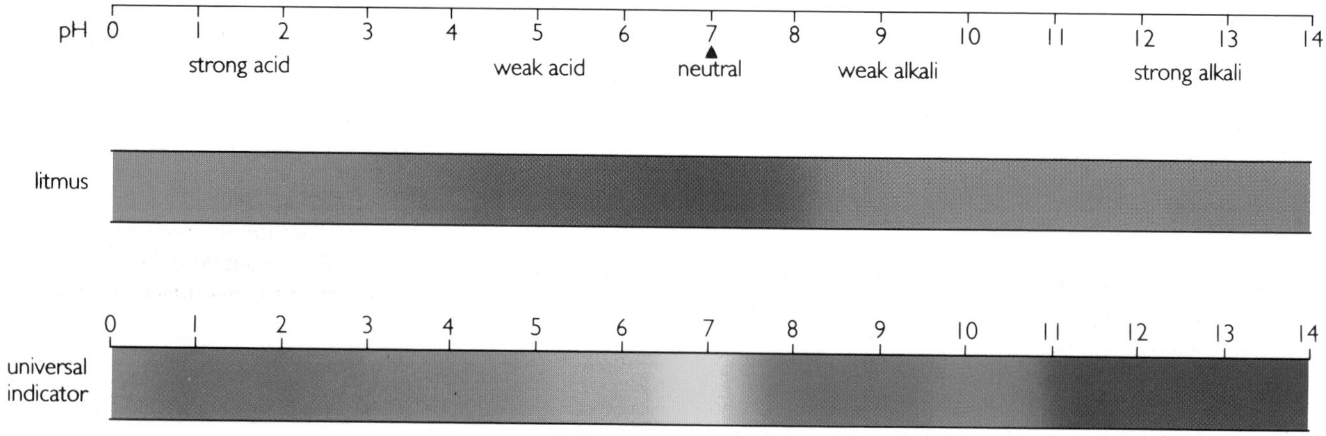

Figure 1
The colours of litmus and universal indicator with solutions of different pH

Making acids

- **From non-metal oxides and water.** Acids are produced when non-metal oxides are added to water. For example:

carbon dioxide	+	water	→	carbonic acid
CO_2	+	H_2O	→	H_2CO_3
sulphur dioxide	+	water	→	sulphurous acid
SO_2	+	H_2O	→	H_2SO_3

Sulphuric acid is manufactured via sulphur dioxide. The raw materials are sulphur, air (oxygen) and water. Most of the sulphur is shipped to the UK from North America.

There are three stages in the manufacture of sulphuric acid.

1 First, sulphur dioxide is made by burning sulphur in air (oxygen).

$$S + O_2 \rightarrow SO_2$$

This photo shows part of a sulphuric acid plant. The hot gas filter removes any unburnt sulphur dust from the sulphur dioxide produced in the first stage of the process. What do you think happens in the converter? The heat exchanger uses the heat produced in the contact process to warm up the reactant gases. Why is this economical?

2 The sulphur dioxide is then converted to sulphur trioxide. This is called **the contact process**. The sulphur dioxide is mixed with air (oxygen) and passed over a catalyst (vanadium(V) oxide) at 450°C.

$$\text{sulphur dioxide} + \text{oxygen} \rightarrow \text{sulphur trioxide}$$
$$2SO_2 + O_2 \rightarrow 2SO_3$$

3 Finally, the sulphur trioxide is converted to sulphuric acid. Sulphur trioxide does not dissolve very easily in water. So sulphur trioxide is dissolved in 98% sulphuric acid and then some water is added.

$$SO_3 + H_2O \rightarrow H_2SO_4$$

- **By combining hydrogen with reactive elements.** The most important acid to be manufactured in this way is hydrochloric acid. Hydrogen chloride is first made by burning hydrogen in chlorine.

$$H_2 + Cl_2 \rightarrow 2HCl$$

The hydrogen chloride is then passed up a tower packed with stones (or some other unreactive substance) over which water trickles. The gas dissolves in the water to form hydrochloric acid.

This photo shows sulphur being mined in an open pit in North America. Most sulphur used in the manufacture of sulphuric acid in the UK is imported from North America

QUESTIONS

1 (a) Describe how you would check the pH of toothpaste.
(b) Why is the pH of toothpaste important?
2 The following table shows the main uses of sulphuric acid:

Use	**Percentage of total**
Fertilisers	30%
Detergents	14%
Paints and pigments	11%
Fibres (rayon, paper, etc.)	8%
Plastics	5%
Cleaning metals	2%
Other uses (explosives, dyes, batteries, insecticides, etc.)	30%

Draw a pie chart to show these main uses. Illustrate your pie chart in an appropriate way.
3 This question is about the manufacture of sulphuric acid.
(a) Write equations for the three stages in the manufacture of sulphuric acid.
(b) What is the contact process?
(c) Suggest two reasons why a catalyst is used in the contact process.
(d) 450°C is an optimum temperature for the contact process. What do you think the term 'optimum temperature' means?
(e) Sulphur trioxide from the contact process is not reacted directly with water to make sulphuric acid. Explain why.
(f) Most of the sulphuric acid plants in the UK are situated near large ports. Suggest two reasons for this.

2 ACIDS IN EVERYDAY LIFE

Carbonic acid in drinks like cola, lemonade and soda water makes them 'fizz'

Acid rain, caused by fumes from industry has had drastic effects on the natural environment of Czechoslovakia. Trees have been stunted or killed and the decrease in the pH of lakes has wiped out whole food chains

Acids are important in everyday life and in industry. They are used to make foods, clothes and medicines.

Acids in our foods

Many foods and drinks contain acids. Citrus fruits – oranges, lemons, pineapples, grapefruit – contain citric acid. Tomato sauce, brown sauce and mint sauce get their sharp taste from vinegar which contains acetic acid.

One of the simplest and cheapest drinks is soda water. This is made by dissolving carbon dioxide in water under pressure. Some of the carbon dioxide reacts with the water to form carbonic acid. When soda water is poured from the bottle it fizzes because the pressure is lower, which makes carbon dioxide form as bubbles in the drink. Other fizzy drinks, like cola and lemonade, also contain carbonic acid.

Acids in our bodies

Our stomach walls produce hydrochloric acid. This gives a pH of about 2 in the stomach. These acidic conditions help us to **digest** (break down) foods, particularly proteins. Proteins are broken down into smaller molecules, called peptides and amino acids. These small molecules can pass through the lining of our intestines into the bloodstream. The blood carries these small molecules round the body to where they are needed.

The pH of our blood is 7.4 (slightly alkaline). Most reactions in our bodies can only take place within a narrow range of pH. Change in the pH in your stomach might give you indigestion, but a pH change of just 0.5 in the blood would probably kill you. In order to prevent changes in pH, our bodies contain substances that counteract the effects of acids and alkalis.

Water pollution in the Sea of Galilee has killed these fish

Acids in the air – acid rain

When fuels burn, sulphur dioxide and carbon dioxide are produced. Because of this, city air may contain five times as much sulphur dioxide as fresh air. Sulphur dioxide and carbon dioxide react with water vapour and rain in the air to form sulphurous acid, H_2SO_3, and carbonic acid, H_2CO_3.

$$SO_2 + H_2O \rightarrow H_2SO_3$$

$$CO_2 + H_2O \rightarrow H_2CO_3$$

These acids make the rain acidic, which is why we use the term 'acid rain'. (See section B, unit 5.)

Many foods especially sauces and pickles, contain acetic acid from vinegar

Most heather grows best in acid soil

These beech trees grow best in slightly alkaline soil

Acids in the soil

The pH of soil can vary from about 4 to 8, but most soils have a pH of between 6.5 and 7.5. In chalk or limestone areas, the soil is usually alkaline, but in moorland, sandstone and forested areas it is generally acidic. Peat bogs and clay soils are normally acidic. For general gardening and farming purposes, the best results are obtained from a neutral or slightly acidic soil of pH 6.5 to 7.0. Only a few plants, including rhododendrons and azaleas, can grow well in soils which are acidic. In areas where the soil is too acidic, it can be improved by treatment with powdered limestone (calcium carbonate) or slaked lime (calcium hydroxide). These substances react with acids in the soil and raise the pH to the right value.

QUESTIONS

1 (a) Describe how you would check the pH of soil samples.
(b) Why do you think that forested and peaty areas have acid soils?
(c) Why is it important for gardeners to know the pH of the soil?

2 Make a list of foodstuffs which are acidic. Try to find out what acids they contain.

3 (a) Why is soda water both fizzy and acidic?
(b) Would you expect mineral waters to be (i) fizzy, (ii) acidic? Explain your answers.
(c) Design an experiment to compare the fizziness and pH of different soda waters and mineral waters.

4 Find out more about acid rain. Here are some questions to consider.
(a) When did acid rain first become a problem?
(b) Why did this problem not arise earlier?
(c) What are the major factors causing acid rain?
(d) What steps might be taken to reduce acid rain?
Prepare a short report for your class.

3 NEUTRALISATION

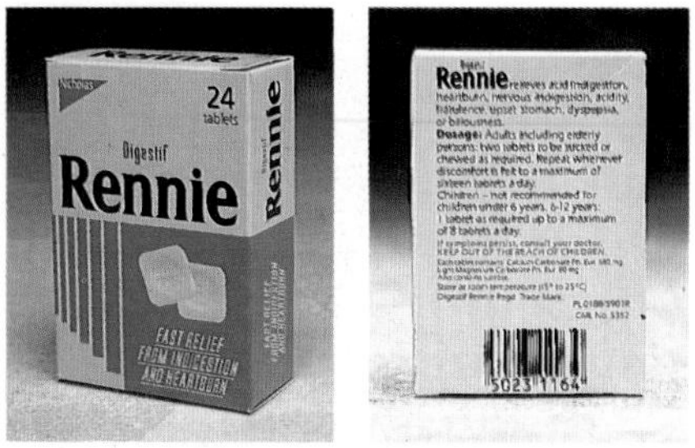

What substances in 'Rennies' *help to neutralise acids in the stomach?*

Adverts for indigestion cures usually talk of 'acid stomach' and 'acid indigestion'. Medicines which ease stomach ache (such as Milk of Magnesia, Setlers and Rennies) are called antacids (anti-acids) because they neutralise excess acid in the stomach.

- Substances which neutralise acids are called **bases**. Bases which are soluble in water are called **alkalis**. The bases in indigestion tablets include magnesium hydroxide and calcium carbonate.
- The reactions between acids and bases are called **neutralisations**.

As we saw in the last section, farmers use slaked lime to neutralise acid soils. The main base in slaked lime is calcium hydroxide.

Neutralisation is also important in dental care. We use toothpaste to neutralise the acids produced when food breaks down in the mouth (Figure 1).

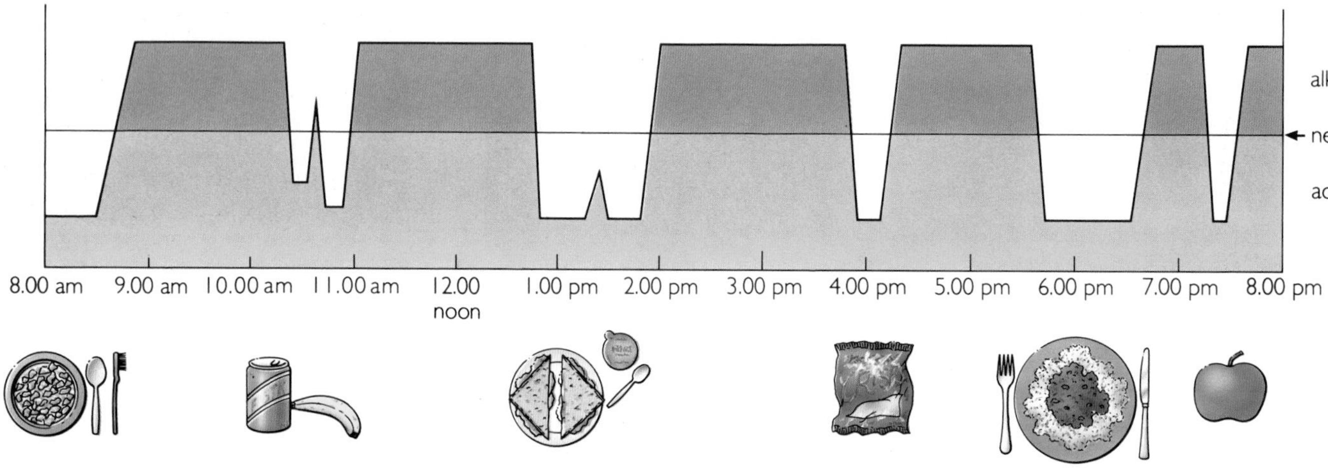

Figure 1
This shows how the pH in your mouth changes during the day. Notice how the pH becomes acidic during and just after meals. This is because sugar and other chemicals in food are being broken down into acids. These acids cause tooth decay

Neutralisation is used to treat splashes of acid or alkali on people's skin or clothes. The stings from many plants and animals contain acids or bases so they can also be treated by neutralisation. For example, nettle stings and wasp strings contain bases like **histamine**. They are treated with acidic substances called **anti-histamines**. Unlike wasp stings, those caused by bees are acidic and should be treated with bases.

Neutralisation is used to remove carbon dioxide from the air in air-conditioned buildings. Carbon dioxide is an acidic oxide. So the air is passed over soda lime, a mixture of two bases (sodium hydroxide and calcium hydroxide). These react with the carbon dioxide and neutralise the air.

One of the most important applications of neutralisation is in making fertilisers like ammonium sulphate ($(NH_4)_2SO_4$) and ammonium nitrate (NH_4NO_3). Ammonium sulphate is manufactured by neutralising sulphuric acid (H_2SO_4) with ammonia (NH_3). Ammonium nitrate is manufactured by neutralising nitric acid (HNO_3) with ammonia (NH_3).

Investigating neutralisation

When acids are neutralised by bases, the pH of the solution changes. You can follow the neutralisation and the pH change when a base is added to an acid, in the following way.

Put about 10 cm^3 of dilute hydrochloric acid in a conical flask. Add 2 or 3 drops of universal indicator and note the colour and pH. Now add dilute sodium hydroxide to the mixture, 2 cm^3 at a time, until the solution becomes alkaline. Record the colour and the pH after each addition.

Notice that the base (sodium hydroxide) has neutralised the hydrochloric acid, and by the end is in excess.

We can write an equation for the reaction as:

sodium hydroxide	+	hydrochloric acid	→	sodium chloride	+	water
base	+	acid	→	salt	+	water
NaOH	+	HCl	→	NaCl	+	H_2O

Substances such as sodium chloride, which are produced when acids react with bases, are called **salts**. Sodium chloride is usually known as common salt. Other important salts include copper sulphate ($CuSO_4$), ammonium nitrate (NH_4NO_3) and potassium chloride (KCl).

'Nitram' *fertiliser is ammonium nitrate. It is manufactured by a neutralisation reaction*

Naming salts

Look at Table 1. This shows how the name of a salt comes from its parent base and acid. So magnesium oxide and nitric acid would produce magnesium nitrate.

Notice also in Table 1 that

- acids with names ending in '-ic' form salts with names ending in '-ate'. (Hydrochloric acid is an exception – its salts are called chlorides.)
- acids with names ending in '-ous' form salts with names ending in '-ite'.

Base	**Name of salt**
Magnesium oxide	Magnesium –
Potassium hydroxide	Potassium –
Zinc oxide	Zinc –
Acid	**Name of salt**
Sulphuric	– sulphate
Sulphurous	– sulphite
Nitric	– nitrate
Nitrous	– nitrite
Carbonic	– carbonate
Hydrochloric	– chloride

Table 1 The names of salts

QUESTIONS

1 Design an experiment to compare the effectiveness of indigestion tablets. (Which tablets are best at neutralising acid? Which tablets are the best buy?)

2 Get into a small group with 2 or 3 others. Discuss the following questions.

(a) What causes indigestion?

(b) Why can calcium carbonate be classified as both a salt and a base?

(c) Why is neutralisation important?

3 (a) Why is ammonium nitrate (NH_4NO_3) used as a fertiliser?

(b) What substances are used to manufacture ammonium nitrate?

(c) Write an equation for the reaction to make ammonium nitrate.

4 Complete the following.

(a) Magnesium oxide + sulphuric acid → __?__ + __?__

(b) Potassium hydroxide + __?__ → __?__ nitrate + water

(c) __?__ + hydrochloric acid → zinc __?__ + __?__

4 THE PROPERTIES OF ACIDS

Car batteries contain sulphuric acid. H^+ ions and SO_4^{2-} ions in the acid react with lead plates in the battery in a way which makes the battery produce electricity. This electricity starts the engine, and powers things like the radio, lights and windscreen wipers

Many things we eat and drink contain acid. Cola contains phosphoric acid and rhubarb contains oxalic acid. What acids do grapefruits and vinegar have in them?

We know a lot about acids already.

1 Acids are soluble in water. Their solutions have a pH below 7.
2 Acids give characteristic colours with indicators. They give a red colour with litmus and an orange or red colour with Universal Indicator.
3 Acids react with metals above copper in the reactivity series to form a salt and hydrogen.

$$\text{metal} + \text{acid} \rightarrow \text{salt} + \text{hydrogen}$$

e.g. $Mg + H_2SO_4 \rightarrow MgSO_4 + H_2$

Nearly all salts contain a metal and at least one non-metal. They are ionic compounds with a cation (e.g. Cu^{2+}, Mg^{2+}) and an anion (Cl^-, SO_4^{2-}). Sodium chloride (Na^+Cl^-) is known as common salt.

4 Acids react with bases (such as metal oxides and hydroxides) to form a salt and water. These reactions are called neutralisations.

$$\text{base} + \text{acid} \rightarrow \text{salt} + \text{water}$$

For example, when black copper oxide is added to warm dilute sulphuric acid, the black solid disappears and a blue solution of copper sulphate forms.

$$CuO + H_2SO_4 \rightarrow CuSO_4 + H_2O$$

5 Acids react with carbonates to give a salt plus carbon dioxide and water.

$$\text{carbonate} + \text{acid} \rightarrow \text{salt} + CO_2 + H_2O$$

For example, when sodium carbonate is added to hydrochloric acid, sodium chloride forms and bubbles of carbon dioxide are produced.

$$Na_2CO_3 + 2HCl \rightarrow 2NaCl + CO_2 + H_2O$$

6 Acid solutions conduct electricity and are decomposed by it. This shows that solutions of acids consist of ions. All acids produce hydrogen at the cathode during electrolysis. This shows that all acids contain H^+ ions (Table 1).

Name of acid	Formula	Ions in the acid
Acetic acid	CH_3COOH	$H^+ + CH_3COO^-$
Carbonic acid	H_2CO_3	$2H^+ + CO_3^{2-}$
Hydrochloric acid	HCl	$H^+ + Cl^-$
Nitric acid	HNO_3	$H^+ + NO_3^-$
Sulphuric acid	H_2SO_4	$2H^+ + SO_4^{2-}$
Sulphurous acid	H_2SO_3	$2H^+ + SO_3^{2-}$

Table 1 Some acids and the ions which they form

Care with acids

The photo on the right shows the label from a bottle of conc. sulphuric acid.

Concentrated sulphuric acid contains about 98% H_2SO_4 and only 2% water. *It must be used very carefully, and eye protection MUST be worn.* It is an oily liquid which reacts with water producing a lot of heat. Because of this, *always add concentrated H_2SO_4 to a large volume of water when mixing the two. NEVER add water to acid.* If water is added to concentrated H_2SO_4, the heat produced can boil the water and spit out drops of acid.

Pure H_2SO_4 and pure water are both poor conductors of electricity. They contain simple molecules, *not* ions. But a solution of sulphuric acid in water is a good conductor of electricity, so it *must* contain ions. The ions are formed when the H_2SO_4 reacts with water.

$$H_2SO_4(l) \xrightarrow{water} 2H^+(aq) + SO_4^{2-}(aq)$$

Notice that each molecule of H_2SO_4 produces two H^+ ions in water.

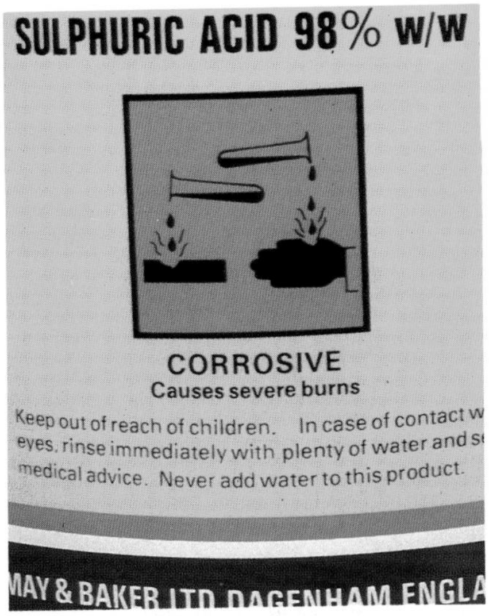

Concentrated sulphuric acid attacks skin, clothing, metal and other materials. Because of these hazards containers for concentrated sulphuric acid must carry the 'corrosive' warning sign (above)

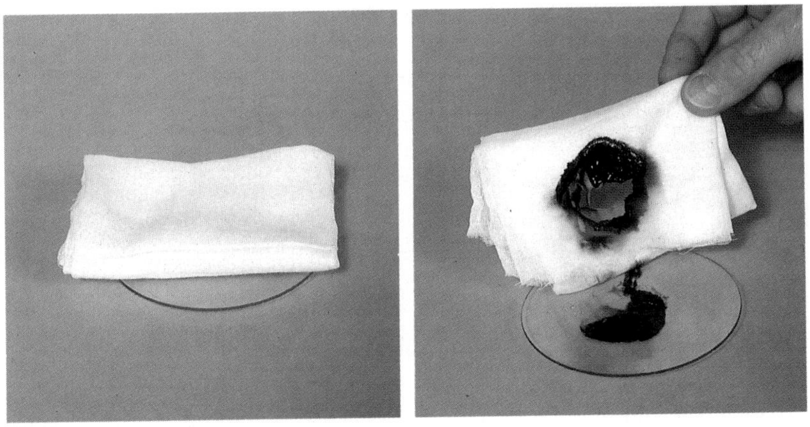

Concentrated sulphuric acid will burn through cotton cloth as it rapidly removes water from the carbohydrate molecules in the cotton fibres (left)

Concentrated sulphuric acid removes water fom sugar and leaves a black mass of carbon

Concentrated sulphuric acid will absorb water very rapidly. It is a **dehydrating agent**. Because of this, it can be used to dry gases. The concentrated acid also removes water from hydrated salts such as blue copper(II) sulphate ($CuSO_4.5H_2O$), from carbohydrates such as sugar ($C_{12}H_{22}O_{11}$) and from compounds containing hydrogen and oxygen in clothes and skin (as shown in the photos). This is why it burns and chars clothing and skin. Dilute sulphuric acid does *not* react as a dehydrating agent in this way.

QUESTIONS

1 Discuss the following questions and then write your answers.

(a) Why can vinegar be used to descale kettles?

(b) Accurate clocks sometimes have silica crystals (a dehydrating agent) placed near their working parts. Why is this?

(c) Why is concentrated sulphuric acid a poorer conductor of electricity than dilute sulphuric acid?

(d) Why does concentrated sulphuric acid burn the skin but dilute sulphuric acid does not?

(e) When a thermometer bulb dipped in water is placed in a gas jar of dry hydrogen chloride, the temperature rises by 10°C. Why is this?

2 Write word equations and balanced equations for the following reactions:

sodium oxide + sulphuric acid
zinc oxide + hydrochloric acid
copper oxide + nitric acid
iron + sulphuric acid
aluminium + hydrochloric acid
copper carbonate + hydrochloric acid

3 (a) Write the formula for blue copper(II) sulphate crystals.

(b) What colour will these crystals become when concentrated sulphuric acid is added?

(c) Write an equation for the reaction.

5 BASES AND ALKALIS

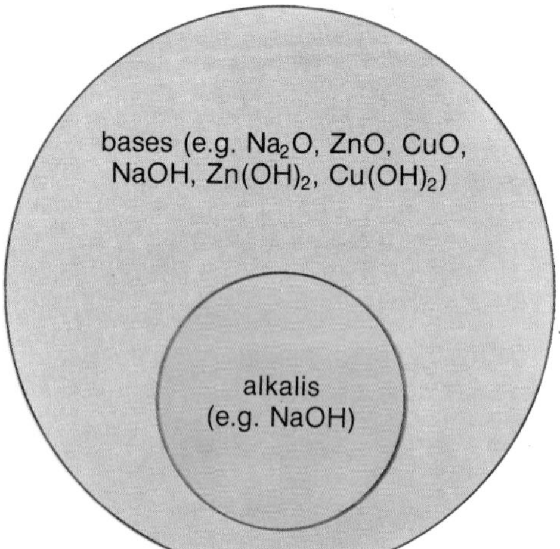

Figure 1
This Venn diagram shows the relationship between bases and alkalis. All alkalis are bases, but not all bases are alkalis

Bases are substances which neutralise acids. They are the chemical opposites to acids. The largest group of bases are metal oxides and metal hydroxides such as sodium oxide, copper oxide, sodium hydroxide, and copper hydroxide.

A special class of bases are called **alkalis**. *Alkalis are bases which are soluble in water.* Their solutions have a pH above 7. They turn litmus blue and give a green, blue or purple colour with Universal Indicator.

The most common alkalis are sodium hydroxide (NaOH), calcium hydroxide ($Ca(OH)_2$) and ammonia (NH_3). Calcium hydroxide is much less soluble than sodium hydroxide. A solution of calcium hydroxide in water is often called 'limewater'. Sodium oxide (Na_2O), potassium oxide (K_2O) and calcium oxide (CaO) react with water to form their hydroxides. So the reaction of these three metal oxides with water produces alkalis. For example

$$Na_2O(s) + H_2O(l) \rightarrow 2NaOH(aq)$$

$$CaO(s) + H_2O(l) \rightarrow Ca(OH)_2(aq)$$

Most other metal oxides and hydroxides are insoluble in water. These insoluble metal oxides and hydroxides are bases but *not* alkalis. The relationship between bases and alkalis is shown in a Venn diagram in Figure 1.

Explaining neutralisation in terms of ions

Bases react with acids to form a salt and water. For example

copper oxide	+	sulphuric acid	$\rightarrow$	copper sulphate	+	water
$CuO(s)$	+	$H_2SO_4(aq)$	$\rightarrow$	$CuSO_4(aq)$	+	$H_2O(l)$
sodium hydroxide	+	hydrochloric acid	$\rightarrow$	sodium chloride	+	water
$NaOH(aq)$	+	$HCl(aq)$	$\rightarrow$	$NaCl(aq)$	+	$H_2O(l)$

These bases contain either oxide ions (O^{2-}) like copper oxide, or hydroxide ions (OH^-) like sodium hydroxide. During neutralisation these ions react with H^+ ions in the acids to form water.

$$2H^+ + O^{2-} \rightarrow H_2O$$

$$H^+ + OH^- \rightarrow H_2O$$

These equations will help you to understand why bases are the chemical opposites to acids. Bases *take* H^+ ions, whereas acids *donate* them.

Alkalis in industry

A 'noodling' machine, part of the soap making process

The most important industrial alkalis are sodium hydroxide (caustic soda) and calcium hydroxide (slaked lime). Calcium hydroxide is made by adding water to quicklime (calcium oxide). It is used in treating acid soils, in cement and in the manufacture of sodium hydroxide and bleaching powder. Large amounts of sodium hydroxide are used to make soap, paper, rayon and cellulose fibres.

Paper, rayon and cellulose fibres are all made from wood. The wood is made into pulp and soaked in sodium hydroxide solution. This removes gums and resins and leaves the natural fibres of cellulose. The cellulose fibres are then squashed into thin white sheets which look like blotting paper. This purified pulp can now be used to make paper, rayon or cellulose acetate (Tricel).

Soaps and soap powders are made by boiling fats and oils with sodium hydroxide in large vats (Figure 2).

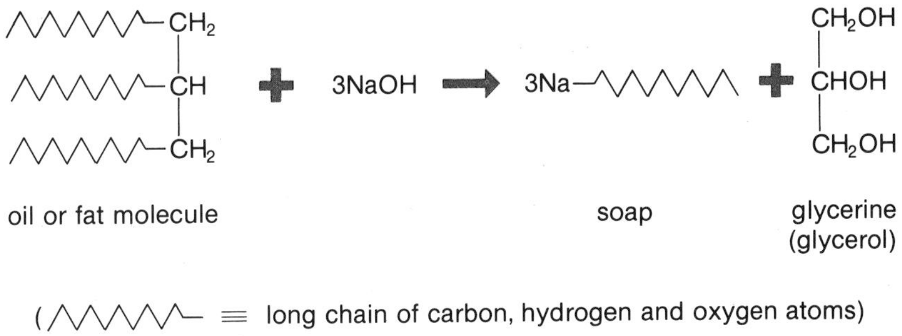

Figure 2

In the manufacture of paper, wood pulp (cellulose) is purified by soaking in concentrated sodium hydroxide solution

Rayon is made by forcing a solution containing cellulose (wood pulp) through tiny holes into dilute acid. This forms filaments of yarn

QUESTIONS

1 Suppose your best friend has missed the last few lessons, and asks you for help. How would you explain to him or her what the difference is between (i) acids and bases, (ii) bases and alkalis, (iii) bases and salts?

2 Explain the following:

(a) All alkalis are bases, but all bases are *not* alkalis.

(b) Sodium hydroxide is used in oven cleaners to remove fat and grease.

(c) Limestone, slaked lime and lime are all bases, but only slaked lime is an alkali.

3 Write word equations and then balanced equations for the following reactions:

(a) the action of heat on limestone (calcium carbonate),

(b) the reaction of lime (calcium oxide) with water,

(c) the reaction of zinc oxide with hydrochloric acid.

4 (a) Why is it important to recycle paper?

(b) What are the main stages in recycling paper?

(c) What are the problems in recycling paper?

(d) Paper can be made from certain rags. Why is this?

LIME AND LIMESTONE

A limestone quarry. The mining of limestone is big business, because it has so many commercially valuable uses and products

Limestone for industry

Limestone is one of the most important raw materials for industry. The main substance in limestone rock is calcium carbonate ($CaCO_3$). Quicklime (calcium oxide, CaO) and slaked lime (calcium hydroxide, $Ca(OH)_2$) are easily manufactured from limestone, which makes it even more important. The uses of limestone and its products are shown in Figure 1. Notice that

1 the main uses of limestone itself are in making iron and steel (section B, unit 10), in neutralising acid soil and in fertilisers such as nitro-chalk.
2 the main substances made from limestone are calcium oxide (quicklime), calcium hydroxide (slaked lime), cement and glass (section H, unit 9).

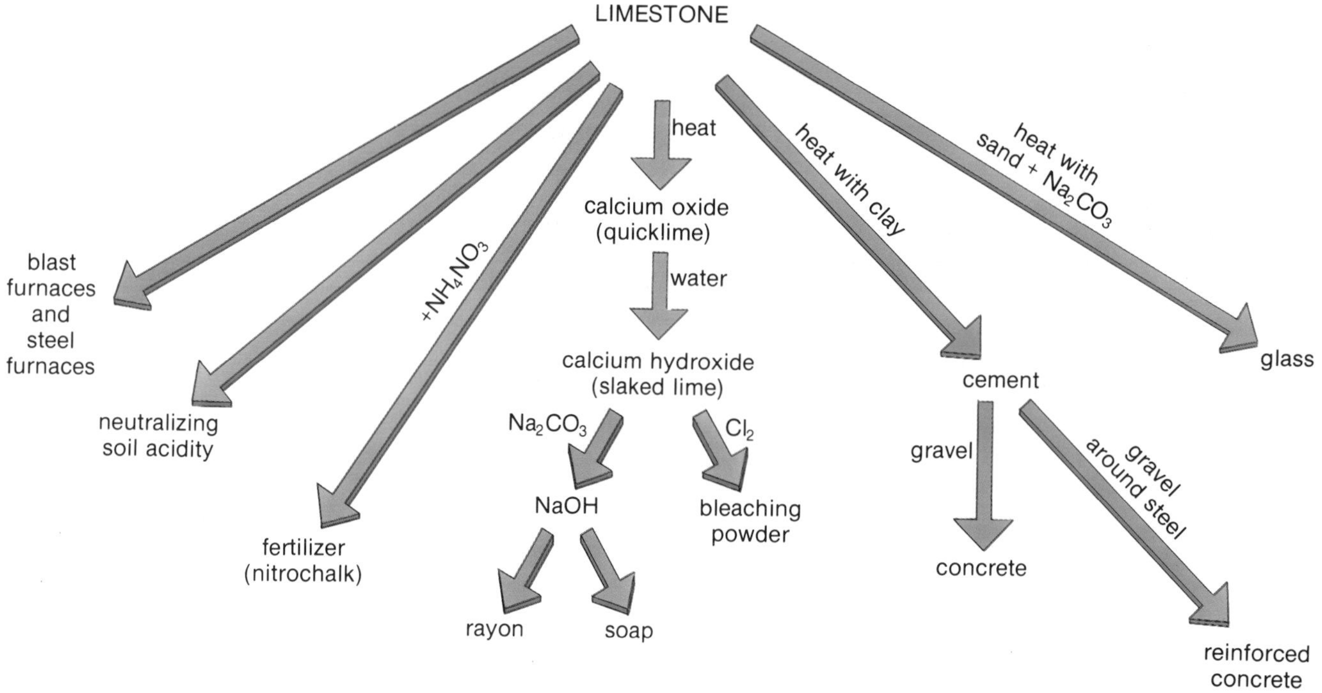

Figure 1
Important uses and products of limestone

- **Calcium oxide (quicklime)** is made by heating limestone in lime kilns (Figure 2). Calcium carbonate in the limestone is decomposed to calcium oxide (quicklime).

$$CaCO_3(s) \rightarrow CaO(s) + CO_2(g)$$

The quicklime is then reacted with water to make calcium hydroxide (slaked lime).

$$CaO(s) + H_2O(l) \rightarrow Ca(OH)_2(s)$$

- **Calcium hydroxide (slaked lime)** is the cheapest industrial alkali. It is used to make sodium hydroxide and bleaching powder. Powdered calcium hydroxide is only slightly soluble in water. The dilute alkaline solution which it forms is called **limewater**.
- **Cement** is made by heating limestone with clay. It contains a mixture of calcium silicate and aluminium silicate. This mixture reacts with water to form hard, interlocking crystals as the cement sets. When cement is used, it is normally mixed with two or three times as much sand as well as water.
- **Concrete** is a mixture of cement, sand and water with gravel (aggregate), broken stones or bricks.

Reinforced concrete is made by allowing concrete to set around a steel framework. It is used in building large structures like office blocks and bridges. Reinforced concrete is an example of a **composite material**. Composite materials contain two different materials, but the particles in the materials do not mix. Reinforced concrete contains concrete and steel. Because of this, it has the hardness of concrete and the flexibility and tensile strength of steel.

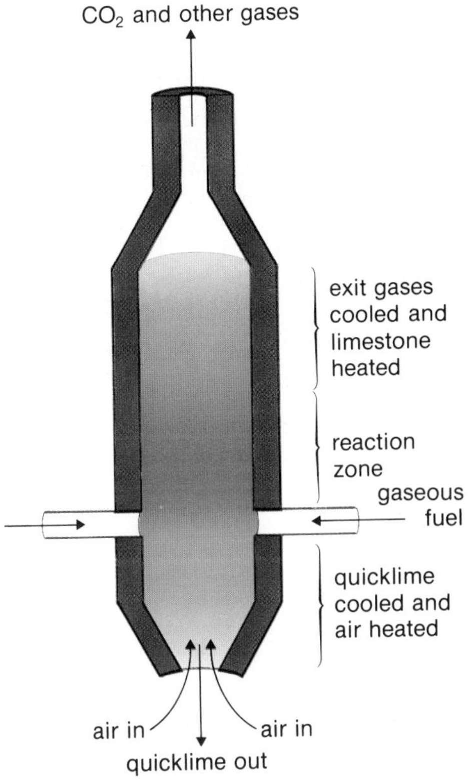

Figure 2
A gas-fuelled lime kiln

Chalk, limestone and marble

Calcium carbonate occurs naturally as chalk and marble, as well as limestone. In fact, calcium carbonate is the second most abundant material in the Earth's crust after silicates like clay, sand and sandstone.

Chalk is the softest form of calcium carbonate. Deposits of chalk have formed from the shells of dead sea creatures that lived millions of years ago. In some places, the chalk was covered with other rocks and put under great pressure. This changed the soft chalk into a harder rock – limestone. In other places the chalk was under pressure and heat. This changed the soft chalk into marble, the hardest form of calcium carbonate.

QUESTIONS

1 Chalk, marble and limestone are different natural forms of calcium carbonate.
(a) Why do you think there are three different forms?
(b) Find out about the classification of rocks as 'sedimentary', 'metamorphic' and 'igneous'. Which classes of rocks do chalk, marble and limestone belong to?

2 Various mixtures of sand and cement (by volume) are used in building. Plan an experiment to find out which mixture (ratio of sand to cement 2 : 1, 2.5 : 1, 3 : 1 or 3.5 : 1) gives the *hardest* product.

3 Try to explain the following. Write balanced equations where appropriate.
(a) Finely ground limestone is used to neutralise acids in the soil.
(b) Lime water is used to test for carbon dioxide.

4 'Wattle and daub' might be described as the medieval equivalent of reinforced concrete as a composite material. Find out about 'wattle and daub'. Why is it a composite material? What was the advantage of 'wattle and daub' over just 'wattle' or just 'daub'?

5 (a) What problems are associated with the large-scale quarrying of chalk and limestone?
(b) What steps can be taken to overcome or reduce these problems?

7 HARD WATER

At Malham Cove in the Yorkshire Dales, water once poured from the top of the rocks. Over many years carbonic acid in the water has reacted with the limestine and formed caves. The water now flows through these caves and finally emerges at the bottom of the cove.

Look at this photo. Notice the scum which has formed in the water. Notice also that there is no lather

In chalk and limestone areas, soap does not lather easily with the water. The water also forms a **scum** when it is mixed with soap. This kind of water is known as **hard water**.

Which ions cause hard water?

Table 1 shows what happens when 10 cm^3 of each of the solutions listed are shaken with 1 cm^3 of soap solution.

Solution used	Ions present	Reaction with soap
Sodium chloride	Na^+, Cl^-	no scum, lots of lather
Calcium chloride	Ca^{2+}, Cl^-	lots of scum, little lather
Potassium nitrate	K^+, NO_3^-	no scum, lots of lather
Magnesium nitrate	Mg^{2+}, NO_3^-	lots of scum, little lather
Sodium sulphate	Na^+, SO_4^{2-}	no scum, lots of lather
Iron(II) sulphate	Fe^{2+}, SO_4^{2-}	lots of scum, little lather

Table 1 The reactions of some solutions with soap

Look closely at the table.

1 Which solutions act like hard water?
2 Which of the following ions cause hard water?

$$Na^+, Cl^-, Ca^{2+}, K^+, NO_3^-, Mg^{2+}, SO_4^{2-}, Fe^{2+}$$

3 Which ions are most likely to cause hard water in the UK?

Why does scum form?

Hard water in the UK is caused mainly by calcium ions, Ca^{2+}. Soaps contain salts such as sodium palmitate and sodium stearate. When hard water is mixed with soap, Ca^{2+} ions in the hard water react with palmitate and stearate ions in the soap forming insoluble calcium stearate and calcium palmitate. This insoluble solid is scum. Scum is an example of a **precipitate** – a solid which forms in a liquid during a reaction.

$Ca^{2+}(aq)$	+	$2X^-(aq)$	→	$CaX_2(s)$
calcium ions in hard water		stearate/palmitate ions in soap		scum

Soaps and detergents are both used for cleaning, but they differ in *one* important way. Detergents, like washing-up liquid, do *not* give a scum with hard water. Unlike soaps they do not contain ions which react with Ca^{2+} ions in hard water to form a precipitate.

In limestone areas, caves have formed as carbonic acid in rain water has reacted with the limestone rock. These caves are at Castleton in Derbyshire

How does hard water form?

When rain falls, it reacts with carbon dioxide in the air to form carbonic acid.

$$H_2O(l) + CO_2(g) \rightarrow H_2CO_3(aq)$$

When this dilute solution of carbonic acid flows over limestone or chalk, it reacts with calcium carbonate in the rocks to form calcium hydrogencarbonate.

$CaCO_3(s)$	+	$H_2CO_3(aq)$	→	$Ca(HCO_3)_2(aq)$
in limestone		in rain water		in hard water

Unlike calcium carbonate, calcium hydrogencarbonate is soluble in water and the calcium ions make the water hard.

Calcium carbonate in chalk and limestone is the main cause of hard water. In some areas, calcium sulphate which occurs as gypsum ($CaSO_4.2H_2O$) and anhydrite ($CaSO_4$) also causes hardness. Calcium sulphate is only slightly soluble in water but enough will dissolve to make the water hard.

Carbonic acid in rain water has helped to dissolve the limestone in the cracks of this limestone 'pavement'

QUESTIONS

1 Design and make a poster to show some of the geological features in limestone areas. Illustrate your poster with diagrams, drawings or photos. On the poster, explain how the geological features have formed.

2 (a) Why does scum form?
(b) Write an equation for the reaction involved.
(c) Why is scum a nuisance?

3 What is the main advantage of detergents over soaps?

4 Plan an experiment to compare the hardness of rain water and tap water.

8 SOFTENING HARD WATER

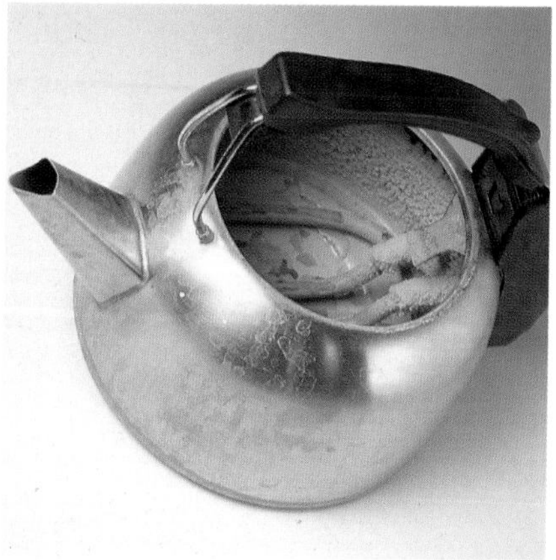

'Fur' inside a kettle is a deposit of calcium carbonate from hard water

Hard water usually tastes better than soft water. The dissolved substances in hard water also help to produce strong teeth and bones, which contain calcium carbonate and calcium phosphate. But hard water has several disadvantages compared to soft water.

1 It uses more soap than soft water.
2 It produces scum, which looks unsightly.
3 It results in the formation of 'scale' in water pipes and 'fur' in kettles. The 'scale' may block pipes and 'fur' will reduce the efficiency of a kettle.
4 It is necessary to remove the hardness from the water in some areas and this causes extra expense.

When hard water is warmed up or boiled, the calcium hydrogencarbonate in it decomposes to calcium carbonate, water and carbon dioxide.

$$Ca(HCO_3)_2(aq) \rightarrow CaCO_3(s) + H_2O(l) + CO_2(g)$$

The calcium carbonate is insoluble and forms a deposit inside the pipe or kettle. This reaction is the reverse of that which forms hard water. The reaction also explains the formation of stalagmites and stalactites in limestone caves. The temperature inside the cave is just warm enough for some of the hard water to decompose and leave a tiny deposit of calcium carbonate. More water drips down and the deposit gets larger. A deposit also forms where the drops hit the floor. After hundreds of years, the deposits grow into large stalagmites and stalactites.

Stalagmites and stalactites in Cox's Cave, Cheddar, Somerset

How is hard water softened?

In some areas, substances that cause hardness in the water must be removed. This is called **water softening**. In order to soften hard water we must take out the Ca^{2+} ions. We can do this in various ways.

- **By boiling**. Boiling decomposes calcium hydrogencarbonate forming insoluble calcium carbonate.

$$Ca(HCO_3)_2(aq) \rightarrow CaCO_3(s) + H_2O(l) + CO_2(g)$$

  This removes the hardness caused by calcium hydrogencarbonate, but boiling does not remove the hardness caused by calcium sulphate. Because of this, the hardness from calcium sulphate is called **permanent hardness**. The hardness caused by calcium hydrogencarbonate (which is removed by boiling) is called **temporary hardness**.
- **By adding washing soda or bath salts**. Washing soda and bath salts contain sodium carbonate (Na_2CO_3). Adding these to hard water removes all the Ca^{2+} ions as a precipitate of calcium carbonate, leaving the water 'soft'.

$$\underset{\text{in hard water}}{Ca^{2+}(aq)} + \underset{\text{in washing soda}}{CO_3^{2-}(aq)} \rightarrow CaCO_3(s)$$

- **By ion-exchange**. The most convenient way of softening water is to use an ion-exchange column. The water passes through a column containing a special substance called a **resin** (Figure 1). The resin contains sodium ions which are displaced by calcium ions as hard water passes through the column. Na^+ ions do not cause hardness, so the water is now 'soft'.

$$\underset{\text{in hard water}}{Ca^{2+}(aq)} + \underset{\text{on resin}}{2Na^+(s)} \rightarrow \underset{\text{on resin}}{Ca^{2+}(s)} + \underset{\text{in the water}}{2Na^+(aq)}$$

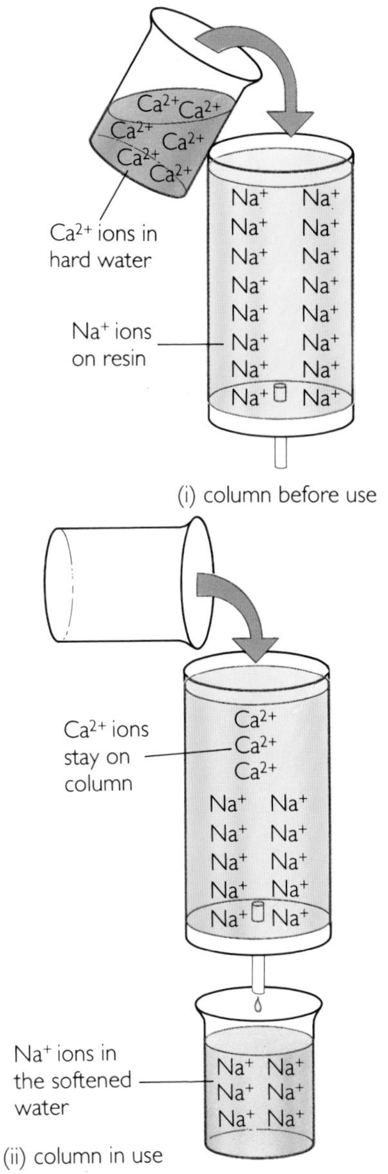

Figure 1 Using an ion-exchange column

QUESTIONS

1 Read the section on permanent hardness and temporary hardness. Design an experiment to compare the amounts of permanent hardness and temporary hardness in a sample of hard water.
(a) Say what you would do.
(b) List the measurements you would make.
(c) Say what methods and precautions you would take to get an accurate result.

2 (a) What are the advantages of hard water?
(b) What are the disadvantages of hard water?

3 Get into a small group with two or three others. Try to explain the following.
(a) Tap water in London is much harder than tap water in Manchester.
(b) After installing a water softener, some people have a separate tap delivering hard water.
(c) During the 1950s, tooth decay was worse in the North West of England than in the South East.
(d) Limescale forms around the taps in hard water areas. Usually, there is more scale around the hot tap than the cold tap.

4 Design a poster for a company which sells water-softening equipment.

SECTION G: THINGS TO DO

Activities

1 Making fertilisers

The flow diagram below shows the raw materials which are used to make the two most important fertilisers:

- N fertiliser ('straight' fertiliser) containing ammonium nitrate,
- NPK fertiliser ('compound' fertiliser) containing ammonium nitrate, ammonium phospate and potassium chloride.

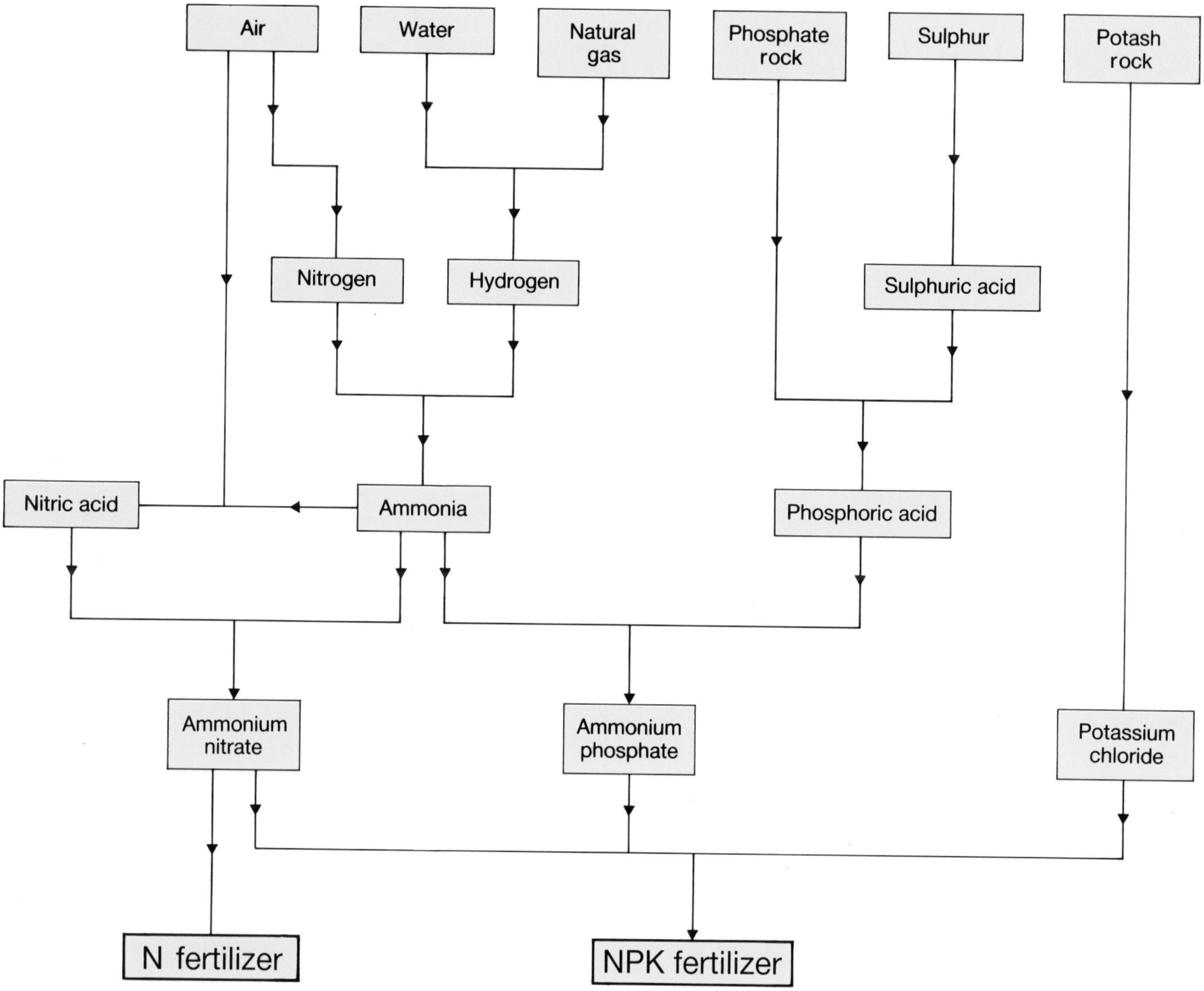

1 What raw materials besides air and water are used to make these fertilisers?

2 Where do the raw materials in question 1 come from to manufacture these fertilisers in the UK?

3 The reaction of ammonia (NH_3) with phosphoric acid (H_3PO_4) to form ammonium phosphate involves neutralisation.
(a) What is neutralisation?
(b) Write a word equation for this reaction.
(c) Use chemical symbols to write a balanced equation for the reaction.

4 One other reaction in the flow diagram also involves neutralisation. Write a word equation and a balanced symbolic equation for this reaction too.

5 Phosphoric acid is manufactured from phosphate rock which contains calcium phosphate ($Ca_3(PO_4)_2$).
The equation for the reaction is

$$Ca_3(PO_4)_2 + 3H_2SO_4 \rightarrow 2H_3PO_4 + 3CaSO_4$$

$$(H = 1, 0 = 16, P = 31, Ca = 40)$$

(a) What is the relative formula mass of calcium phosphate?
(b) What is the relative formula mass of phosphoric acid?
(c) According to the equation above, how many grams of phosphoric acid are produced from 310 g of calcium phosphate?
(d) How much calcium phosphate is needed to produce 100 tonnes of phosphoric acid?

Most of the ammonia produced at this plant is used to make fertilisers

2 pH changes after eating a sweet

Using a special instrument called a pH meter, Dr Razell measured the pH in Lee's mouth after he had eaten a sugary sweet.
The results are shown in the table on the right.

1 Open up a new spreadsheet on your computer.

2 Enter the times and pHs as in the table above.

3 From the spreadsheet, plot a line graph of pH (y-axis) against time (x-axis). (If you cannot plot graphs directly from the spreadsheet, draw the graph by hand.)

4 Print a copy of your line graph.
Sugary foods are converted to acids in our mouths. These acids attack the enamel on our teeth and cause tooth decay. Tooth decay occurs when the pH is lower than 5.5.

5 What does pH measure?

6 Describe how the pH changes in Lee's mouth after eating the sweet.

7 How many minutes after eating the sweet does tooth decay start?

8 How many minutes after eating the sweet does tooth decay stop?

9 How many minutes after eating the sweet is tooth decay at its worst? Explain your answer.

10 Use your graph to explain why dentists advise against eating sweets between meals.

11 What should the pH of toothpaste be? Explain your answer.

12 Dental decay can be prevented by strengthening the enamel of teeth. This is often done using a chemical in drinking water or in toothpaste. Find out more about this.

Time after eating sweet/minutes	pH
0	6.6
2	5.6
4	5.2
6	5.0
8	4.9
10	4.9
12	5.1
14	5.3
16	5.5
18	5.6
20	5.8

Table 1 How pH changed in Lee's mouth after eating a sweet

Study Questions

1 Question **1** concerns acids.
(a) Name *one* indicator that you could use to test for an acid.
(b) What would happen to the indicator you have named in part (a) if you placed it in an acidic solution?
(c) Name the acid that is used in a car battery.
(d) Name *one* substance, usually found around the house, that you could use to neutralise some spilt battery acid.
(e) Bath salts contain crystals of sodium carbonate. What gas would be given off if vinegar (acetic acid) were dropped onto some of these crystals?

2 Copper sulphate crystals ($CuSO_4.5H_2O$) can be made by adding excess copper oxide (CuO), which is insoluble in water, to dilute sulphuric acid.
(a) Why is copper oxide added in excess?
(b) The following apparatus could be used to separate the excess copper oxide from the solution. What words should be used for the labels *A* to *D*?

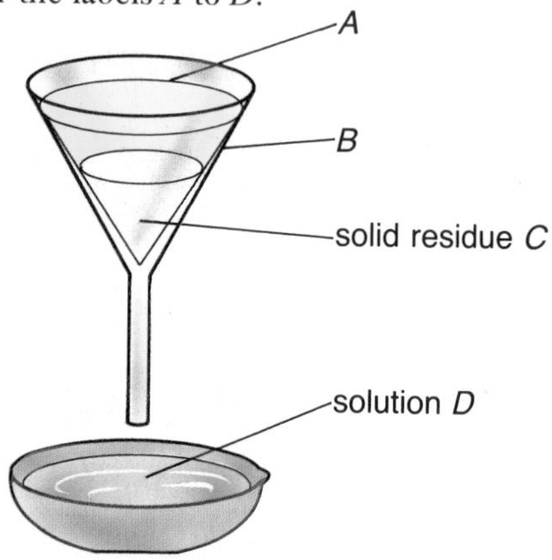

(c) After the excess copper oxide has been removed, the solution is partly evaporated and set aside. Some days later the sample is filtered again and the product is washed and finally dried.
(i) Why is the solution partly evaporated?
(ii) Which substance is removed by the *second* filtering?
(iii) How could the product be dried?
(iv) What would happen to the product if it was then heated strongly?

3 'Growmore' is a general garden fertiliser. It contains ammonium sulphate, calcium phosphate and potassium sulphate.
(a) What *elements* does Growmore contain?
(b) Which three elements in Growmore do plants need the most?
(c) Why is it important that fertilisers contain soluble substances?
(d) What is the formula of potassium sulphate?
(e) What substances would you use to prepare a sample of potassium sulphate for use as a fertiliser? Write an equation for the reaction involved.
(f) Suppose 100 g of Growmore contains 20 g of potassium sulphate. What is the percentage of potassium in Growmore? Explain your working.

4 Study the following reaction scheme shown below:
(a) Give the names of *P* to *W*.
(b) (i) What type of reaction takes place when hydrochloric acid reacts with solution *S*?
(ii) Describe a test that you could use to identify gas *W*.

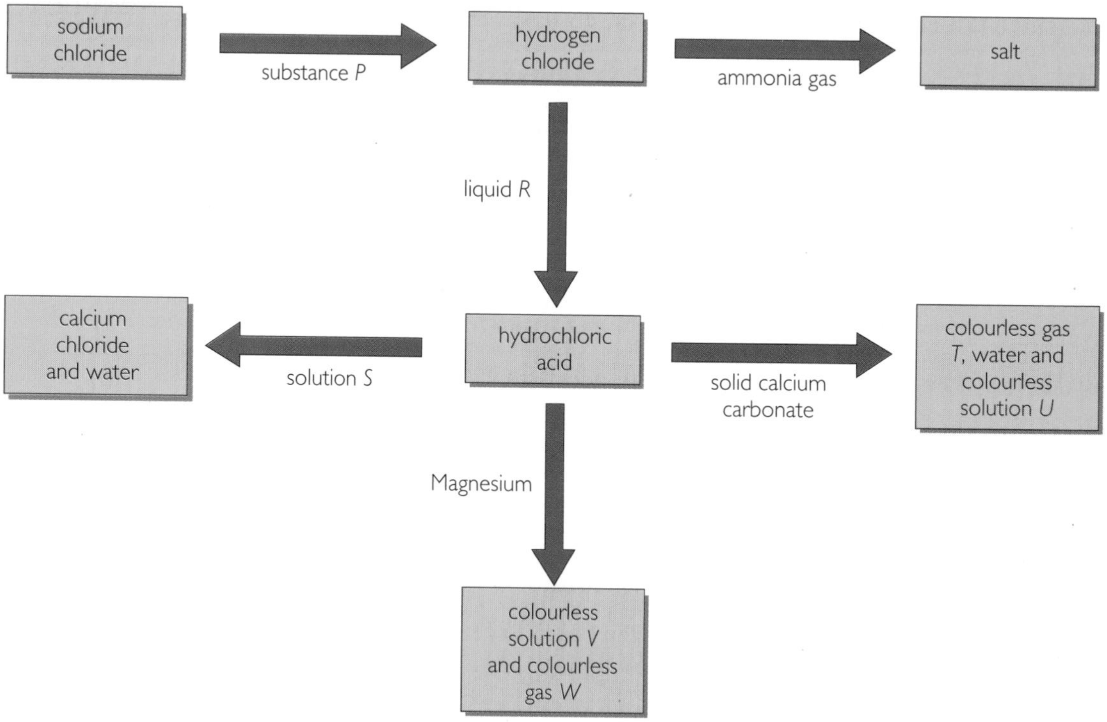

SECTION H

The structure of materials

What are the properties of the materials used in the construction of this building that make them suitable?

1 STUDYING STRUCTURES

White quartz crystals on impure dolomite. Quartz is silicon dioxide. Dolomite is a mixture of calcium carbonate and magnesium carbonate

Crystals of impure rock salt. Even in this impure sample you can see the regular shape of sodium chloride crystals

Look at the crystals of rock salt and quartz in the photographs above. What do you notice about all the salt crystals? What do you notice about all the quartz crystals? All the salt crystals are roughly the same cubic shape. All the quartz crystals have pointed tops. Further studies show that *all the crystals of one substance have similar shapes*. This suggests that the particles in the crystal are always packed in a regular fashion to give the same overall shape. Sometimes crystals grow unevenly and their shapes become distorted. Even so, it is usually easy to see their general shape. Solid substances which have a regular packing of particles are described as **crystalline**. Figure 1 shows how cubic crystals and hexagonal crystals can form. If the particles are always placed in parallel lines or at 90° to each other, the crystal will be cubic. If the particles are placed at 120° in the shape of a hexagon, the final crystal will be hexagonal.

We can compare the way in which a crystal grows to the way in which a bricklayer lays bricks. If the bricklayer always places the bricks in parallel lines or at 90° to each other, then the final buildings will be like cubes or boxes. However, if some bricks are laid at 120° to make hexagons, then the final buildings will be hexagonal.

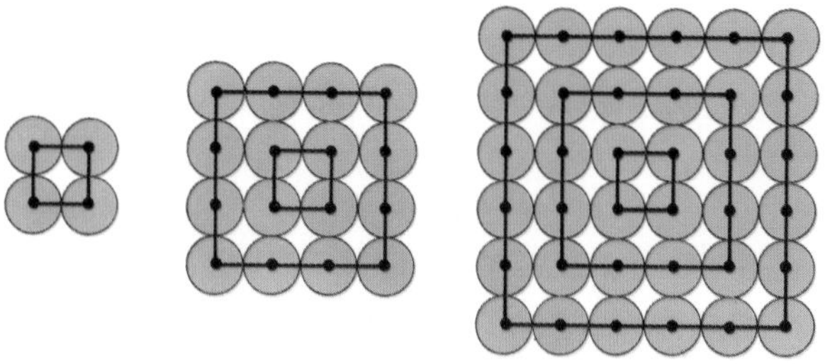

When the particles are arranged in a cubic fashion the final crystal will be cubic

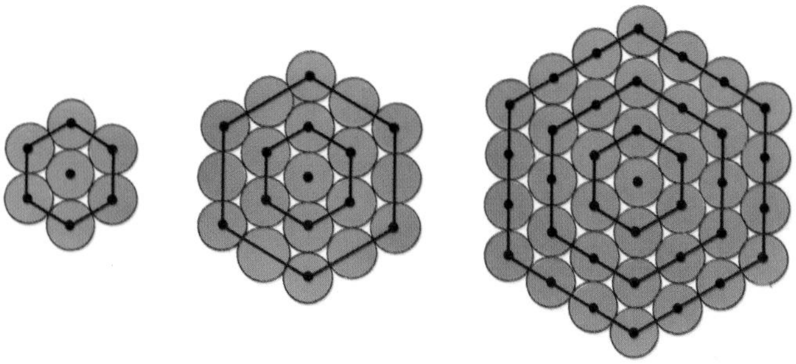

When the particles are arranged in a hexagonal fashion the final crystal will be hexagonal

Figure 1

The overall shape of a crystal can only give a *clue* to the way in which the particles are arranged. X-rays give much better *evidence*.

An X-ray diffraction photo of silicon

Using X-rays to study crystals

Look through a piece of thin stretched cloth at a small bright light. The pattern you see is due to the deflection of the light as it passes through the regularly spaced threads of the fabric. This deflection of light is called **diffraction** and the patterns produced are **diffraction patterns**. If the cloth is pulled so that the threads in the fabric get closer, then the pattern spreads further out. From the diffraction pattern which we *can* see, we can work out the pattern of the threads in the fabric which we *cannot* see. The same idea is used to work out the arrangement of particles in a crystal.

A narrow beam of X-rays is directed at a well-formed crystal (Figure 2). Some of the X-rays are diffracted by particles in the crystal onto X-ray film. When the film is developed, a regular pattern of spots appears. This is the diffraction pattern for the crystal. From the diffraction pattern which we *can* see, it is possible to work out the pattern of particles in the crystal which we *cannot* see. A regular arrangement of spots on the film indicates a regular arrangement of particles in the crystal. This regular arrangement of particles in the crystal is called a **lattice**.

X-rays have been used to study thousands of different crystals. An X-ray diffraction photograph is shown in the top corner of this page.

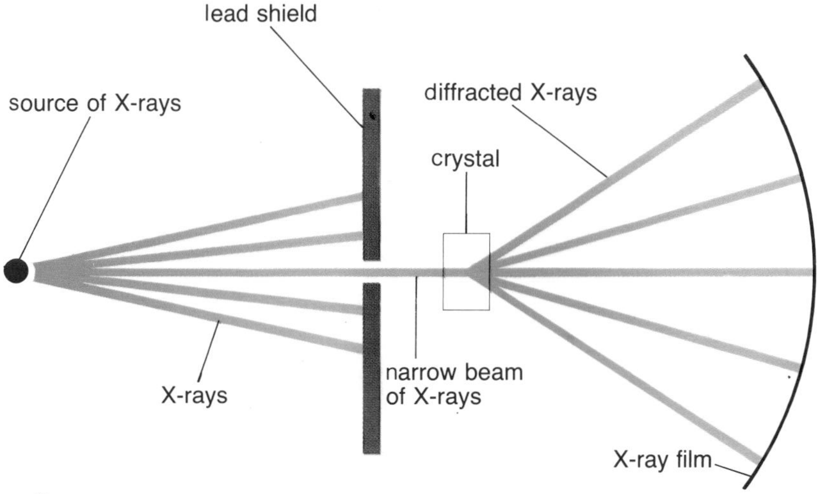

Figure 2

Beautiful, delicate snowflake crystals

QUESTIONS

1 Look through a piece of thin cloth at a small bright light. Describe what happens to the diffraction pattern when the cloth is (i) pulled and stretched, (ii) tilted, (iii) turned, (iv) moved closer to the light, (v) moved further from the light.

2 Look at the snowflake crystals in the photo above.
(a) What substance are snowflakes made of?
(b) What particles do snowflakes contain?
(c) How do you think the particles are arranged in snowflakes?

3 How are X-rays used to give evidence for the arrangement of particles in a crystal?

4 Why do all crystals of one substance have roughly the same shape?

2 THE STRUCTURE OF SUBSTANCES

What properties must the material in nappies have? What do you think the structure of nappy material is like?

Why was metal used to make suits of armour in the Middle Ages?

What properties of clay make it useful for making pots and crockery? What do you think the structure of clay is like?

The uses of materials depend on their properties. For example, copper is used for electrical wires and cables because it can be drawn into wires and it is a good conductor of electricity. All substances are made up of particles. If we know how these particles are arranged (the **structure**) and how the particles are held together (the **bonding**), then we can explain the properties of substances. For example, copper is a good conductor because its metallic bonding allows electrons to move through the structure when it is connected to a battery. It can be drawn into wires because copper atoms can slide over each other in the close-packed structure.

> Notice that the structure and bonding of a substance determine its properties and, in turn, the properties determine its uses.

The links from structure and bonding to uses help us to explain why metals are used as conductors, why graphite is used in pencils and why clay is used to make bricks.

Using X-ray analysis, we can find out how the particles are arranged in a substance (its structure), but it is more difficult to study the forces between these particles (its bonding).

From previous sections, you will know that

> all substances are made up of only three different types of particle – atoms, ions and molecules.

These three particles give rise to four different solid structures.

- metallic structures
- simple molecular structures
- giant molecular structures
- ionic structures

Table 1 shows the particles in these structures, the types of substances formed and examples of these substances.

Type of structure	Particles in the structure	Types of substance	Examples
Metallic (see also section E)	atoms	metals	Na, Fe, Cu
Simple molecular	small molecules containing a few atoms	non-metals, or non-metal compounds	I_2 (iodine) S_8 (sulphur) H_2O (water) CO_2 (carbon dioxide)
Giant molecular	very large molecules containing thousands of atoms	non-metals, or non-metal compounds	diamond, graphite, polythene, sand
Ionic	ions	compounds of metals with non-metals	Na^+Cl^- (salt) $Ca^{2+}O^{2-}$ (lime)

Table 1 The four types of solid structure and the particles they contain

A large natural diamond embedded in an ore, side by side with a cut diamond. How do the structure and properties of diamond lead to its use in jewellery?

Lead sulphide (galena) crystals on a sample of iron(II) carbonate (siderite). What particles will lead sulphide contain? How do you think the particles are arranged?

QUESTIONS

1 Get into a small group with 2 or 3 others. Look at the photos in this unit and discuss the questions in the captions.

2 (a) What are the particles in metal structures?
(b) Why are most metal structures described as close-packed?
(c) How are the particles arranged in most metal structures?

3 What type of structure will the following substances have?
chlorine, limestone (calcium carbonate), silver, air, rubber, polyvinylchloride (PVC), brass, wood

4 Conductivity tests can give evidence for the particles, bonding and structure of substances.
Describe (i) the tests you would carry out, (ii) the results you would expect, (iii) the conclusions you would make from your results. (Hint: See section D, units 3 and 4.)

CARBON AS DIAMOND

Charcoal is a really useful artists' material. It can be used to sketch an image in bold dark lines, or smudged gently with a fingertip, to provide a soft shaded effect

Diamond, graphite and charcoal are all made of the same pure element – carbon. But these three solids have different properties and different uses. Diamond is hard and clear, whereas graphite and charcoal are soft and black. Diamonds are used to cut and engrave glass, but graphite and charcoal are used by artists to get a soft, shaded effect. Diamond, graphite and charcoal have different properties and different uses because *they have different structures*. Their atoms are not packed in the same way.

These different forms of solid carbon are called **allotropes**. A few other elements also have allotropes. Oxygen has two allotropes – oxygen (O_2) and ozone (O_3). Sulphur has three allotropes – rhombic sulphur, monoclinic sulphur and plastic sulphur. *Allotropes are different forms of the same element in the same state.*

Diamond

In diamond, each carbon atom is joined to four others, as in the model shown on the left. Every carbon atom forms a strong covalent bond (see units 5 and 6) with each of its four neighbours. The covalent bonds extend through the whole diamond, forming a continuous, three-dimensional structure. Thus, a diamond is a single **giant molecule** or **macromolecule**.

Carborundum is a compound of carbon and silica, with a structure like diamond. It is widely used as an industrial abrasive

An engraving wheel has to be extremely hard so that the cuts it makes in glass are absolutely clean and straight. Diamond is the best material to use to toughen the wheel because it is so hard

Only a small number of atoms are shown in the model in the photo. In a real diamond, this arrangement of carbon atoms is repeated millions and millions of times.

A model showing how carbon atoms are bonded together in diamond

Properties and uses of diamond

- **Diamonds are very hard.** Carbon atoms in diamond are linked by very strong covalent bonds. This makes diamond hard. Another reason for its hardness is that the atoms are not arranged in layers so they cannot slide over one another like the atoms in metals. Diamond is one of the hardest known substances. Most of its industrial uses depend on this hardness. Diamonds which are not good enough for gems are used in glass cutters and in diamond studded saws. Powdered diamonds are also used as abrasives for smoothing very hard materials.

- **Diamond has a very high melting point.** Carbon atoms in diamond are held in the crystal structure by very strong covalent bonds. This means that the atoms cannot vibrate fast enough to break away from their neighbours until very high temperatures are reached. So the melting point of diamond is very high.

- **Diamond does not conduct electricity.** Diamond does *not* conduct electricity, unlike metals and graphite. In metals and graphite, some of the *outer* electrons of each atom are not strongly attached to any nucleus. They move towards the positive terminal when metals and graphite are connected to a battery. In diamond, however,, the outer electrons of each carbon atom are held firmly in covalent bonds. So diamond does not conduct electricity.

QUESTIONS

1 Why is diamond called a 'giant molecule'?

2 Design a poster to illustrate the uses of diamond and the way in which its uses depend on its properties and structure.

3 The largest diamond which has ever been found is the Cullinan Diamond. This weighs about 600 g.
(a) How many moles of carbon does it contain? (C = 12)
(b) How many atoms of carbon does it contain?

4 'Diamonds are a girl's best friend'. Is this true? What do you think? What do your friends think?

5 **Abrasive carborundum**
Carborundum was one of the first abrasives to be made. Before the discovery of carborundum, powdered diamond was the most common abrasive. When the first carborundum factory was opened, carborundum cost £4.40 per kilogram. After only a few years, its price had fallen to 7p per kilogram.

Carborundum is a compound of carbon and silicon. It has a structure like diamond. Its chemical name is silicon carbide, SiC. Carborundum is manufactured by heating coke (carbon) with sand (silicon(IV) oxide) at 2500°C in a furnace.

(a) What is an abrasive?
(b) What are abrasives used for?
(c) Suggest three important properties of an abrasive.
(d) Why do you think carborundum replaced diamond as the most widely-used abrasive?
(e) Write a word equation and a balanced symbolic equation for the reaction of coke with sand to form carborundum.
(f) Draw a diagram to show the structure of carborundum.
(g) It has been suggested that carborundum made the Industrial Revolution possible, during the nineteenth century. Why do you think this is?

4 CARBON AS GRAPHITE

Graphite has a layered structure. The layers slip and slide over one another very easily. This makes graphite an excellent material to add to oils like the one in the photo, to improve their lubricating properties

Graphite is the second important allotrope of carbon. The photo below shows a model of part of the structure of graphite. Notice that the carbon atoms are arranged in parallel layers. Each layer contains millions and millions of carbon atoms arranged in hexagons. Each carbon atom is held strongly in its layer by covalent bonds, so every layer is a **giant molecule**. The distance between neighbouring carbon atoms in the same layer is only 0.14 nm, but the distance between the layers is 0.34 nm. X-ray studies show that charcoal contains tiny crystals with a similar structure to graphite.

A model showing how atoms are bonded together in graphite. You can see the layers of hexagons one on top of the other

Properties and uses of graphite

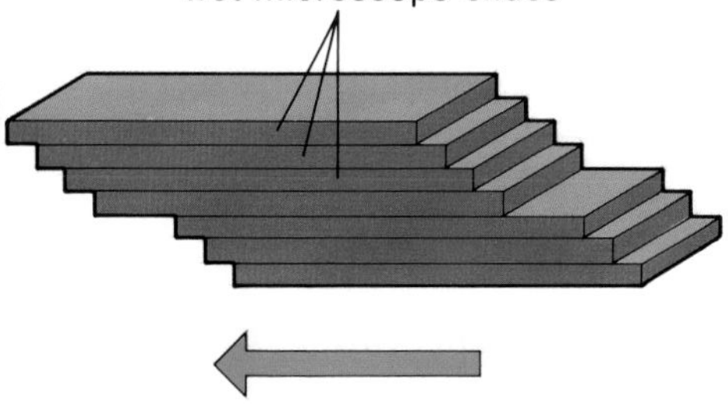

Figure 1
The layers in graphite slide over one another like a pile of wet microscope slides. The wet slides stick together and it is difficult to pull them apart, but a force parallel to the slides pushes them over each other easily

- **Graphite is a lubricant.** In graphite, each carbon atom is linked by strong covalent bonds to three other atoms in its layer. But the layers are 2.5 times further apart than carbon atoms in the same layer. This means that the forces between the layers are weak. If you rub graphite, the layers slide over each other and onto your fingers. This property has led to the use of graphite as the 'lead' in pencils and as a lubricant. The layers of graphite slide over each other like a pile of wet microscope slides (Figure 1).
- **Graphite has a very high melting point.** Although the layers of graphite move over each other easily, it is difficult to break the bonds between carbon atoms within one layer. Because of this, graphite does not melt until 3730°C and it does not boil until 4830°C. So it is used to make crucibles for molten metals.
- **Graphite conducts electricity.** The bonds *between* the layers of graphite are fairly weak. The electrons in these bonds move along the layers from one atom to the next when graphite is connected to a battery. So graphite will conduct electricity, unlike diamond and other non-metals. Because of this unusual property, graphite is used as electrodes in industry and as the positive terminals in dry cells (like the batteries you might put in a personal stereo).

This large graphite anode is being prepared for the electrolytic extraction of aluminium. Why is graphite used for this?

Pole vaulting in the 1920s. Notice how stiff the bamboo pole is. Notice also that there is no crashmat!

- **Graphite fibres are strong.** The bonds between carbon atoms within the layers of graphite are incredibly strong. But because the layers can slide over each other fairly easily, graphite will *bend* when it is made into parallel fibres. So graphite fibres are both strong *and* flexible. These properties make it ideal for reinforcing metals, broken bones and the shafts of tennis and squash racquets.

  The long poles used in pole vaulting are now made from a composite called carbon fibre. This consists of fibres of graphite embedded in flexible plastic. The graphite makes the pole strong so it won't break under the weight of the athlete. The plastic makes the pole very flexible, to help catapult the athlete over the bar. Over the years, world records for pole vaulting have increased dramatically, as the material used for the poles has improved. In 1991 the world record for the men's outdoor event was 6.1 m. In 1960, when the poles were made of aluminium, it was 4.8 m, and in 1920, when they used bamboo, it was 4.2 m!

QUESTIONS

1 Make a list of the important uses of graphite.

2 Make a table to show the similarities and differences between diamond and graphite.

3 (a) What is a 'composite'?
(b) Why do you think aluminium poles were better than bamboo, for pole vaulting?
(c) Before carbon fibre, a material called glass fibre was used for vaulting poles. This was made of fibres of glass embedded in plastic.
(i) Why do you think glass fibre was better than aluminium?
(ii) Why do you think carbon fibre is better than glass fibre?

4 (a) Why might someone rub a zip-fastener with a soft pencil to make it move more freely?
(b) Why would they use a pencil rather than oil?
(c) Why is graphite better than oil for lubricating the moving parts of hot machinery?

5 (a) Graphite powder is mixed with clay to make pencil 'lead'. How does the hardness of a pencil depend on the amounts of graphite and clay in it?
(b) Suppose you are the chief chemist at a company producing high quality pencils. What simple tests or experiments could you use to check the percentage of graphite in a batch of pencils?

5 SIMPLE MOLECULAR SUBSTANCES

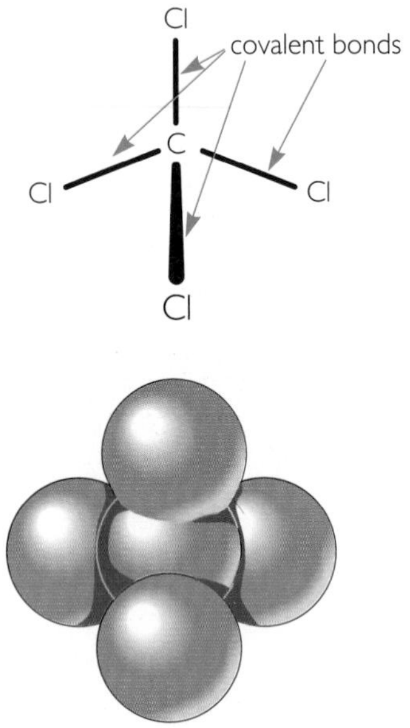

Figure 1
Tetrachloromethane (carbon tetrachloride) is a simple molecular substance. In tetrachloromethane, the carbon atom and four chlorine atoms are held together by strong covalent bonds

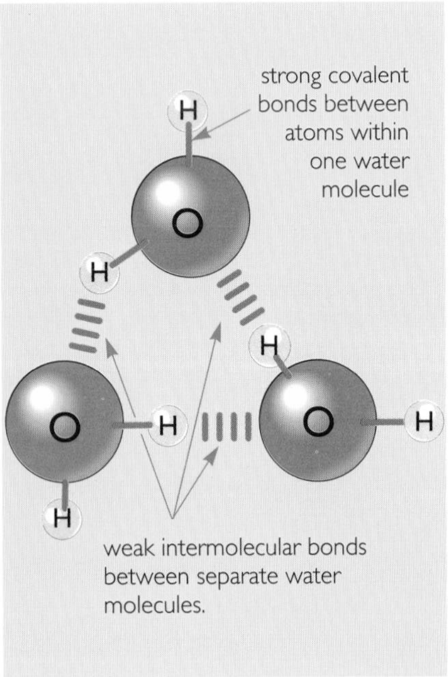

Figure 2
Covalent bonds and intermolecular bonds in water

Oxygen and water are good examples of simple molecular substances. They are made of simple molecules (each containing a few atoms). Their formulas and structures are shown near the top of Table 1. Most other non-metals and non-metal compounds are also made of simple molecules. For example, hydrogen is H_2, oxygen is O_2, iodine is I_2, carbon dioxide is CO_2 and tetrachloromethane is CCl_4. Sugar ($C_{12}H_{22}O_{11}$) has much larger molecules than these substances, but it still counts as a simple molecule.

In these simple molecular substances, the atoms are held together in each molecule by strong covalent bonds (Figure 1). The formulas and structures of a few simple molecular substances are shown in Table 1.

Name	**Molecular formula**	**Model of structure**
Hydrogen	H_2	
Oxygen	O_2	
Water	H_2O	
Methane	CH_4	
Hydrogen chloride	HCl	
Iodine	I_2	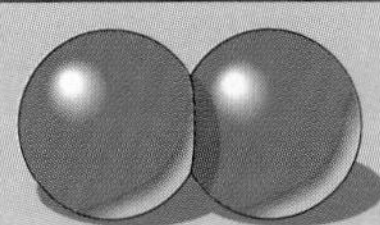
Carbon dioxide	CO_2	

Table 1 Formulas and structures of some simple molecular substances

Properties	Iodine	Sugar
Appearance	soft dark grey crystals	soft white crystals
Smell	sharp disinfectant smell	faint sweet smell
What happens on gentle heating?	vaporises easily forming a purple vapour	melts easily to a clear liquid
Does the solid conduct electricity?	no	no
Does the liquid conduct electricity?	no	no

Table 2 Properties of iodine and sugar

This butcher is using 'dry ice' (solid carbon dioxide) to keep meat cool and bacteria-free during mincing. After mincing, the 'dry ice', which is a simple molecular substance, will evaporate rapidly without spoiling the meat

Properties of simple molecular substances

Simple molecular substances have similar properties. These properties are shown by iodine (a non-metal element) and sugar (a non-metal compound) in Table 2.

Look carefully at Table 2. What properties do iodine and sugar have in common?

The properties of simple molecular substances can be explained in terms of their structure. The molecules in these substances have no electrical charge (like ions in ionic compounds). So there are no electrical forces holding them together. But as molecular substances do form liquids and solids, there must be some forces holding their molecules together. These weak forces between the separate molecules are called **intermolecular bonds** or **van der Waals' bonds** (Figure 2).

- **Simple molecular substances are soft.** The separate molecules in simple molecular substances, such as I_2, are usually further apart than atoms in metal structures and further apart than ions in ionic structures. The forces between the molecules are only weak and the molecules are easy to separate. Because of this, crystals of these substances are usually soft.
- **Simple molecular substances have low melting points and boiling points.** It takes less energy to separate the molecules in simple molecular substances than to separate ions in ionic compounds, or atoms in metals. So simple molecular compounds have lower melting points and lower boiling points than ionic compounds and metals.
- **Simple molecular substances do not conduct electricity.** Simple molecular substances have no mobile electrons like metals. They do not have any ions either. This means that they cannot conduct electricity as solids, as liquids or in aqueous solution.

QUESTIONS

1 Use your own words to explain the following:
covalent bond, intermolecular bond, simple molecule

2 (a) Summarise the main properties of simple molecular substances.
(b) Make a list of ten everyday substances which would be classed as simple molecular.

3 Simple molecular substances often have a smell, but metals do not. Why is this?

4 What properties does butter have that show that it contains simple molecular substances?

5 A substance is a poor conductor of electricity in the solid state. It melts at 217°C and boils at 685°C. Could this substance be (i) a metal, (ii) a non-metal, (iii) a giant molecule, (iv) an ionic solid, (v) a simple molecular solid?

6 MOLECULAR COMPOUNDS

If the nucleus of an atom was magnified a million, million times it would be as big as a pea and the total volume of the atom in which the electrons move would be as large as Westminster Abbey!

Metals can be mixed to form alloys, but they *never* react with each other to form compounds. For example, zinc and copper will form the alloy brass, but the two metals cannot react chemically because they both want to lose electrons and form positive ions.

Unlike metals, non-metals can react with each other to form a compound. These *non-metal compounds* are composed of simple molecules as we saw in the last unit. They include water (H_2O), carbon dioxide (CO_2), sugar ($C_{12}H_{22}O_{11}$), and ammonia (NH_3).

Forming molecular compounds – electron sharing

All atoms have a small positive centre called a **nucleus**, surrounded by a larger region of negative charge. The negative charge consists of electrons. It is balanced by positive charge in the nucleus, so that the whole atom is neutral. Almost all the mass of the atom is concentrated in the nucleus (Figure 1). Different atoms have different numbers of electrons. Hydrogen atoms are the smallest with only one electron, helium atoms have two electrons and oxygen atoms have eight electrons.

When two non-metals react to form a molecule, the regions of electrons in their atoms overlap so that each atom gains negative charge. The positive nuclei of both atoms attract the electrons in the region of overlap and this holds the atoms together (Figure 2). This type of bond formed by *electron sharing* between non-metals is a **covalent bond**. Notice that covalent bonding, like ionic bonding, involves attraction between opposite charges. Covalent bonds hold the atoms together *within* a molecule, but there are also **intermolecular bonds** holding the separate molecules together in liquids like water, and in solids like sugar.

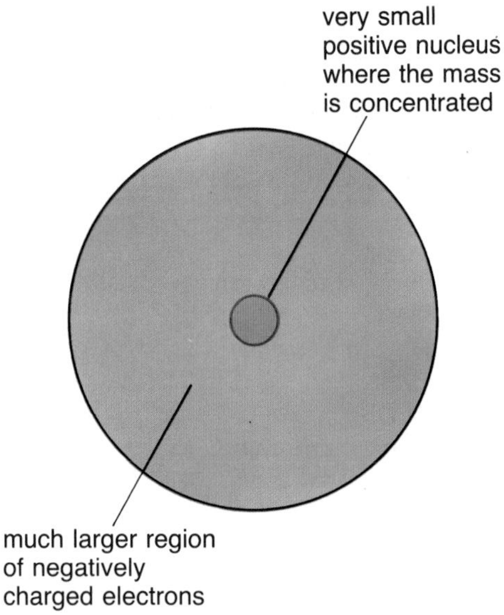

Figure 1
A simple picture of atomic structure

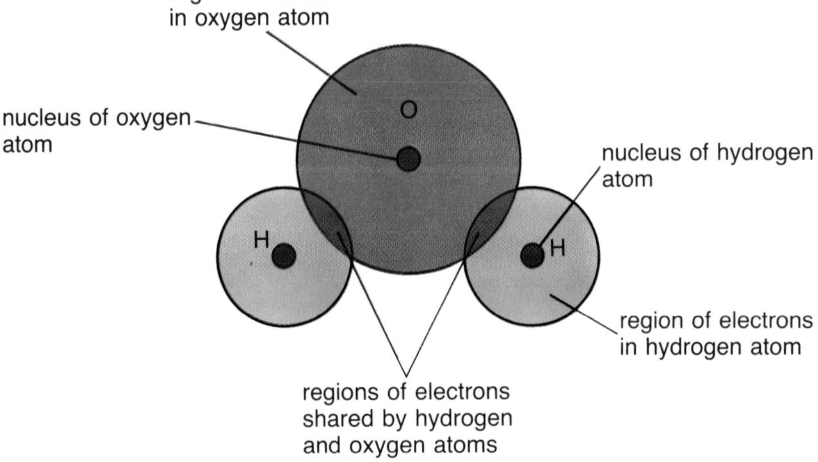

Figure 2
The simple structure of a molecule of water

Formulas of molecular compounds

Table 1 shows the formulas and structures of some well-known molecular compounds. The structures are written so that the number of covalent bonds (drawn as a line —) to each atom can be seen. Each bond is shown as a single line, so two lines show that there is a double bond between the atoms. Notice that hydrogen can form 1 bond with other atoms (H—), so its **combining power**, or **valency**, is 1. The combining powers of chlorine and bromine are also 1. Oxygen and sulphur both form 2 bonds with other atoms (—O— and —S—). Their combining power is therefore 2. Nitrogen atoms form 3 bonds and carbon atoms form 4 bonds, so their valencies are 3 and 4 respectively.

Although we can predict the formulas of molecular compounds from the number of bonds which the atoms form, the only sure way of knowing a formula is by chemical analysis. For example, carbon forms 4 bonds and oxygen forms 2 bonds, so we would predict that carbon and oxygen will form a compound O=C=O. This compound (carbon dioxide) does exist, but so does carbon monoxide, CO, which we would not predict.

<table>
<tr><th>Compound</th><th>Formula</th><th>Structure</th></tr>
<tr><td>Hydrogen chloride</td><td>HCl</td><td>H—Cl</td></tr>
<tr><td>Hydrogen bromide</td><td>HBr</td><td>H—Br</td></tr>
<tr><td>Hydrogen iodide</td><td>HI</td><td>H—I</td></tr>
<tr><td>Water (hydrogen oxide)</td><td>H_2O</td><td>H—O—H</td></tr>
<tr><td>Hydrogen sulphide</td><td>H_2S</td><td>H—S—H</td></tr>
<tr><td>Ammonia (hydrogen nitride)</td><td>NH_3</td><td>H　H
\ /
N
|
H</td></tr>
<tr><td>Methane</td><td>CH_4</td><td>H
|
H—C—H
|
H</td></tr>
<tr><td>Carbon dioxide</td><td>CO_2</td><td>O=C=O</td></tr>
<tr><td>Carbon disulphide</td><td>CS_2</td><td>S=C=S</td></tr>
<tr><td>Tetrachloromethane
(carbon tetrachloride)</td><td>CCl_4</td><td>Cl
|
Cl—C—Cl
|
Cl</td></tr>
</table>

Table 1 The formulas and structures of some well-known molecular compounds

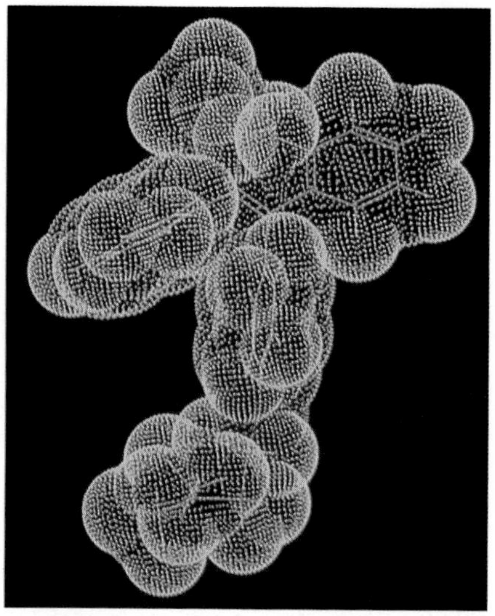

This model of a molecule was drawn using a computer to analyse the results obtained during the investigation of a complex organic substance

QUESTIONS

1 Explain the following:
non-metal compound, molecular compound, covalent bond, nucleus, intermolecular bond

2 Assuming the usual combining powers of the elements, draw the structures of the following compounds. (Show each bond as a line —.)
hydrogen sulphide (H_2S), hydrogen bromide (HBr), ammonia (NH_3), sulphur dichloride (SCl_2), dichlorine oxide (Cl_2O), tetrabromomethane (carbon tetrabromide (CBr_4), nitrogen triiodide (NI_3), hydrogen peroxide (H_2O_2), ethane (C_2H_6), ethene (C_2H_4)

3 Suppose you are helping to look after a group of ten-year-old children at a holiday centre. During one session, they are playing with jam jars, water and steel spoons. The following questions come up while they're playing.
(i) Jam jars and spoons are solid. What is a solid?
(ii) Water is liquid. What is a liquid?
(iii) Why is steel a solid and water a liquid?
How would you answer their questions?

7 IONIC COMPOUNDS

Chalk cliffs are composed of an ionic compound containing calcium ions (Ca^{2+}) and carbonate ions (CO_3^{2-})

Forming ionic compounds – electron transfer

Ionic compounds form when metals react with non-metals. For example, when sodium burns in chlorine, sodium chloride is formed. This contains sodium ions (Na^+) and chloride ions (Cl^-).

These ions form by *transfer of electrons*. During the reaction, each sodium atom gives up one electron and forms a sodium ion (Figure 1). The electron is taken by a chlorine atom to form a chloride ion.

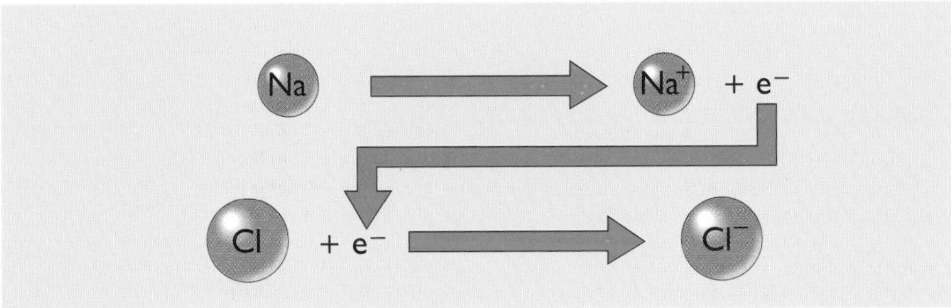

Figure 1

When ionic compounds form, metal atoms *lose* electrons and form *positive* ions, whereas non-metal atoms *gain* electrons and form *negative* ions. This transfer of electrons from metals to non-metals explains the formation of ionic compounds.

Bonding and properties of ionic compounds

In solid ionic compounds, the ions are held together in a **giant structure** by the attraction between positive ions and negative ions. This force of attraction between oppositely-charged ions is called an **ionic bond**.

Figure 2 shows how the ions are arranged in one layer of sodium chloride. Strong ionic bonds hold the ions together very firmly. This explains why ionic compounds

- are solids at room temperature, with high melting points,
- are hard substances,
- cannot conduct when solid because the ions cannot move freely,
- conduct electricity when molten or aqueous.

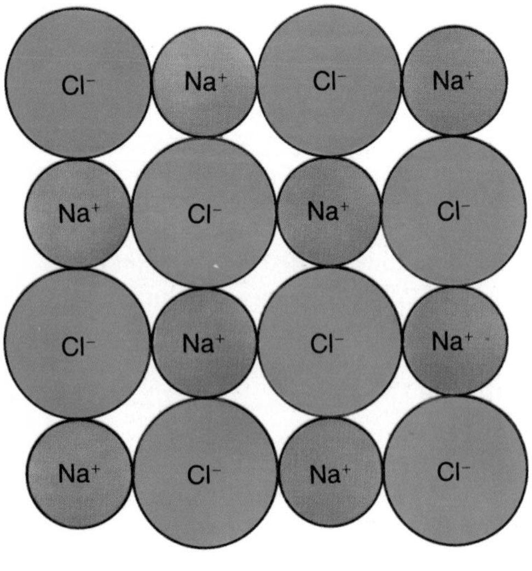

Figure 2

When molten sodium chloride conducts electricity it is decomposed into sodium and chlorine. The decomposition is caused by electrical energy in the current.

During electrolysis, sodium ions (Na^+) are attracted to the cathode. Here they pick up an electron and form sodium atoms (Na).

Cathode (−) $Na^+ + e^- \rightarrow Na$

At the same time, chloride ions (Cl^-) in the electrolyte are attracted to the anode. These ions give up an electron to the anode and form chlorine atoms (Cl).

$$\textbf{Anode (+)} \qquad Cl^- \rightarrow e^- + Cl$$

The chlorine atoms immediately join up in pairs to form molecules of chlorine gas (Cl_2)

An underground salt mine.
The ionic structure of salt means sodium and chlorine can be extracted by electrolysis

Formulas of ionic compounds – balancing charges

The formulas of ionic compounds, like sodium chloride (NaCl) and calcium oxide (CaO), can be obtained by balancing the charges on the positive ions with those on the negative ions. For example, the formula of calcium chloride is $Ca^{2+}(Cl^-)_2$ or simply $CaCl_2$. Here, two Cl^- ions balance the charge on one Ca^{2+} ion. The formula $CaCl_2$ has a small 2 after the Cl to show that two Cl^- ions are needed for one Ca^{2+} ion. These formulas show the ratio of the numbers of ions present in the ionic compound.

Can you see that the number of charges on an ion is a measure of its combining power, or valency? Na^+ has a combining power of 1, whereas Ca^{2+} has a combining power of 2. Na^+ can combine with only one Cl^- to form Na^+Cl^-, whereas Ca^{2+} can combine with two Cl^- ions to form $Ca^{2+}(Cl^-)_2$.

Elements such as iron, which have two different ions (Fe^{2+} and Fe^{3+}), have two valencies. Thus, iron can form two different compounds with chlorine – iron(II) chloride, $FeCl_2$, and iron(III) chloride, $FeCl_3$.

Table 1 shows the names and formulas of some salts. Notice that the formula of calcium nitrate is $Ca(NO_3)_2$. The brackets around NO_3^- show that it is a single unit containing one nitrogen and three oxygen atoms with one negative charge. Thus, two NO_3^- ions balance one Ca^{2+} ion. Other ions like SO_4^{2-}, CO_3^{2-} and OH^- must also be regarded as single units and put in brackets when there are 2 or 3 of them in a formula.

Name of salt	Formula
Calcium nitrate	$Ca^{2+}(NO_3^-)_2$ or $Ca(NO_3)_2$
Zinc sulphate	$Zn^{2+}SO_4^{2-}$ or $ZnSO_4$
Iron(III) chloride	$Fe^{3+}(Cl^-)_3$ or $FeCl_3$
Copper(II) bromide	$Cu^{2+}(Br^-)_2$ or $CuBr_2$
Sodium carbonate	$(Na^+)_2CO_3^{2-}$ or Na_2CO_3
Potassium iodide	K^+I^- or KI

Table 1 The names and formulas of some salts

QUESTIONS

1 Which of the following substances conduct electricity (i) when liquid, (ii) when solid?
diamond, potassium chloride, copper, carbon disulphide, sulphur

2 Write the symbols for the ions and the formulas of the following compounds:
potassium hydroxide, iron(III) nitrate, barium chloride, sodium carbonate, silver sulphate, calcium hydrogencarbonate, aluminium oxide, zinc bromide, copper(II) nitrate, magnesium sulphide

3 Use polystyrene balls to make a model of one layer in the structure of sodium chloride, as in Figure 2. Put your model with those of 2 or 3 other groups to make a three-dimensional model.
(a) How many Cl^- ions surround one Na^+ ion in the three-dimensional structure?
(b) How many Na^+ ions surround one Cl^- ion in the three-dimensional structure?

4 Sodium fluoride and magnesium oxide have the same crystal structure and similar distances between ions. The melting point of NaF is 992°C, but that of MgO is 2640°C. Why is there such a big difference in their melting points?

5 Suppose you have a new student in your class who has only recently moved to your school. She knows about the way atoms and ions are made up, but because chemistry was taught in a different order in her last school, she hasn't done the structure of materials yet. How would you explain to her what we mean by the following terms?
giant structure, ionic bond, covalent bond, intermolecular bond

8 SALT – AN IMPORTANT IONIC COMPOUND

Ionic compounds are present in the sea and in the Earth's crust. Many rocks contain ionic compounds. These include rock salt (Na^+Cl^-), limestone ($Ca^{2+}CO_3^{2-}$) and iron ore ($(Fe^{3+})_2(O^{2-})_3$). Clay, sandstone and granite also contain ionic compounds.

Tennis players and people who work in hot places usually take salt tablets to replace the salt they lose by sweating. Sweat is mainly salt solution

Salt in our diet

Salt (sodium chloride) is one of the most important ionic compounds. It is an essential mineral in our diet. Most foods contain salt but some foods are saltier than others. Our diet must contain the right amount of salt. Too much salt may cause high blood pressure. Too little salt causes sharp pains ('cramp') in our muscles.

Cattle licking a salt block. The salt block contains various salts to supplement the diet of the cattle

Properties of ionic compounds

Ionic compounds

- are hard
- have high melting points
- have high boiling points
- do not conduct when solid
- conduct when liquid and in aqueous solution

The box above shows the important properties of ionic compounds. These properties can be explained in terms of their giant structure of ions.

- **Hardness.** In order to cut a lump of sodium chloride, the ions must be separated. This is very difficult because each ion is held in the crystal lattice by strong attractions from the ions of opposite charge around it.
- **Melting and boiling.** In ionic solids, the ions vibrate about fixed positions. As the temperature rises, the ions vibrate more and more. Eventually, the

ions vibrate so much that they 'escape' from their places in the crystal and slide freely around each other. When this happens, the solid is melting. The forces between ions of opposite charge are so strong that ionic compounds have high melting points and even higher boiling points.

Some ionic compounds, such as magnesium oxide and aluminium oxide, have *extremely* high melting points. Because of this, they can be used as a lining for furnaces. They are called **refractory materials**.

- **Conductivity.** Solid ionic compounds cannot conduct electricity. The ions are held in the crystal and cannot move towards the electrodes. When the solid melts, and when it is dissolved in water, the ions are free to move. So molten and aqueous ionic compounds *will* conduct.

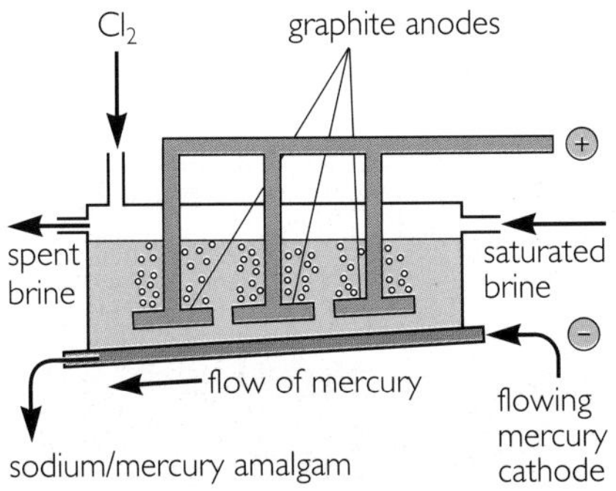

Figure 1
The electrolysis of saturated sodium chloride solution (brine)

Salt – an important source of other substances

The electrolysis of *molten* sodium chloride is used to produce sodium and chlorine (section D, unit 4). The electrolysis of saturated sodium chloride *solution* (brine) is used to manufacture chlorine, hydrogen and sodium hydroxide (Figure 1).

During electrolysis, chloride ions in the brine are attracted to the graphite anodes. At the anodes, they are converted into chlorine gas.

$$\textbf{Anode } (+) \qquad 2Cl^-(aq) \rightarrow 2e^- + Cl_2(g)$$

At the same time, sodium ions in the brine are attracted to the flowing mercury cathode. The sodium ions take electrons from the cathode and form sodium metal. This dissolves in the mercury to form an amalgam (alloy).

$$\textbf{Cathode } (-) \qquad 2Na^+(aq) + 2e^- \rightarrow 2Na(s)$$

The sodium/mercury amalgam flows out of the cell and into a second tank. Here, it reacts with water to produce hydrogen and sodium hydroxide.

$$2Na/Hg(l) + 2H_2O(l) \rightarrow 2NaOH(aq) + H_2(g) + 2Hg(l)$$

Large amounts of impure sodium chloride (crushed rock salt) are used for gritting and de-icing roads. The salt mixes with the ice and lowers its melting point. Mixtures of ice and salt will melt at temperatures down to −22°C. This means that the temperatures can be well below 0°C and the roads still do not ice up

QUESTIONS

1 Draw a diagram or design a poster to illustrate the uses of sodium chloride (salt) and the materials which are obtained from it.

2 Solid sodium chloride does not conduct electricity, but liquid sodium chloride conducts well.
(a) Explain this statement.
(b) Write equations for the processes at the electrodes when liquid NaCl conducts.

3 Give three important uses of sodium chloride as pure salt, rock salt or brine. Explain why sodium chloride has these uses.

4 Substance X melts at a high temperature. Liquid X conducts electricity.
(a) Which of the following could be X?
calcium chloride, starch, copper, sulphur, polythene, bronze, carbon disulphide, zinc oxide
(b) Explain your answers to part (a).

5 Get into a small group with 2 or 3 others. Discuss and explain the following points.
(a) Cattle require a salt lick during hot summer months.
(b) The use of crushed rock salt on icy roads does have a drawback for motorists.
(c) Salt is very unreactive, but sodium and chlorine, from which it is made, are both very reactive.

9 FROM SAND TO GLASS

Different kinds of glass

We use different kinds of glass for different purposes. Some of these are illustrated in the photos below.

Bottles and windows are made from cheap soda lime glass

Glass dishes and ornaments use lead glass which is harder and shinier

Laboratory glassware and glass ovenware are made from borosilicate glass which is heat resistant

Making glass

Glass is usually made by heating a mixture of metal oxides and metal carbonates with pure sand (silicon dioxide, SiO_2) in a furnace. At the high temperatures involved, any carbonates decompose to oxides and bubbles of carbon dioxide form in the mixture. The glass is heated to make sure that all the bubbles escape and this produces a runny, molten liquid. The liquid is cooled until it is thick enough to be moulded or blown into different shapes. It is then allowed to cool further until it sets solid.

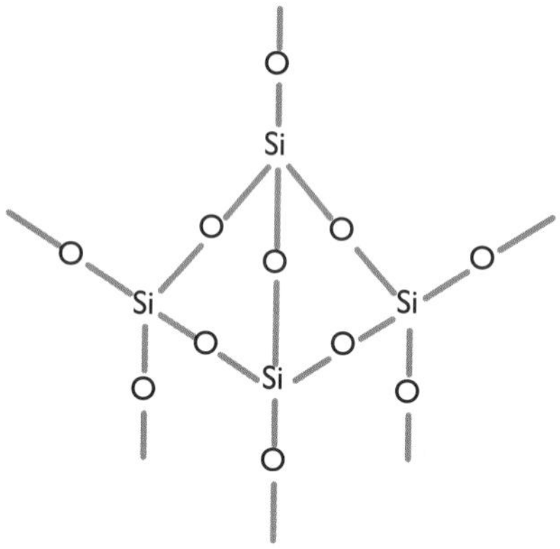

Figure 1
The structure of pure silicon dioxide glass. (For simplicity, the structure is not drawn in three dimensions)

The structure of glass

Glass can be made from pure silicon dioxide. This melts at 1700°C to give a thick, viscous liquid. On cooling, the liquid forms a glassy, transparent solid. This glass has a giant structure of atoms. Every silicon atom is bonded to four oxygen atoms (Figure 1).

This three-dimensional structure with strong covalent bonds between silicon and oxygen atoms makes the glass very hard. Unfortunately, the high melting point of this glass makes it expensive to manufacture and difficult to mould. Because of this, most glasses are made from a mixture of silicon dioxide with metal oxides. The metal oxides lower the melting point. For example, soda lime glass is made from sand (silicon dioxide), limestone (calcium carbonate) and sodium carbonate. The carbonates decompose to oxides on heating. So the glass has a giant structure of atoms *and* ions (Figure 2).

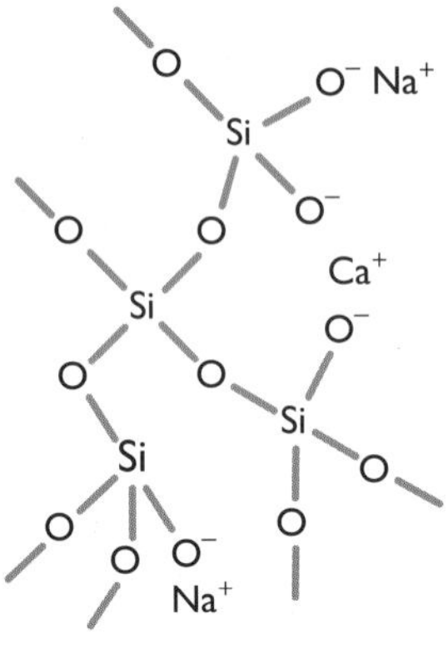

Figure 2
The structure of soda-lime glass. (For simplicity, the structure is not drawn in three dimensions)

The properties of glass

Glass is
- hard
- chemically unreactive
- easy to clean
- heat resistant
- an insulator
- transparent
- fairly cheap
- easily moulded

Glass is sometimes used to encase nuclear waste before it is transported. Why do you think glass is used for this purpose?

Recycling glass

More than six thousand million glass bottles and jars are sold in Britain each year. Most of these are thrown away. A small proportion (perhaps one tenth) are recycled. This is disappointing for a number of reasons.

- Glass containers are relatively easy to clean, sterilise and refill.
- Glass can easily be crushed and added to furnaces to make new containers.
- Glass is made from materials such as sand, limestone and salt which have to be extracted from the earth by processes which spoil the landscape and use up our energy supplies.
- Re-using and recycling glass could save huge amounts of energy.

QUESTIONS

1 Get into a group with 2 or 3 others. Look at the properties and the structure of glass described in this unit.
(a) In your group, discuss how the structure of glass results in its properties.
(b) Why do you think glass cracks and breaks if it is dropped or if the temperature suddenly changes?

2 Glass insulators are often used on electricity pylons. Make a list of the properties of glass which make it particularly useful for this purpose.

3 Look at the energy costs involved in recycling glass, in this table.

Recycling process	Energy cost per kg of recycled product/kJ
Collecting waste glass	350
Processing waste glass	100
Delivering waste glass to glass furnace	50

(a) What is the total energy cost of collecting, processing and delivering enough waste glass to the furnace to produce one kilogram of recycled glass?
(b) The energy cost of extracting, processing and delivering enough raw materials to the furnace to produce one kilogram of *new* glass, is 4500 kJ. How much energy is saved, per kilogram, if glass is recycled?
(c) How much energy is saved, per *tonne*, if glass is recycled? (1 tonne = 1000 kg)
(d) Suppose that all the fuels used to provide this energy (petrol, diesel, fuel oil) produce 40 000 kJ per litre. How many litres of fuel are saved, per tonne, if glass is recycled?
(e) Suppose that all the fuels used in glass manufacture cost 50p per litre. How much money is saved, per tonne, if glass is recycled?

4 (a) Why do you think so few glass containers are re-used at present?
(b) What proportion of glass containers do you think could be re-used or recycled? Explain your answer.
(c) How could we encourage people to re-use or recycle glass?

10 FROM CLAY TO CERAMICS

What are ceramics?

The word 'ceramic' comes from a Greek word meaning 'pottery' or 'burnt clay'. Clay has been dug from the ground and heated to make pots, bricks and ornaments since prehistoric times.

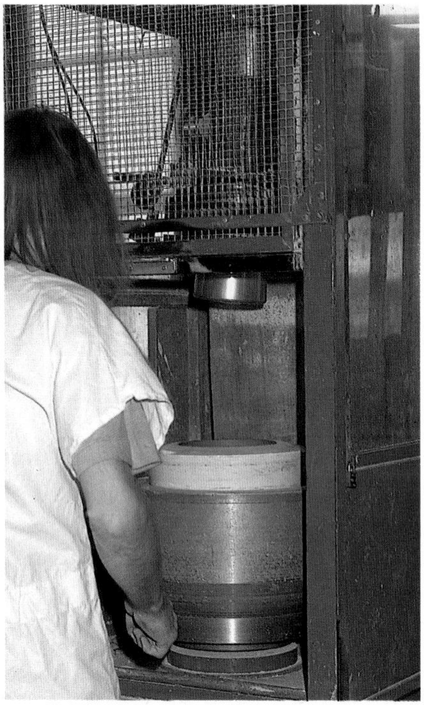

This piece of pottery is being made using a machine

This photo shows where china clay is extracted. China clay is the best in quality clay for pots and ornaments

Finished pieces of pottery after glazing and firing

What is clay?

Clay is a mixture of several materials. The best quality clay (china clay), which is used to make crockery and ornaments, contains a high proportion of kaolinite. Kaolinite contains silicon, oxygen and aluminium atoms. The silicon and oxygen atoms are joined by strong covalent bonds and there are ionic bonds between the oxygen and aluminium. These atoms are linked together in separate flat layers in a giant structure. These flat layers are described as **two-dimensional structures**.

Figure 1 gives a very simplified picture of the bonding and structure in raw clay.

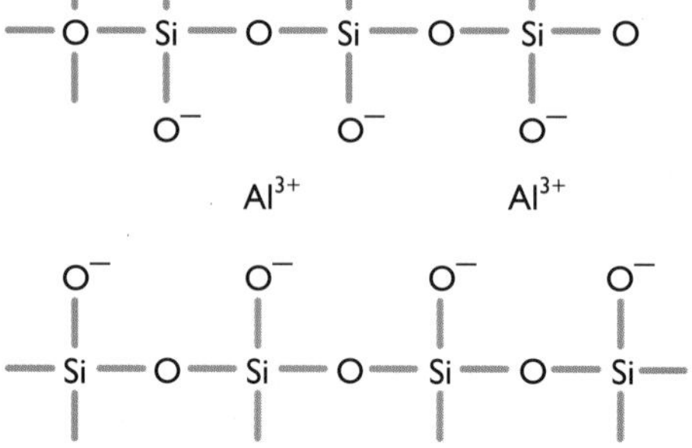

Figure 1
A simplified diagram to show the bonding in clay

Notice the similarity between the structure of raw clay and those of sand and glass. There is, however, one important difference. Clay has a flat, two-dimensional structure whereas the structures of sand and glass are three-dimensional.

Moulding and firing clay

When clay is wet, water molecules get between the layers and allow them to slide over one another. This is why wet clay is soft and slippery and can easily be moulded into different shapes. The water acts as a lubricant. When clay is left to dry, most of the water between the layers evaporates. The clay also shrinks and the layers get closer so that they cannot move over each other so easily.

When clay is heated in a furnace at about 1000°C (**fired**), it forms a ceramic. This is hard, gritty and rigid. During firing, all the water molecules are driven out of the clay and complicated chemical changes take place. Atoms in one layer form bonds with those in the layers above and below. So the flat, two-dimensional structure in clay changes to a cross-linked, three-dimensional structure in the ceramic. The structure of ceramics is therefore very similar to that of glass.

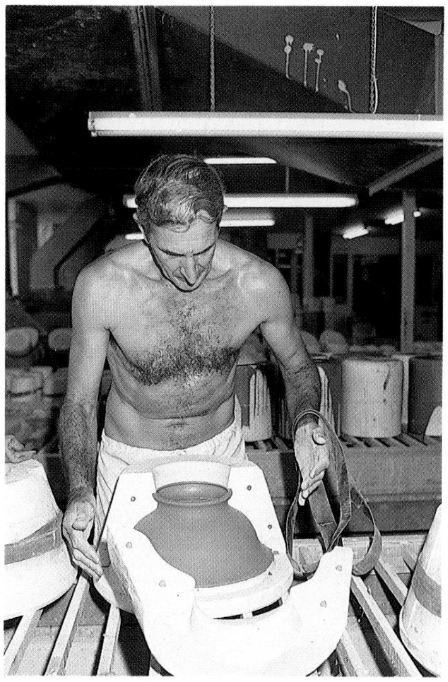

A dried clay article being removed from its mould before firing

Using ceramics

Ceramics and glasses are very important in our homes and industries. They have high melting points and are chemically unreactive. They contain elements like silicon and aluminium already combined with oxygen. So they do not burn when heated in air. In this respect, they differ from metals and plastics.

Because of their heat resistance, ceramics are used to line most industrial furnaces. Any other material would either melt or burn. They are used in furnaces for the manufacture of iron, steel, glass and cement. The bricks used for these furnaces are made from fireclay. This contains a higher proportion of aluminium than other clays and it produces bricks which are more heat resistant and chemically inert.

Probably the most important use of ceramics is for crockery. At one time, crockery was shaped by hand on a potter's wheel, but now most of it is shaped and moulded by machines, as shown in the photo on the opposite page.

Hollow articles like teapots and jugs are made by pouring runny clay into moulds. The moulds are made of porous plaster. This absorbs water from the clay and a solid lining forms on the inside of the mould. When this lining is sufficiently thick, the rest of the runny clay is poured out. The article can then be removed from the mould and dried.

Bricks and tiles are made by pressing dryish clay into steel moulds. Once the clay has been shaped or moulded, it must be dried before firing. Otherwise, the water will evaporate and shatter the clay during firing.

Apart from bricks and roof tiles, most ceramic articles are covered with a **glaze**. The glaze gives the ceramic a smooth, glassy, waterproof surface. After the first firing, the pottery is dipped in a glaze and then fired again. The glaze consists of finely powdered oxides of silicon and metals suspended in water. When the glazed pottery is fired a second time, the glaze is converted to glass.

QUESTIONS

1 Get into a group with 2 or 3 others. Discuss the following points and try to explain them.
(a) Clay is soft and slippery, but ceramic is hard and gritty.
(b) Dry, crumbly clay can be made soft and easily moulded by adding water.
(c) Ceramic cannot be converted back to clay by adding water to it.

2 (a) Make a list of the similarities between glasses and ceramics.
(b) Make a list of the differences between glasses and ceramics.
(Hint: You might like to look at the properties of glass described in the last unit.)

3 (a) Draw a flow diagram to show the stages in making a teapot from china clay.
(b) Why are the moulds used for teapots made of porous material?
(c) Why must clay be dried before it is fired?
(d) What is the advantage of coating articles like teapots with a glaze?

SECTION H: THINGS TO DO

Activities

1 ChlorChem

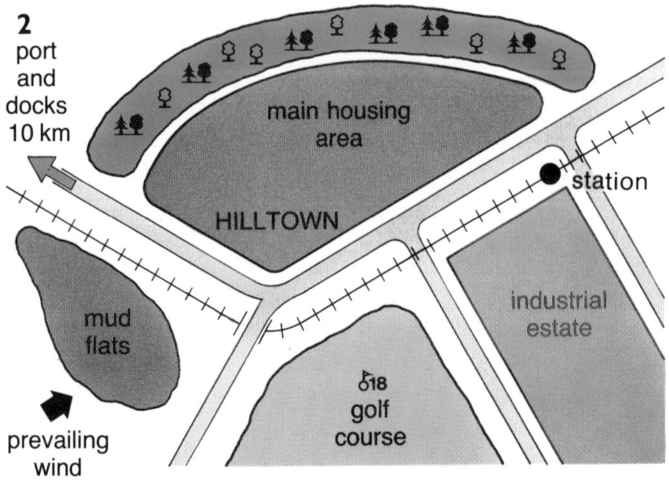

A chemical company (ChlorChem) wants to build a new factory near Hilltown in order to manufacture chlorine and sodium hydroxide using brine.

(a) What problems does this pose?

(b) Where would you site this new factory?

(c) What are the reasons for your choice of site?

(d) The Council at Hilltown wants ChlorChem to build their factory on the mud flats. They have offered ChlorChem financial help if the factory is sited on the mud flats. What disadvantage will this site have for (i) ChlorChem, (ii) Hilltown?

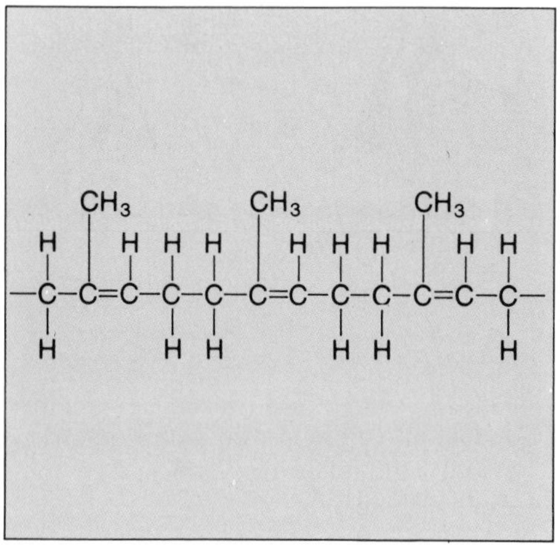

Figure 1
Part of a molecule of natural rubber

2 Explaining the structure of rubber

Natural rubber is a **hydrocarbon**. It contains long chains of carbon atoms with hydrogen atoms attached to them (Figure 1). Each rubber molecule contains between 10 000 and 50 000 carbon atoms. Notice that every fourth bond between the carbon atoms is a double bond. In natural rubber, millions of these long chains are tangled together like tangled pieces of wire.

Unfortunately, natural rubber is only partly elastic. When it is pulled or deformed the molecules move over one another and do not return to their original position when the force is removed (Figure 2). Pulling a tangle of wires will have the same effect.

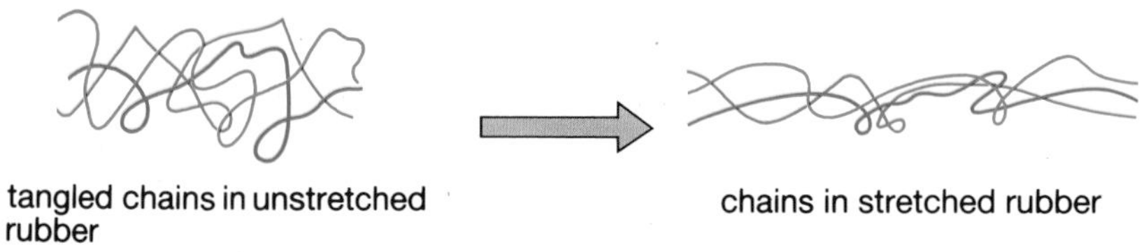

Figure 2
Stretching natural rubber

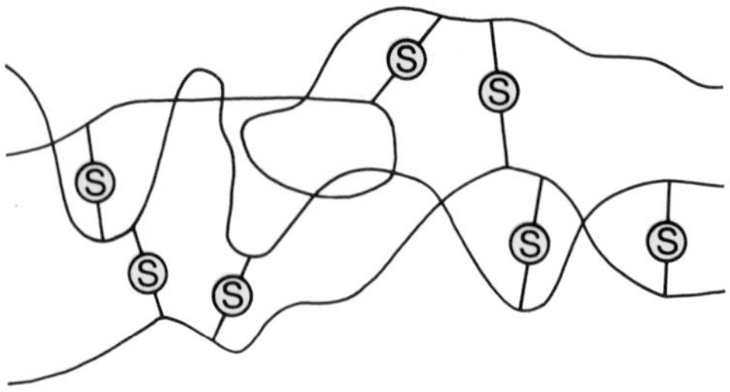

Figure 3
Sulphur links between molecules of vulcanised rubber

The structure of natural rubber must be changed slightly to make it more elastic. If we use the wire model again, it is clear that the sliding and slipping can be prevented if pairs of wires are fixed together where they cross. But how can this be done with natural rubber?

The answer lies in the carbon-carbon double bonds. If sulphur is heated with natural rubber, the carbon-carbon double bonds break open and sulphur atoms form links between the long carbon chains (Figure 3). This process is called **vulcanisation**.

The links between the rubber molecules stop the chains sliding over one another too much, yet the rubber stays elastic.

1 Make a model of the structure of rubber using four or five tangled pieces of wire.

2 Does your model behave like natural rubber when it is stretched?

3 Fix pairs of wire together using paper clips or short pieces of wire. How does this affect the stretching and elastic properties of your wire model?

4 What do the following terms mean?
hydrocarbon, double bond, elastic, vulcanisation.

5 How do you think the elasticity of rubber will change as the amount of sulphur used for vulcanisation increases?

6 If possible, use ball and spoke molecular models to build a model of part of a molecule of natural rubber like that shown in Figure 1.

7 Use your molecular model from part 6 to explore the way in which sulphur atoms can form crosslinks between the rubber molecules.

A vulcanised rubber tyre coming off the production line

3 Using X-rays to study materials

Most of our ideas about the structure of materials has come from X-ray crystallography. One of the most important pieces of X-ray crystallography involved finding the structure of penicillin. During the Second World War increased quantities of penicillin were needed for the treatment of wounds incurred by our fighting troups.

Scientists were asked to find ways of producing larger quantities of penicillin synthetically. But, before they could do this, they needed to know its structure. Dorothy Hodgkin used X-ray crystallography to find the structure of penicillin. In 1964, she received the Nobel Prize for her work on the structure of penicillin and other biological substances.

1 Find out what you can about the use of X-rays to study the structure of materials. (You could start by reading unit 1 in section H of this book.)

2 Design and prepare one page of notes, diagrams and photos, etc. to help you understand how X-rays are used to study the structure of materials.

3 Use a word processor to prepare a neat version of your notes. Try to use all the facilities of your word processor including different sized letters, diagrams, shading, brushwork, etc.

4 Print a copy of your notes. Add in any photos and diagrams that are necessary and then take a photocopy of the final version.

Dorothy Hodgkin

Study Questions

1 In order to find the formula of a compound of tin and chlorine, a stream of dry chlorine was passed over 3.0 g of tin in the apparatus below. When the reaction was complete, the liquid tin chloride was distilled into the receiver.

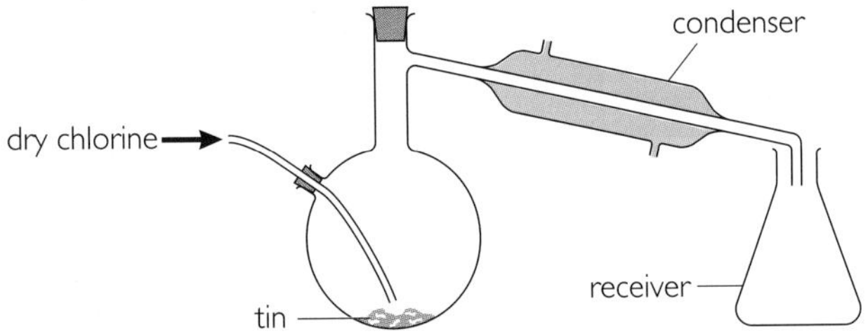

(a) Why is a condenser used?
(b) Make a neat sketch of the diagram.
(i) Show where the water should enter and leave the condenser.
(ii) The thermometer has been left out. Draw it in your sketch in the right place.
(c) Why should this experiment be done in a fume cupboard?
(d) Write down the weighings you would make to find the mass of tin chloride collected.
(e) In the experiment, 6.5 g of tin chloride should form.
(i) What mass of chlorine reacts with 3.0 g of tin?
(ii) How many moles of chlorine atoms (Cl = 35) combine with 3.0 g of tin?
(iii) How many moles of tin (Sn = 120) are there in 3.0 g?
(iv) How many moles of chlorine atoms combine with 1 mole of tin?
(v) What is the formula of the tin chloride?
(f) How would you check the purity of the tin chloride by a simple method?

2 Anaesthetics are compounds which cause temporary loss of consciousness and remove all feeling of pain. They have been known since the early 1800s when Humphry Davy suggested the possible use of dinitrogen oxide for operations. By the middle of the 19th century, both ether and chloroform were being used as anaesthetics. However, both these substances had drawbacks. Ether is highly flammable and chloroform is very toxic (poisonous). There was no real understanding of anaesthetics until the 20th century. In 1939, J. Ferguson suggested that all volatile compounds could act as anaesthetics, as long as they were not toxic.

This started the hunt for the perfect anaesthetic. It would form vapour easily, but not *too* easily. It would show no toxic effects. It would also be good if the compound was chemically unreactive. This would allow it to be stored for some time without going off. It was particularly important that it did not react with soda lime, because this is used in anaesthetic apparatus in hospitals. Soda lime absorbs carbon dioxide and allows the recycling of the expensive anaesthetic compounds. The table below shows a number of compounds with good anaesthetic properties. None of these substances reacts with soda lime.
After many trials, fluothane, $CF_3CHBrCl$, was found to be the best. It is a colourless liquid with a pleasant smell. It is now the most common anaesthetic in Britain.
(a) Write *one* sentence about the contribution of the two scientists mentioned in the passage.
(b) List *five* characteristics of a good anaesthetic.
(c) Write down *two* reasons why soda lime is used by anaesthetists.
(d) The active compounds in soda lime are sodium hydroxide and calcium hydroxide. Write a balanced equation for the reaction of carbon dioxide with *one* of these compounds.
(e) What do the non-flammable compounds in the table, apart from N_2O, have in common?
(f) The structural formula for $CF_2ClCHFBr$ is

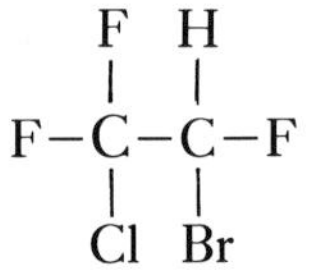

Make a similar drawing for fluothane.
(g) How do fluothane and $CF_2ClCHFBr$ differ?

Compound	Relative molecular mass	Boiling point /°C	Flammability	Toxicity
Dinitrogen oxide, N_2O	44	−88	not flammable	safe
Cyclopropane, C_3H_6	42	−33	very flammable	safe
Ether, $(C_2H_5)_2O$	74	35	very flammable	safe
Fluothane, $CF_3CHBrCl$	197.5	50	not flammable	safe
$CF_2ClCHFBr$	197.5	50	not flammable	toxic
Chloroform, $CHCl_3$	119.5	75	not flammable	toxic

SECTION I

Energy and fuels

Look carefully at this diagram. What examples of energy sources can you see in the diagram?

1 ENERGY IN EVERYDAY LIFE

Everyone needs energy for living. We use energy from electricity or from burning fuels to keep us warm at home and at school. The African bush people in this photo warm themselves with the energy from an open wood fire

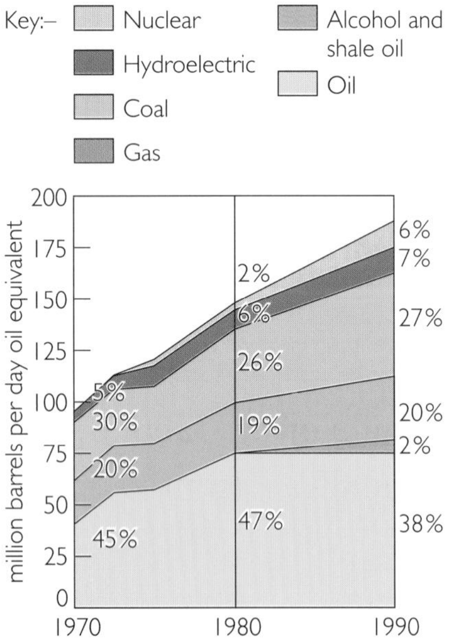

Figure 1
The percentage contributions that different fuels have made to the world's energy requirements since 1970

Everyday we use energy in our homes, schools and industries in thousands of different ways. We need energy to warm a room, cook a meal or light a torch.

Although we use a lot of energy in our homes, industry uses far more. Energy is needed to turn raw materials like clay into useful things like bricks. It is also needed to mine coal and other minerals and to generate electricity. In fact,

Energy is essential to our society.

Transferring energy to or from chemicals is the basis of chemical changes. Some chemical changes, such as the decomposition of limestone to lime, need heat to make them happen.

$$\underset{\text{limestone}}{CaCO_3} \xrightarrow{\text{heat}} \underset{\text{lime}}{CaO} + \underset{\text{carbon dioxide}}{CO_2}$$

Chemical reactions like this, which *take in* heat, are described as **endothermic**. Other chemical reactions *give out* heat. In these reactions, chemical energy is converted into heat. This happens when fuels such as coal and natural gas are burnt.

$$\underset{\text{natural gas}}{CH_4} + 2O_2 \rightarrow CO_2 + 2H_2O + \text{heat}$$

Chemical reactions like this, which *give out* heat are described as **exothermic**.

Everyday, the world needs energy equivalent to about 180 million barrels of oil. This energy is provided by a range of fuels. Figure 1 shows how the world's energy requirements have been provided by different fuels since 1970.

Most coal in Britain is used to generate electricity. That is why most power stations have huge coal heaps nearby. This photo shows the power station at Didcot in Oxfordshire, with its huge supply of coal in the foreground

Foods such as fats and carbohydrates are biological fuels. They are important stores of energy. When foods are broken down (**metabolised**) in our bodies, the chemical energy in the food is changed into

- *thermal energy* (heat) to keep us warm.
- *mechanical energy*, to help us move around and keep our heart and breathing muscles working,
- *electrical energy* in our nerves, to help us to think and respond to stimuli like heat or pain.

QUESTIONS

1 (a) Give one example in each case of a food which contains a high proportion of (i) fat, (ii) carbohydrate.
(b) What elements do fats and carbohydrates contain?
(c) Name two carbohydrates.
(d) What substances are produced when carbohydrates are broken down (metabolised)? (Hint: the carbohydrates act like fuels.)

2 Look closely at Figure 1.
(a) Draw a pie chart showing the percentages that different fuels made to the world's energy requirements in 1990.
(b) Which fuel showed
(i) the largest increase in percentage between 1970 and 1990,
(ii) the largest decrease in percentage between 1970 and 1990,
(iii) the same percentage in both 1970 and 1990?
(c) Why do you think the world energy requirements have increased steadily since 1970?

3 (a) Describe what happens when
(i) water is heated in a saucepan,
(ii) ice cubes are made in a freezer,
(iii) coal burns,
(iv) margarine is heated in a saucepan,
(v) a steel poker is heated in a fire.
(b) Make a summary of the types of change which can occur when substances are heated. For example, one type of change is that solids melt.

2 FOSSIL FUELS

Fossils in coal show the plants from which the coal was formed

The most commonly used fuels are coal, oil and natural gas. These fuels are called **fossil fuels** because they have formed from the remains of dead animals and plants.

How did fossil fuels form?

When plants and animals die and decay they rot away completely. During this process, compounds containing carbon, hydrogen and oxygen in the decaying material react with oxygen in the air. The products are carbon dioxide and water.

Sometimes, the plants and animals cannot react with oxygen when they die. If this happens, the carbon compounds in the decaying material turn into energy rich substances like coal and oil. Three hundred million years ago the Earth was covered in forests and the sea was full of tiny organisms. When some of these living things died, they were covered with mud and protected from oxidation. Over millions of years these deposits were changed by bacteria and compressed by the layers of earth, and the sea above them. Coal has formed mainly from plants that grew on land. Oil and gas have formed mainly from plants and animals that lived in the sea.

Obtaining fossil fuels

A mineshaft is sunk through layers of earth and rock to the coal seam so that the coal can be dug out

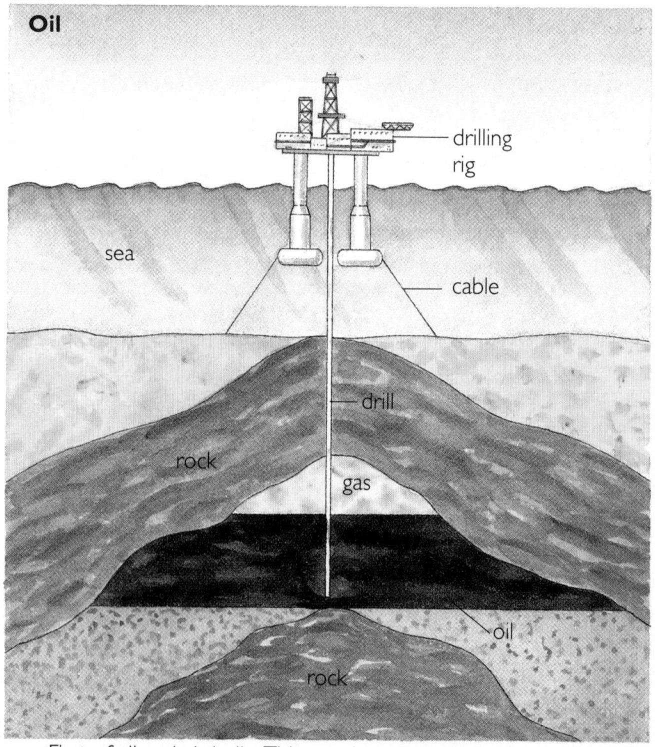

First of all, a rig is built. This may be on land, on the sea bed or anchored to the sea bed. Long drills are then used to bore through rock layers to the gas and oil. The pressure of the gas forces oil through pipes to the surface

Conserving fossil fuels

Fossil fuels are concentrated stores of chemical energy. When they burn, some of their chemical energy is released as heat. For 150 years, industrial countries have relied on fossil fuels for energy. Vast amounts of coal, oil and natural gas have been used for heating, for transport and to generate electricity. This is still true today, but fossil fuels cannot last forever. The Earth's resources of coal, oil and gas are limited. Eventually, they will run out. Figure 1 shows how long these fuels will last if we continue to use them as we do now.

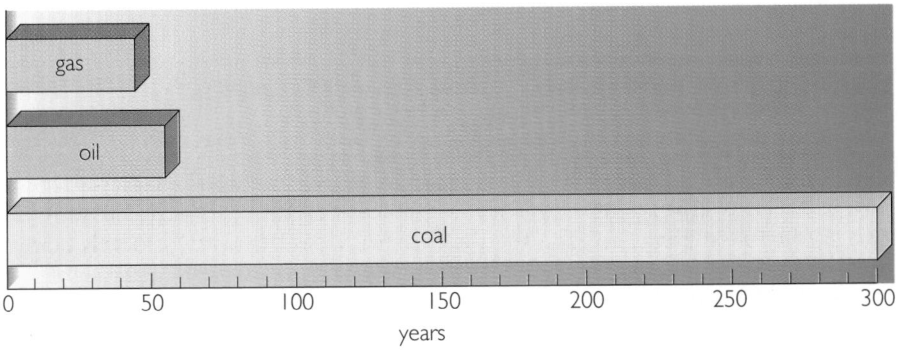

Figure 1
How long will the world's reserves of fossil fuels last?

Although coal is plentiful, oil and natural gas will probably start to run out in your lifetime. Because of this, it is important to conserve fossil fuels and avoid wasting energy. Our attempts to conserve fuels have led to

- more economic and more efficient use of fuels,
- manufacture of special fuels for some uses,
- research into alternative energy sources to fossil fuels,
- use of better materials and better methods for insulation.

QUESTIONS

1 (a) Draw a flow diagram to show how coal formed.
(b) Why is coal described as a fossil fuel?
(c) Why is good ventilation important when coal burns?
2 Make a table showing how people have tried to conserve (save) fossil fuels. The table has been started for you.

Action taken	How it saves fossil fuels
Driving smaller cars	Uses less petrol
...	...

3 What problems does coal mining pose for (i) the miners, (ii) the environment?

3 ALTERNATIVE ENERGY SOURCES

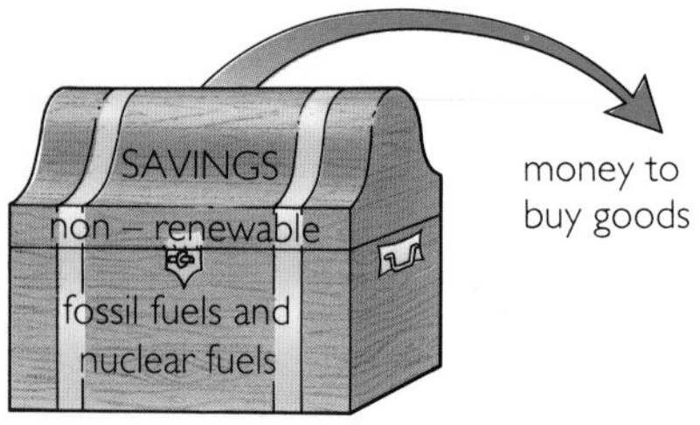

Figure 1
Non-renewable energy sources are like life savings – once they are spent, they can never be replaced

At one time, almost all the energy we used in our homes and industries came from fossil fuels. We still use vast amounts of coal, oil and natural gas, but these fossil fuels will not last forever. Because of this, scientists and engineers are looking for and developing **alternative** (other) **energy sources**.

- **Nuclear power.** Nuclear power is being used more and more to generate electricity. Uranium is used as the fuel. Nuclear energy has many advantages, but some people think that it is unsafe. Nuclear energy is discussed further in section K.
- **Tidal power.** A tidal power station traps the high tides behind a barrage across a river estuary. When the tide falls, the water is made to flow through turbines which generate electricity.

Only a few tidal power stations have been built. This photo shows a tidal barrage in Nova Scotia, Canada

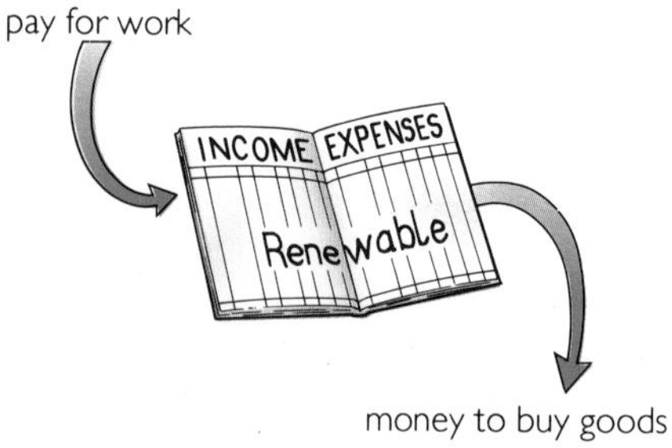

Figure 2
Renewable energy sources are like a regular income from a job – you can keep spending and the money is continually replaced by your wages

- **Hydro-electric power.** Falling water can be used to drive turbines which generate electricity. Once the power station has been built, the cost of hydro-electricity is fairly cheap. But because the process relies on water falling from a height, the use of hydro-electric generators is limited.
- **Solar power.** The Sun provides us with energy in two forms – heat and light. The Sun's thermal energy (heat) can be used to heat water in solar panels, or it can be used in solar furnaces. Light from the Sun can be used to make electricity using photo-cells.
- **Wind power.** Windmills, like water mills, have been used as a source of power for hundreds of years. Some scientists believe that giant windmills (or wind turbines as they are called) could be used to generate electricity on a large scale in the future.

Energy resources – savings and income

The Earth's energy resources can be compared to money. Fossil fuels and nuclear fuels are like *savings in the bank* (Figure 1). They represent energy stored (saved) over millions of years. They are sometimes called **non-renewable energy sources**. Once used, they are gone forever. We cannot get them back and use them again.

Fortunately, there are some **renewable energy sources**. Every day the Sun shines (somewhere!), the wind blows, rain falls and tides come in. So solar power, wind power, hydro-electric power and tidal power are renewable energy sources. We can use them all the time, but they keep on being replaced. They are like an *income from a job* (Figure 2)). In the future, we may have to use these renewable energy sources more and more as our fossil fuel savings run out.

Solar water-heating panels on the roof of a house

The CEGB's first wind turbine at the Carmarthen Bay Power Station in South Wales. The 24 m high machine can supply up to 200 kW of electricity – enough to operate 200 one-bar electric fires. (An average-sized power station can supply about 200 MW of electricity)

QUESTIONS

1 Draw a large sketch map of the British Isles. (You might like to copy or trace from an atlas.) Mark on your map (with a letter T) those places where tidal power might be used. Use the letters H, S and W to indicate areas where hydro-electric, solar and wind power might be used.

2 Get into a group with 2 or 3 others. Try to imagine how people in the UK might obtain the energy they need between now and the year 3000. Make a timetable with dates and important energy sources.

3 There are economic and environmental reasons for choosing different energy sources. What are the advantages and disadvantages of the following energy sources?
coal, tidal power, wind power

4 (a) What is meant by (i) renewable energy sources, (ii) non-renewable energy sources?
(b) Is wood a renewable or a non-renewable energy source? Explain your answer.
(c) Is wave power a renewable or non-renewable energy source? Explain your answer.

5 Why is electricity much cheaper in Norway than in Britain?

4 ENERGY TRANSFERS

The energy that Sally Gunnell needs to hurdle is provided by food

Most of the energy that we need for heating, for transport and for generating electricity comes from the chemical energy stored in materials.

Figure 1 summarises the most usual ways in which chemical energy is converted into other forms of energy.

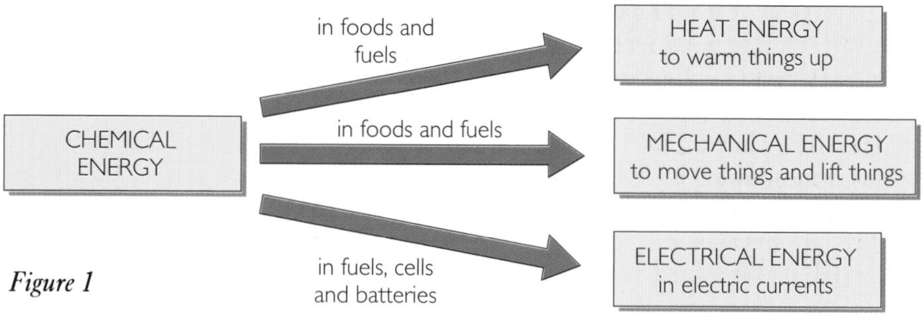

Figure 1

When you turn on a gas fire, digest a meal or use a torch, chemical energy changes into other forms of energy. But when you do these things, *no energy is ever lost*. It may, however, be changed into a less useful form.

> Energy cannot be destroyed. It can only be changed from one form to another. This is called the **Law of Conservation of Energy**.

Energy is most useful when it is concentrated. This is why fuels and foods are so useful. They store energy in a concentrated form until it is needed. Once the fuel is burnt or the food is eaten, the energy is released.

When energy is transferred from one form to another, some of it always ends up as heat. Unfortunately, heat spreads out very easily and cannot be stored. For example, we burn fuels and use electricity to keep our homes warm, but the heat spreads out too easily. It leaks through the roof, the windows, the walls and the floor. This heat energy which leaks out of our homes is useless. It is too spread out (**dissipated**) to be any use. So the overall effect of most energy transfers is to dissipate the useful energy stored in foods and fuels. The energy is dissipated as heat which raises the temperature of the surroundings.

Energy from the Sun

Light energy from the Sun is trapped by plants. The light energy is absorbed by a green substance called **chlorophyll** in the leaves of plants. The light energy is then used to join together molecules of carbon dioxide and water to form carbohydrates (e.g. $C_6H_{12}O_6$.) Oxygen is also produced.

$$\text{carbon dioxide} + \text{water} + \underset{\text{in sunlight}}{\text{energy}} \longrightarrow \underset{\text{(food)}}{\text{carbohydrate}} + \text{oxygen}$$

$$6CO_2 + 6H_2O \xrightarrow{\text{energy}} C_6H_{12}O_6 + 6O_2$$

This endothermic process is called **photosynthesis** (Figure 2). The carbohydrates produced by photosynthesis include substances like starch and glucose. They are important sources of energy in foods such as rice, bread, pasta, potatoes and soya beans.

When plants and animals die, the chemical energy which they contain becomes part of the decaying maaterial. Millions of years ago some of this decaying material became fossil fuels. Heat from the Sun also makes water evaporate and results in winds, rain and waves. This leads to renewable energy supplies from wind power, hydroelectric power and wave power (which uses the energy of buffeting waves to generate electricity). So, without the Sun we would not have

- heat to keep the Earth warm,
- food from plants,
- food from animals which live on plants,
- fossil fuels
- various renewable fuels.

This means that the Sun is our major source of energy. Without the Sun, life on Earth would be impossible.

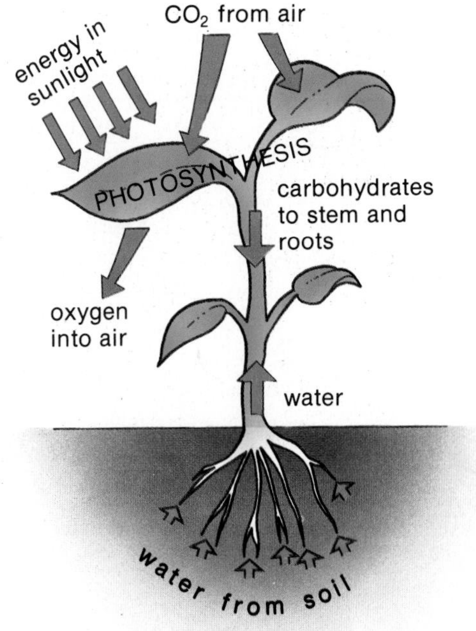

Figure 2

Ancient Britons appreciated that life on Earth relies completely on the Sun. In fact they saw the Sun as a life-giving god. They worshipped and made sacrifices, in order to please the god and make sure the Sun continued to rise each day

QUESTIONS

1 What energy change occurs when (i) a match is struck, (ii) a torch is used, (iii) a gas cooker is used?

2 Generating electricity
Look at Figure 3.
(a) Draw a flow chart to show the energy transfers when coal or uranium is used to produce electricity for your home.
(b) What are the good points about electricity as a power supply?
(c) What are the bad points about electricity as a power supply?
(d) Only a small proportion of the energy stored in the fuel gets converted into electricity.
(i) Where does the rest of the energy go?
(ii) Why is so much energy lost?
(iii) Why is the lost energy less useful?

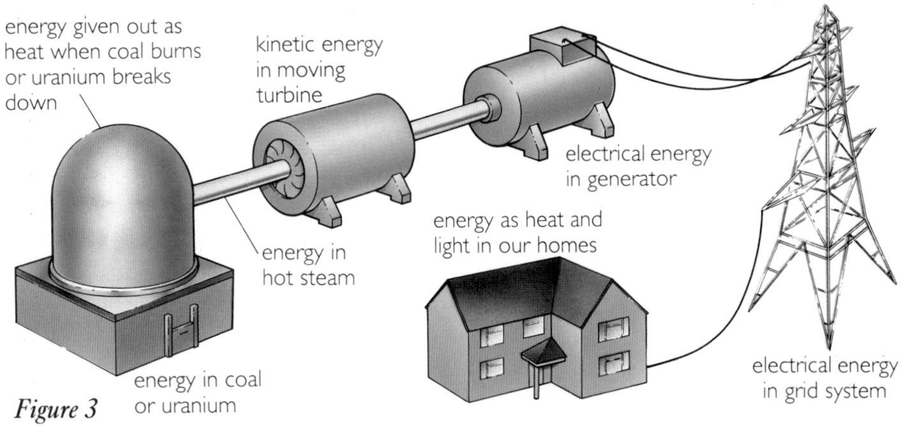

Figure 3

3 (a) How do our bodies use the energy given out during respiration?
(b) Why are animals unable to photosynthesise?
(c) In what ways is photosynthesis the opposite of respiration?

4 Get into a group with 2 or 3 others and discuss the following statement.
Without the Sun, life on Earth would be impossible. So we should worship the Sun.
Write a short report of your discussion for the rest of the class.

5 CRUDE OIL

An aerial view of an oil refinery near Stavanger, Norway. Notice the fractionating column, the storage tanks and the jetty at which tanks can berth

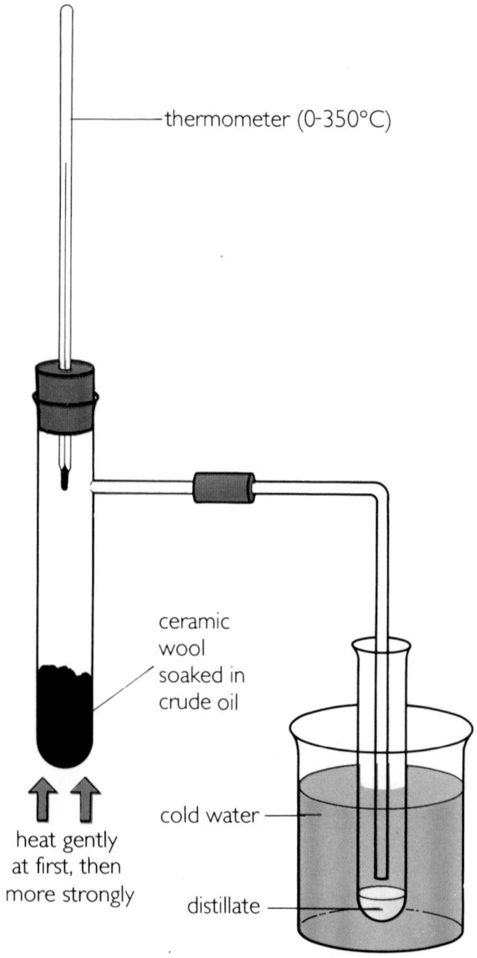

Figure 1
The small-scale fractional distillation of crude oil

Crude oil (petroleum) is the main source of fuel and organic chemicals in the UK. The crude oil comes to our refineries from the North Sea and the Gulf area in the Middle East. It is a sticky, smelly, dark-brown liquid. It contains hundreds of compounds, from simple substances like methane (CH_4) to complicated substances with long chains and rings of carbon atoms. Nearly all the substances in crude oil contain carbon. These carbon compounds are often called **organic compounds**.

A century ago oil was almost unknown. Now we could hardly survive without it. It is almost as important to our lives as air and water. In the UK, 70% of all organic chemicals come from oil. Antifreeze, brake fluid, lipstick, nylon, explosives and paint are all made from it. You may be dressed entirely in oil-based textiles, like Terylene and nylon. Without oil, most transport would come to a standstill and any machine larger than a toy car would seize up from lack of lubricant. We have enough oil to last another 60 years and new reserves are being discovered all the time. Even so, we are using up oil reserves so fast that we need to use crude oil more economically.

Boiling range	Name of fraction	Colour	Viscosity	How does it burn?
20–70°C	petrol	pale yellow	runny	easily, with clear yellow flame
70–120°C	naphtha	yellow	fairly runny	quite easily, yellow flame, some smoke
120–170°C	light paraffin	dark yellow	fairly viscous	harder to burn, quite smoky flame
170–240°C	heavy paraffin	brown	viscous	hard to burn, smoky flame

Table 1 Properties of fractions obtained by the small scale fractional distillation of crude oil

At oil refineries, crude oil is separated into fractions by fractional distillation (section A, unit 9). Then it is treated and purified to produce different fuels and chemicals. Figure 1 shows the small scale fractional distillation of crude oil. Four fractions are collected, with the boiling ranges and properties shown in Table 1. Table 1 also shows the industrial fractions to which our fractions correspond.

Bitumen is mixed with stone chippings and used to surface roads

Constituents and uses of the different fractions

The fractions from crude oil contain mixtures of similar substances. In one fraction, the similar substances have roughly the same number of carbon atoms. Table 2 shows what each fraction contains and what it is used for. The uses of the fractions depend on their properties. Petrol vaporises easily and is very flammable, so it is ideal to use in car engines. Lubricating oil, which is very viscous, is used in lubricants and in central heating. Bitumen, which is solid but easy to melt, is used for waterproofing and asphalting on roads.

Fraction	Boiling range	Number of carbon atoms	Uses
Fuel gas	−160–20°C	1–4 mainly methane (CH_4), ethane (C_2H_6), propane (C_3H_8) and butane (C_4H_{10})	fuels for gas ovens, LPG, GAZ, chemicals
Petrol (gasoline)	20–70°C	5–10 e.g. octane (C_8H_{18})	fuel for vehicles, chemicals
Naphtha	70–120°C	8–12	chemicals
Paraffin (Kerosine)	120–240°C	10–16	fuel for central heating and jet engines, chemicals
Diesel oils and lubricating oils	240–350°C	15–70	fuel for diesel engines, trains and central heating, chemicals, lubricants
Bitumen	above 350°C	more than 70	roofing, waterproofing, asphalt on roads

Table 2 The constituents and uses of fractions from crude oil

QUESTIONS

1 Look at the results in Table 1. How do the properties of the fractions change?

2 (a) Why is crude oil important?
(b) Why should we try to conserve our reserves of crude oil?
(c) How is crude oil separated into various fractions at a refinery?

3 (a) List the main fractions obtained from crude oil.
(b) Give the main uses of each fraction.

4 Discuss the following question with 2 or 3 others.
What should be done to ensure that crude oil is used more economically?

6 NATURAL GAS, CALOR GAS AND GAZ

The main constituents of natural gas are methane (CH_4) and ethane (C_2H_6). These two gases, together with propane (C_3H_8) and butane (C_4H_{10}) are the constituents of fuel gas, the first fraction from crude oil. Propane and butane are the main constituents of 'liquified petroleum gas' (LPG). The best known uses of LPG are 'Calor Gas' and GAZ for camping, caravans and boats. Butane is also used as lighter fuel and in portable hair tongs.

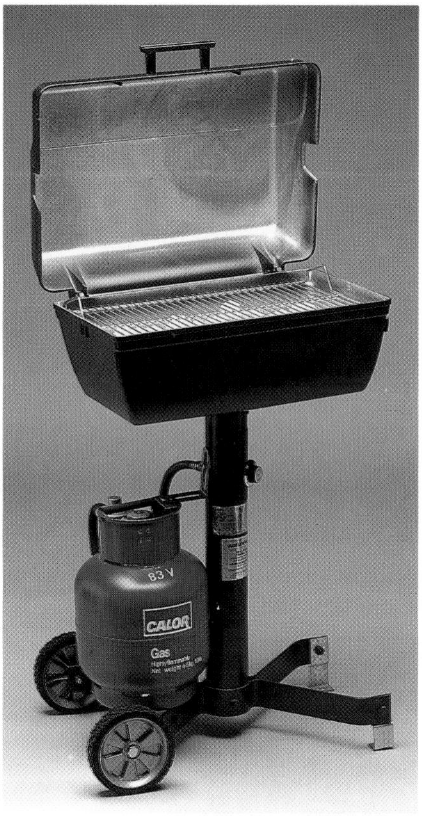

A small blue butane cylinder used to fuel a gas barbecue

Two red propane cylinders stored outside and used to fuel a gas cooker in the kitchen inside

Name	**methane**	**ethane**	**propane**	**butane**
Formula	CH_4	C_2H_6	C_3H_8	C_4H_{10}
Structural formula	H | H–C–H | H	H | H–C–H | H–C–H | H	H | H–C–H | H–C–H | H–C–H | H	H | H–C–H | H–C–H | H–C–H | H–C–H | H
Molecular model	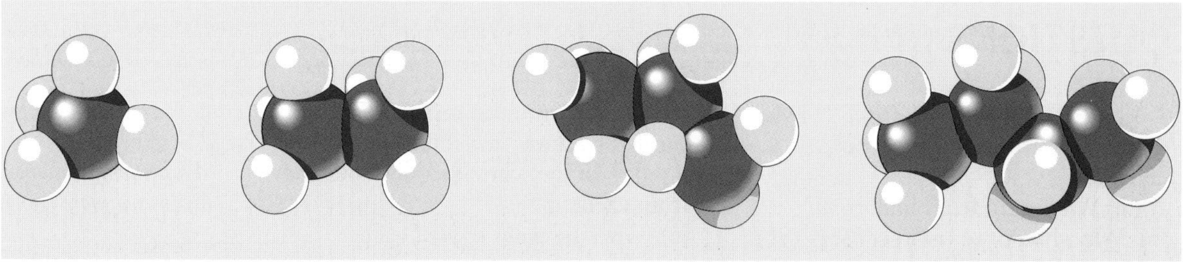			

Table 1 The formulas and molecular models for four alkanes

Methane, ethane, propane and butane are the four simplest hydrocarbons. Table 1 shows the formulas and molecular models for these four hydrocarbons. The structural formulas show which atoms are attached to each other. But they cannot show the three-dimensional structure of the molecules. The 3-D structure is shown in the molecular models. There are four covalent bonds to each carbon atom. These bonds around each carbon atom spread out tetrahedrally, as in diamond (Figure 1).

Methane, ethane, propane and butane are members of a series of compounds called **alkanes**. All other alkanes are named from the number of carbon atoms in one molecule. So C_5H_{12} is *pent*ane, C_6H_{14} is *hex*ane, C_7H_{16} is *hept*ane, and so on. The names of all alkanes end in *-ane*.

Look at the formulas in Table 1. Notice that the difference in carbon and hydrogen atoms between methane and ethane is CH_2. The difference between ethane and propane is CH_2 and the difference between propane and butane is also CH_2. This is an example of a **homologous series** – a series of compounds with similar properties in which the formulas differ by CH_2.

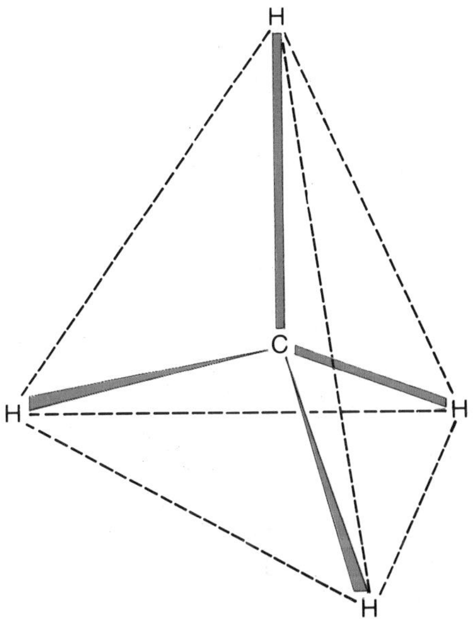

Figure 1
The tetrahedral arrangement of bonds in methane

Properties of alkanes

Alkanes with one to four carbon atoms in each molecule are gases at room temperature. Alkanes with 5 to 17 carbon atoms are liquids at room temperature. Mixtures of these liquids are used in petrol, in paraffin and in lubricating oils. Alkanes with 18 or more carbon atoms per molecule are solids at room temperature. Notice that alkanes become less volatile and change from gases to liquids and then to solids as their molecules get larger.

Alkanes do not contain ions and their C—C and C—H bonds are very strong. So they have very few reactions. They do not react with metals, acids or alkalis. It may surprise you, but petrol will not react with sodium or with concentrated sulphuric acid. Ask your teacher to show you.

The most important reaction of alkanes is **combustion**. They burn in oxygen producing carbon dioxide and water.

$$C_4H_8 + 6O_2 \rightarrow 4CO_2 + 4H_2O$$

butane in GAZ

The combustion reactions of alkanes are very exothermic. So alkanes in natural gas and crude oil are used as fuels. When there is too little oxygen for combustion, carbon (soot) and carbon monoxide form as well as carbon dioxide. Carbon monoxide is very poisonous, so it is dangerous to burn carbon compounds in a poor supply of air.

A small RAF training plane being refuelled with kerosene

QUESTIONS

1 (a) Use molecular models to build a molecule of C_6H_{14}.
(b) What is C_6H_{14} called?
(c) Draw a structural formula for C_6H_{14}.
(d) Draw a formula for C_6H_{14} which shows its 3-D structure.
(e) How many H atoms are there in the alkane with 10 carbon atoms?
(d) If an alkane has n carbon atoms, how many H atoms will it have?

2 A homologous series of compounds can be compared to a group of elements in the periodic table. Why is this?

3 (a) Write an equation for the complete combustion of octane in petrol.
(b) What are the products when petrol burns in a poor supply of oxygen?
(c) Why is it dangerous to allow a car engine to run in a garage with the doors closed?

4 Why do alkanes change from gases to liquids and then to solids as their molecules get larger?

7 PETROL FROM CRUDE OIL

Fraction	Approximate % in	
	crude oil	everyday demand
Fuel gas	2	4
Petrol	6	22
Naphtha	10	5
Kerosine	13	8
Diesel oil	19	23
Fuel oil and bitumen	50	38

Table 1 Relative amounts of different fractions in crude oil and the demand for each fraction

After 1930, the demand for petrol increased much faster than the demand for heavier fractions which make up three quarters of crude oil (Table 1). This meant that refineries were left with large surpluses of the heavier fractions. Fortunately, chemists found ways of converting the heavier fractions into petrol and other useful products.

One method of breaking up the large alkane molecules in these heavier fractions is called **catalytic cracking**.

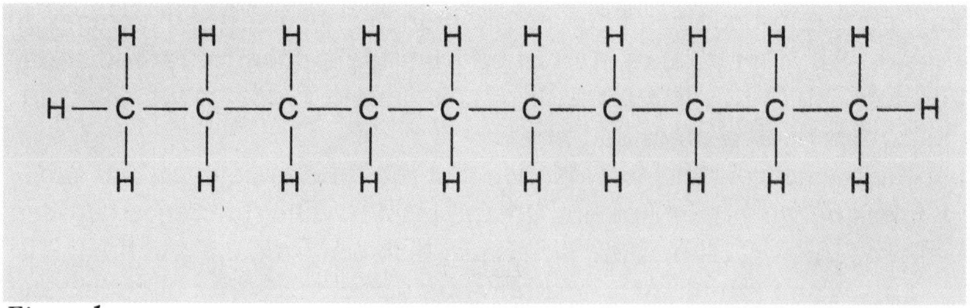

Figure 1
The structural formula of decane

The three-dimensional structure of decane

What are the products of cracking?

Look at the long molecule of decane ($C_{10}H_{22}$) in Figure 1. It has a chain of 10 carbon atoms with 22 hydrogen atoms. Imagine that decane is cracked (broken) between two carbon atoms. This will not produce two smaller alkane molecules because there are not enough hydrogen atoms to go round. But suppose one product is the alkane, octane (C_8H_{18}). If C_8H_{18} is split off from $C_{10}H_{22}$, the molecular formula of the remaining part is C_2H_4 (Figure 2). The chemical name for C_2H_4 is **ethene**. Notice in Figure 2 that ethene has a *double* bond between the two carbon atoms. This double bond allows all the carbon atoms in the product to have four bonds.

Hydrocarbons such as ethene, which contain a double bond (C=C), are known as **alkenes**. Their names come from the alkane with the same number of carbon atoms, using the ending *-ene* rather than *-ane*.

How are molecules cracked?

Unlike distillation, cracking is a chemical process. It involves breaking a strong covalent bond between two carbon atoms. This requires high temperatures and a catalyst. At high temperatures the larger alkane molecules have more energy and they break apart into two or more smaller molecules. The catalyst is finely powdered aluminium oxide and silicon(IV) oxide. These substances do not react with the crude oil fractions but they do provide a hot surface that speeds up the cracking process.

Why is diesel cheaper than petrol?

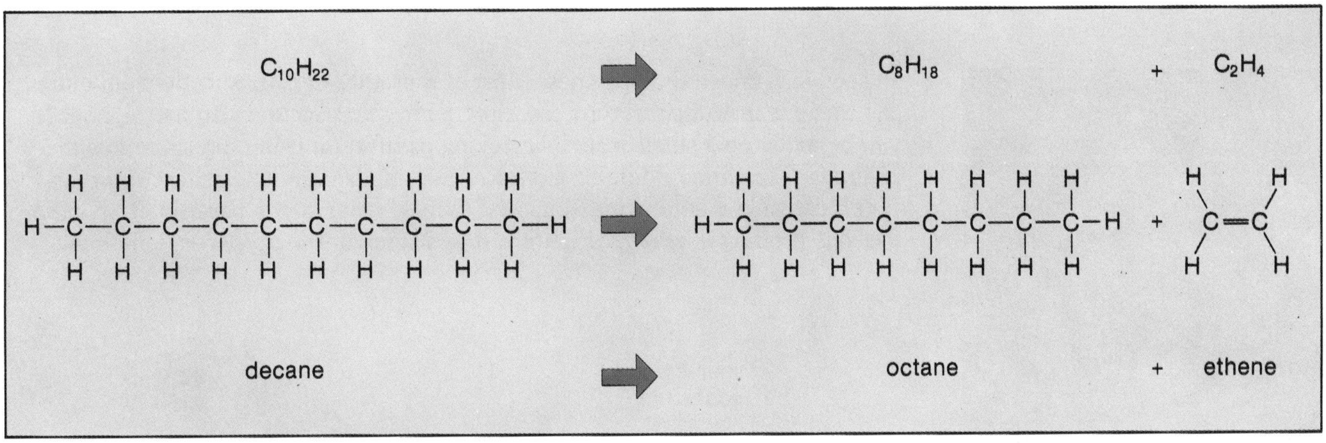

Cracking is important because it helps to produce more petrol. The petrol obtained in this way has a better quality than that obtained by the distillation of crude oil. Cracked petrol is therefore blended with other petrols to improve their quality.

Figure 2
When decane undergoes catalytic cracking, it forms an alkane, like octane, and an alkene, like ethene

The catalytic cracking plant at the Wilmington refinery, Connecticut

QUESTIONS

1 Use molecular models to find out how the bonds are broken and made when decane is cracked as in Figure 2 above.

2 (a) Why is cracking important?
(b) What conditions are used for cracking?
(c) Write an equation for the cracking of octane in which one of the products is pentane.
(d) Draw the structural formulas of the products in the equation in (c).
(e) Use molecular models to picture the cracking process in (c).

3 (a) What is the name of the alkene of formula C_3H_6?
(b) Draw the structure of this alkene.
(c) Why is there no alkene called methene?

4 (a) Design a simple experiment in which you could attempt to crack some kerosine (paraffin) in order to obtain ethene. (Ethene is a gas at room temperature.)
(b) What safety precautions will you take in your experiment?
(c) How does your simple experiment compare with the industrial cracking of crude oil fractions.

5 Discuss the following question in a small group with 2 or 3 others.
What changes do you think there will be in our lives when crude oil begins to run out?

8 PLASTICS AND CHEMICALS FROM CRUDE OIL

Ethene is used in nature as a signal to trigger the ripening of fruit

In section E we studied plastics. Most of these plastics were made from ethene.

Ethene is manufactured by cracking the heavier fractions from crude oil. It can be made on a small scale by cracking paraffin oil using the apparatus in Figure 1. Heat the middle of the tube below the aluminium oxide or porous pot. Heat will be conducted along the tube to vaporise the paraffin. The main gaseous product is ethene. Figure 2 shows some of the properties of ethene.

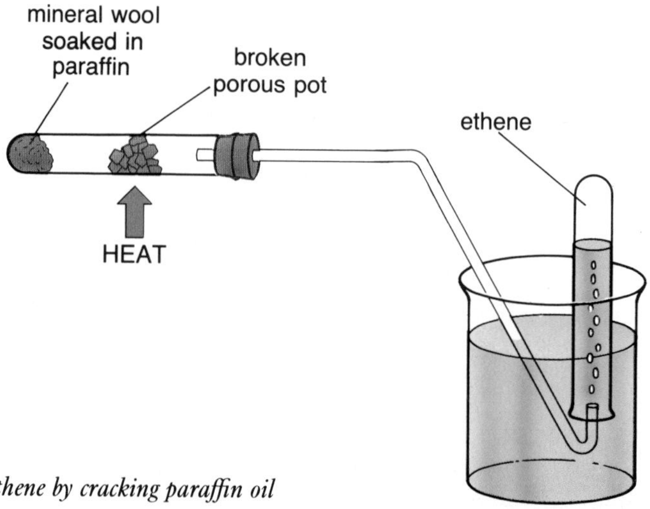

Figure 1
Preparing ethene by cracking paraffin oil

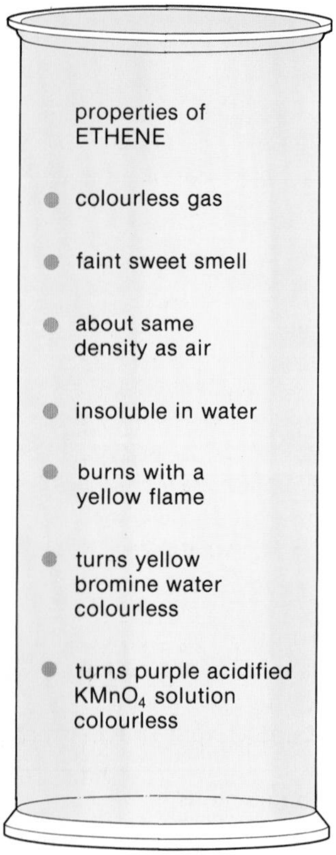

Figure 2

Chemicals from ethene

Alkenes, such as ethene, are much more reactive than alkanes. The most stable arrangement for the four bonds to a carbon atom is a tetrahedral one. This means that a C=C bond is unstable. Other atoms can add across the double bond to make two single bonds. So, alkenes readily undergo **addition reactions**.

This explains why ethene and other alkenes decolorise bromine water and acidified potassium permanganate solution, but alkanes do not. When ethene is shaken with bromine water the yellow colour disappears. The bromine molecules add across the double bond in ethene forming 1,2-dibromoethane. This reaction is used as a test for alkenes.

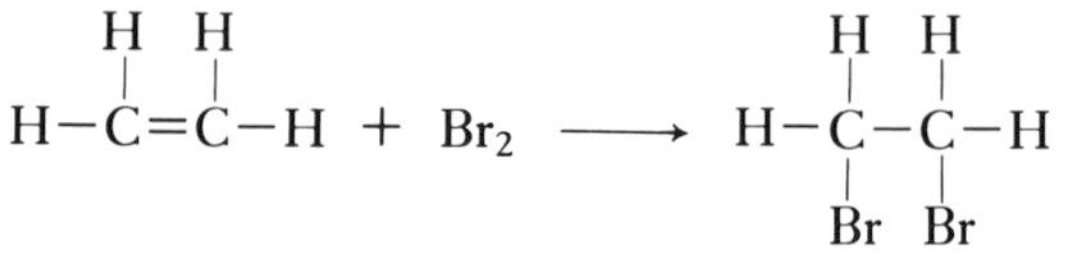

1,2-dibromoethane is an important additive in petrol. Tetraethyllead(IV), $Pb(C_2H_5)_4$, is also added to petrol to help it burn smoothly. Unfortunately the tetraethyllead(IV) causes lead to be deposited in the cylinders and round the sparking plugs. 1,2-dibromoethane reacts with the deposit of lead to form volatile lead bromide. This passes out with the exhaust gases.

Ethene also has an addition reaction with hydrogen at 150°C using a nickel catalyst. The product of this reaction is ethane.

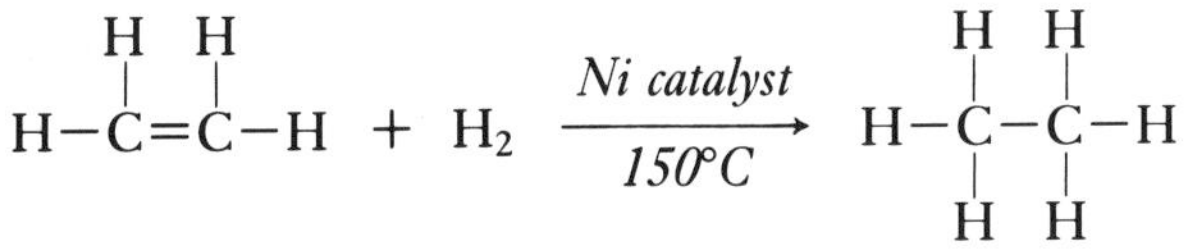

A similar reaction to this occurs when lard and margarine are made from the vegetable oils in palm seeds and sunflower seeds. The vegetable oils are liquids containing alkenes. On reaction with hydrogen, these alkenes are converted to alkanes. This change in structure can turn an oily liquid into a harder, fatty solid that can be used to make margarine.

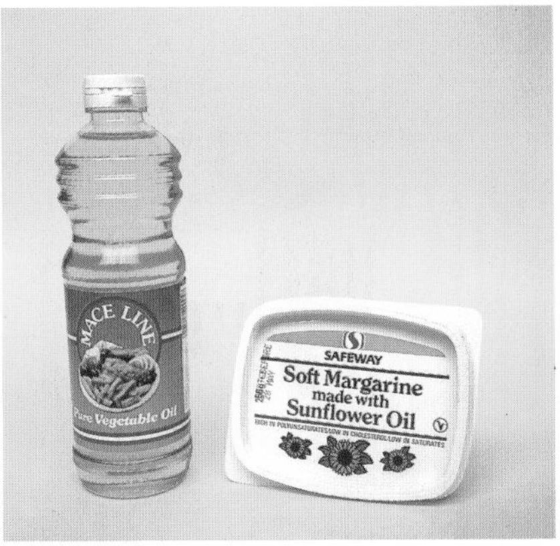

Compounds like ethene, which contain double bonds, are described as **unsaturated**. *Compounds like ethane, which have no double bonds, are* **saturated** *(ie saturated with hydrogen). Which of the products in this photo is the most unsaturated?*

By controlling the amount of hydrogen added to vegetable oils, margarine can be made as hard or as soft as required. Soft margarine is being made in this factory

QUESTIONS

1 Ethanol (alcohol) is manufactured by the addition of water to ethene. The reaction is carried out at 300°C and at very high pressure using a catalyst of phosphoric acid. Ethanol is the main constituent of methylated spirits.
(a) Write a word equation and a balanced formula equation for the manufacture of ethanol.
(b) What conditions are used to enable the ethene and water to react?
(c) Why do you think the temperature used is *not* higher than 300°C?
(d) What is methylated spirits used for?
(e) Pure ethanol is needed for certain uses. Name one of these uses.

2 Make a summary for the manufacture of important products from ethene, following the guidelines below.
(a) Write 'ethene' in the centre of your sheet. Draw arrows from ethene to the various products.
(b) Show the structures as well as the names of these products.
(c) Write the conditions needed to make the products along the arrows.

3 (a) Make a table to show the similarities and differences between ethane and ethene.
(b) Why is ethene much more reactive than ethane?

4 Three cylinders of gas are known to contain ethane, ethene and carbon dioxide but their labels are not clear. What simple tests would you make to show which gas is which?

5 Why do some people object to the addition of tetraethyllead(IV) to petrol?

6 (a) How is margarine made from vegetable oils?
(b) Explain what happens during the reaction in (a).
(c) How can the melting point of the margarine be controlled?

7 (a) Write the structural formula for propene, C_3H_6.
(b) Write equations for the reactions of propene with (i) hydrogen, (ii) bromine.
(c) Draw the structural formulas of the products in part (b).

9 ENERGY FROM FUELS

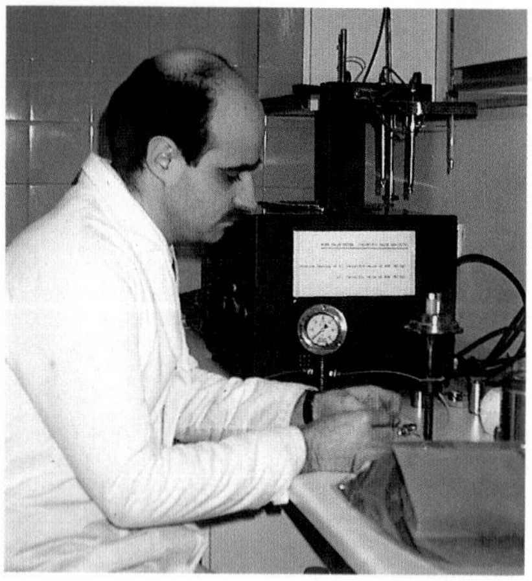

A food scientist using a bomb calorimeter

The amount of energy given out in a chemical reaction can be measured. Normally, we measure the heat given out or taken in when one mole of a substance reacts. This is called the **heat of reaction** and is given the symbol ΔH. For an exothermic reaction, the sign of ΔH is *negative*. This is because the chemicals have *lost* energy to the surroundings. For an endothermic reaction, ΔH is *positive* because the chemicals *gain* energy from the surroundings.

For example, when 1 mole (12 g) of carbon burns, 394 kJ of heat is given out. So we can write

$$C(s) + O_2(g) \rightarrow CO_2(g) \qquad \Delta H = -394 \text{ kJ}$$

It is very useful to know how much energy is produced when a fuel burns. This can help us to decide the best fuel to use for a particular job. Figure 1 shows the apparatus we can use to measure the heat given out when a liquid fuel like meths burns. The heat produced warms the water in the metal can. If we measure the temperature rise of the water, we can work out the heat produced. We can find the mass of meths which is burnt from the loss in weight of the liquid burner. Then we can calculate the heat produced when one gram of the fuel burns.

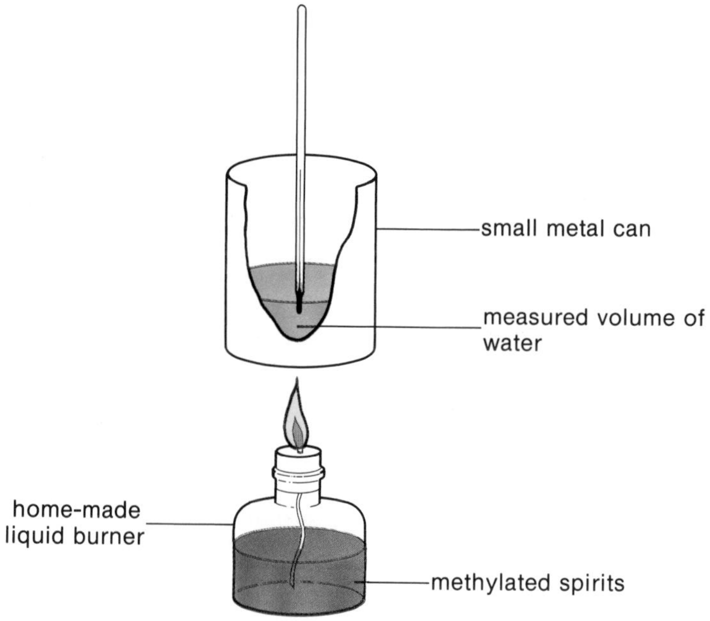

Figure 1

Mass of burner at start of experiment	= 271.8 g
Mass of burner at end of experiment	= 271.3 g
∴ Mass of meths burnt	= 0.5 g
Volume of water in metal can	= 250 cm^3
∴ Mass of water in can	= 250 g
Rise in temp. of water	= 10°C

Table 1 The results of an experiment to measure the heat produced when meths is burnt

The results from one experiment are shown in Table 1. In the experiment, 250 g of water are warmed up by 10°C.

We know that 4.2 joules of heat warm up 1 g of water by 1°C.

So, 250 × 4.2 J warm up 250 g of water by 1°C,
∴ 250 × 4.2 × 10 J warm up 250 g of water by 10°C.

This amount of heat is produced by 0.5 g of meths

0.5 g meths produce 250 × 4.2 × 10 J = 10 500 J
= 10.5 kJ

∴ 1 g of meths produces 21 kJ

Meths is mainly ethanol (alcohol), C_2H_6O. If we assume that meths is pure ethanol (M_r = 46), then

the heat produced when 1 mole of ethanol burns = 46 × 21 kJ
= 966 kJ

So we can write

$$C_2H_6O(l) + 3O_2(g) \rightarrow 2CO_2(g) + 3H_2O(l) \qquad \Delta H = -966 \text{ kJ}$$

Where does the energy come from?

In a reaction, bonds in the reactant chemicals are first broken. This requires energy. New bonds are then formed. This releases energy. If the energy released is greater than the energy required for the initial bond breaking, the reaction is exothermic.

Figure 2 shows the bond breaking and bond forming when methane in natural gas reacts with oxygen.

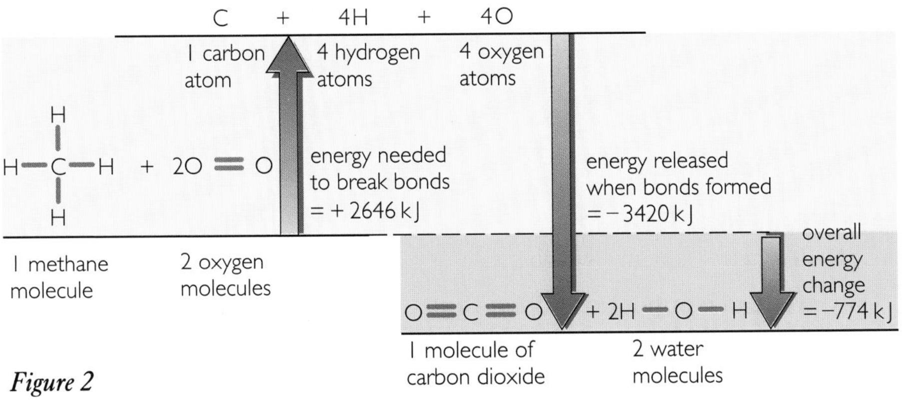

Figure 2

The energy changes in chemical reactions, like those in Figure 2, can be summarised in energy level diagrams. Figure 3 shows the energy level diagram for the reaction of methane with oxygen. This reaction is exothermic. Energy is lost during the reaction, so the products are at a lower energy level than the reactants.

Notice in Figure 2 that

- breaking chemical bonds is an *endothermic* process
- making chemical bonds is an *exothermic* process

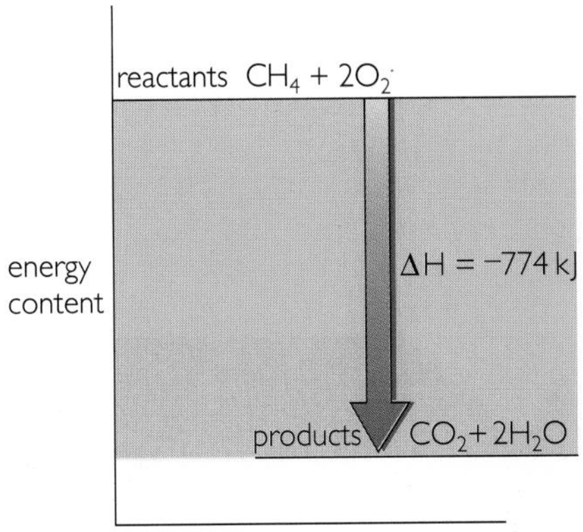

Figure 3

QUESTIONS

1 The simple apparatus in Figure 1 does not give very accurate results. An accurate value for the heat produced when meths burns is about 30 kJ per gram.
(a) Why do you think the apparatus in Figure 1 gives inaccurate results?
(b) How would you modify the apparatus in Figure 1 to obtain more accurate results?

2 Draw a diagram of the apparatus you would use to find the heat produced when 1 g of a firelighter burns.

3 When 100 cm^3 of 1.0 M hydrochloric acid is added to 100 cm^3 of 1.0 M sodium hydroxide, the temperature of the mixture rises by 6.5°C.
(a) What mass of solution is warmed up? (Assume 1 cm^3 of solution has a mass of 1 g).
(b) How many joules warm up the mixture? (Assume 4.2 J warms up 1 g of solution by 1°C.)
(c) How many moles of hydrochloric acid react?
(d) How much heat is produced when 1 mole of hydrochloric acid reacts? This is called the **heat of neutralisation** of hydrochloric acid.
(e) Write a balanced equation for the reaction involved in (d).

4 The energy change for the manufacture of lime (CaO) from limestone ($CaCO_3$) can be written as

$$CaCO_3(s) \rightarrow CaO(s) + CO_2(g) \qquad \Delta H = +178 \text{ kJ}$$

(a) Is the reaction exothermic or endothermic?
(b) Do the products contain more or less energy than the reactants?
(c) Draw an energy level diagram for the reaction.

SECTION I: THINGS TO DO

Activities

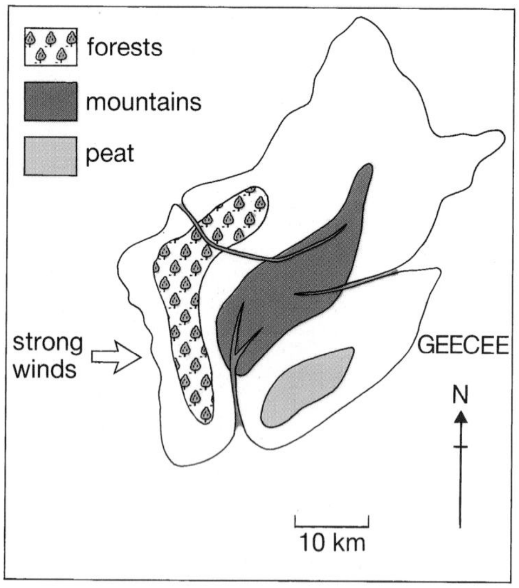

1 Using natural resources

Geecee is an imaginary island off the west coast of Britain. The island has strong winds from the west, high mountains with fast flowing rivers, thick forests along the west coast and peat bogs in the south east. There are no supplies of coal, oil or natural gas. At present no one lives on Geecee.

Suppose you have been sent to study whether people could live on the island and develop a fishing industry.

1 What material would you use to build the first homes on Geecee?

2 Describe two ways of producing heat to cook food.

3 Describe briefly how you would produce electricity to run machinery.

4 Where would you build the port for the fishing fleet? Give two reasons for your choice of site.

5 Where would you build homes on the island? Explain your answer.

6 Which natural resource(s) would you try to conserve? Explain your answer.

2 Unleaded petrol

Why are more motorists using unleaded petrol?

What is unleaded petrol?
Unleaded petrol is petrol which contains no *added* lead compounds.

Why are lead compounds added to petrol?
Small quantities of tetraethyllead(IV) ($Pb(C_2H_5)_4$) are added to petrol to make it burn more easily and more smoothly in car engines. If tetraethyllead(IV) is not added, we must use better quality petrol.

Why should we remove lead compounds from petrol?
When lead compounds are used in petrol, the exhaust fumes contain lead compounds. These substances are very harmful to living things. One of the worst effects of lead compounds is that they affect the brain. They are particularly harmful to young children whose brains are growing quickly.

How are cars modified to use unleaded petrol?
Cars which run on unleaded petrol must be designed so that their exhaust gases contain no unburnt fuel. This means using either special 'lean-burn' engines or special catalysts in the exhaust system. All new petrol-driven cars are now built to run on unleaded petrol. Many older cars can be modified to run on unleaded fuel.

How long will leaded petrol be available?
As yet, there is no date by which *all* cars must use unleaded petrol. Unleaded petrol is being phased in and more and more cars will use it. Unleaded is less expensive than leaded petrol, when you buy it at the petrol station, to encourage more people to use it. But both leaded and unleaded petrol will be available for a long time.

1 Write down the 6 most important points in the text above.

2 Use these 6 points to write a short report for your local newspaper. Your article should be no more than 150 words.

3 Windmills

Harnessing the energy of the wind is not a new idea. Windmills like this were used until the end of the last century to grind cereals to flour

Windmills have been used to drive machinery for hundreds of years. Recently, large windmills (sometimes called wind turbines) have been used to generate electricity. Giant windmills with a blade diameter of 50 metres have been built on the west coasts of Scotland and Wales. The photo on page 193 shows a giant wind turbine at Carmarthen Bay in Wales. Even so, one giant windmill cannot produce much electricity compared to a power station. One way to overcome this problem is to have windfarms with lots of giant windmills.

1 Make a list of the advantages of generating electricity using windmills in a windfarm rather than using a coal-fired power station.

2 Make a list of the disadvantages of generating electricity using windmills in a windfarm.

4 Nuclear energy – yes or no?

One of the most important issues about energy today is whether nuclear energy should be used to produce electricity. In Sweden there is a programme to get rid of nuclear power stations. On the other hand, France gets over 70% of its electricity from nuclear energy and this percentage is likely to increase.

1 Carry out a survey amongst your family and friends. Ask them two important questions.
(a) Should we, in Britain, make more use of nuclear energy than at present?
(b) What are the reasons for their answers in part (a)?

2 (a) Prepare a report from your survey. Show the percentages of people for and against using more nuclear energy and the number of times particular reasons were given.
(b) How do your results compare with those for the rest of the class? What are the overall conclusions for your class?

Modern windmills, or wind turbines are used to generate electricity

Hartlepool nuclear power plant

Study Questions

1 (a) Petroleum is a mixture of *hydrocarbons*. The simplest hydrocarbons in petroleum are a *homologous series* of *alkanes*. When petroleum is subjected to *fractional distillation*, a series of fractions is obtained, some of which are used as *fuels*. Some of the higher fractions are subjected to *catalytic cracking* to make chemical feedstocks including ethene.

Explain the terms in italics.

(b) State and explain what happens in the polymerisation of ethene to polythene.

(c) Polythene is a thermoplastic polymer whereas polyurethane is a thermosetting polymer. Explain, with the aid of simple diagrams, the difference between thermoplastic and thermosetting polymers.

(d) Why is plastic furniture more dangerous in a fire than wooden furniture?

(e) What environmental problems have resulted from the increased use of plastics?

2 This diagram shows some of the reactions of ethene.

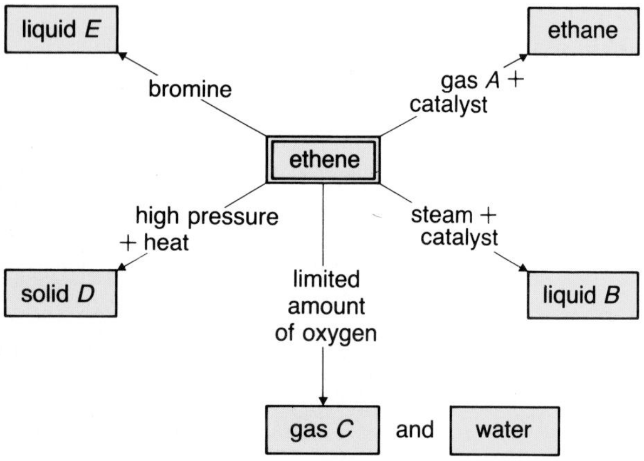

(a) Draw the structural formulas for ethane and ethene.

(b) Give the names for gas *A*, liquid *B*, gas *C*, and solid *D*.

(c) Write a balanced equation for the reaction between ethene and steam to form liquid *B*.

(d) Give the name and draw the structural formula for liquid *E*.

(e) Ethene is an unsaturated compound. By means of gas *A*, ethene is converted into ethane, a saturated compound. Give one commercial application of this type of reaction.

3 This question is about a threatened strike at 'Petrochem', a large chemical company which produces petrol from crude oil. Suppose you are (i) the Chairperson of Petrochem, (ii) the chief chemist at Petrochem, (iii) the Trade Union Representative at Petrochem.

(a) What are your main worries?

(b) What plans would you make in case the strike happens?

4 The major sources of the world's energy supplies are fossil fuels. The chemical energy that they contain can be converted to other useful forms of energy such as heat and electricity. During the last thirty years there has been a considerable increase in demand for electrical energy and other sources of energy are now being investigated as a matter of urgency.

(a) What is a 'fossil fuel'?

(b) Name *three* fossil fuels, each one in a different state at room temperature and pressure.

(c) (i) Which two elements are present in the highest proportion in fossil fuels?

(ii) Write balanced equations for the complete combustion of each of these elements with oxygen.

(d) Both of the reactions represented in part (c) are exothermic. Draw an energy level diagram to show the energy change that takes place when one of the elements is burned in oxygen. Indicate clearly on the diagram the heat of combustion of the element by labelling it ΔH.

(e) Why are urgent efforts being made to find sources of energy as an alternative to the use of fossil fuels?

(f) State one source of energy, other than direct solar energy, which may be used as an alternative to fossil fuels.

5 The apparatus below can be used to add hydrogen to vegetable oils and make margarine.

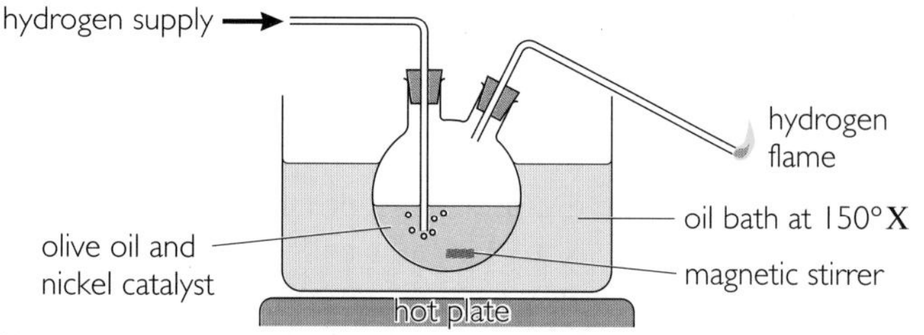

(a) At the start of the experiment, hydrogen is passed through the apparatus for a few minutes before the flame is lit. Why is this done?

(b) Choose a crude oil fraction from the table for use in the oil bath. Explain your choice.

Fraction	**Boiling range in °C**
Gas	below 40
Petrol	40–160
Kerosine	160–250
Diesel oil	250–300
Lubricating oil	300–350

(c) Give one reason for using a hotplate rather than a bunsen for heating.

(d) Each molecule in olive oil contains three carbon-to-carbon double bonds. How many hydrogen atoms will have been added to one molecule when the reaction is complete?

(e) What simple test would you use to show that the reaction had completely saturated the carbon-carbon double bonds in olive oil?

(f) What has been done in the experiment to speed up the reaction?

(g) Some brands of margarine are described as **polyunsaturated**. What does this mean?

SECTION J

Reaction rates

Why do we store food in fridges and freezers?

1 HOW FAST?

Some reactions, like explosions, happen almost instantaneously

Everyday we are concerned about how fast things happen. We want to know how quickly we can get to school, how fast our car is travelling or how soon the post will arrive. We are also interested in how fast some chemical reactions occur, how long we need to boil an egg or bake a cake and how quickly the rust appears on bikes and cars.

Different chemical reactions happen at different rates. Some reactions, like explosions, are so fast they they are almost instantaneous.

A small explosion takes place when you use the 'pop' test for hydrogen. When a burning splint is put into a test tube containing hydrogen, the hydrogen reacts instantly with oxygen in the air forming water vapour.

$$2H_2(g) + O_2(g) \rightarrow 2H_2O(g)$$

Other reactions, like the rusting of steel and the weathering of limestone on buildings, happen so slowly that it may be years or even centuries before we notice their effects.

The copper, used on this domed roof has turned green due to exposure to the elements over time

Most reactions take place at speeds somewhere between those described in the last two paragraphs. The reactions which take place when coal burns and when food is cooked are good examples of reactions which occur at steady rates.

The reactions which take place when food is cooked occur at a steady rate

Why are reaction rates important?

Most of the reactions in our bodies would never happen if the reactants were just mixed together. Fortunately, each reaction is helped along by its own special **catalyst**. Catalysts allow materials to react more easily. The catalysts in our bodies and in all living things are called **enzymes**. Without enzymes the reactions in your body would stop working and you would die.

One enzyme is called amylase. Amylase is present in saliva. It speeds up the first stage in the breakdown of starch in foods such as rice, potatoes and bread. Enzymes in our bodies catalyse the breakdown of our food, as well as reactions which synthesise important chemicals, like proteins.

More and more industrial processes using enzymes are being developed. These processes include the manufacture of fruit juices, beers, yoghurt, vitamins and pharmaceuticals. In some of these processes, enzymes are extracted from living material such as plant extracts, animal tissues, yeast and fungi. Enzyme extracts of this kind are already used in cheese making, food processing and 'biological' washing powders.

Industrial chemists are not usually satisfied with turning one substance into another. They want to carry out reactions faster and cheaper. In industry, speeding up slow reactions makes them more economical because saving time usually saves money. The key reaction in the manufacture of sulphuric acid is the contact process (section G, unit 1). This involves converting sulphur dioxide and oxygen to sulphur trioxide.

$$2SO_2 + O_2 \rightarrow 2SO_3$$

Chemical engineers have found that this reaction takes place quickly at 450°C if a catalyst of vanadium(V) oxide or platinum is used. By using a catalyst, sulphuric acid can be made faster and cheaper. This is important because sulphuric acid is a major industrial chemical.

Fertilisers are not catalysts, but they do improve the growth of plants. How do they do this?

A mechanic fitting a special catalyst section to the exhaust system of a car

QUESTIONS

1 Make a list of the various ways in which gardeners speed up the growth rate of their plants.

2 Find out about the process of fermentation in the manufacture of beer and wine. What part do catalysts play in the process? Prepare a short report of your findings.

3 (a) State *three* ways in which food can be preserved and stored without 'going bad'.
(b) What conditions slow down the rate at which foods deteriorate (go bad)?

4 (a) How do you think a pressure cooker speeds up the rate at which food is cooked?
(b) Why do you think a microwave oven can speed up the rate at which food is cooked?

2 MAKING REACTIONS GO FASTER

It's easier to get a fire started with sticks rather than logs

If you have ever tried to light a fire you will know that it is easier to burn sticks than large logs. The main reason for this is that the sticks have a greater surface area. There is a greater area of contact with the air and so the sticks burn more easily.

> In general, reactions go faster when there is more surface area to react.

Surface area and reaction rate

The reaction between marble chips (calcium carbonate) and dilute hydrochloric acid can be used to study the effect of surface area on reaction rate.

$$CaCO_3(s) + 2HCl(aq) \rightarrow CaCl_2(aq) + H_2O(l) + CO_2(g)$$

During the reaction, carbon dioxide escapes from the reacting mixture so there is a decrease in mass. The results are shown in Figure 1. In experiment I, thirty *small* marble chips (with a total mass of 10 g) reacted with 100 cm^3 of dilute hydrochloric acid. In experiment II, six *large* marble chips (with a total mass of 10 g) reacted with 100 cm^3 of the same hydrochloric acid. There are more than enough marble chips in both experiments, so the acid will be used up first.

1 Why is the overall decrease in mass the same in both experiments?
2 Why do the graphs become flat?
3 Which graph shows the greater decrease in mass per minute at the start of the experiment?
4 Which experiment begins at the faster rate?
5 Why is the reaction rate different in the two experiments?

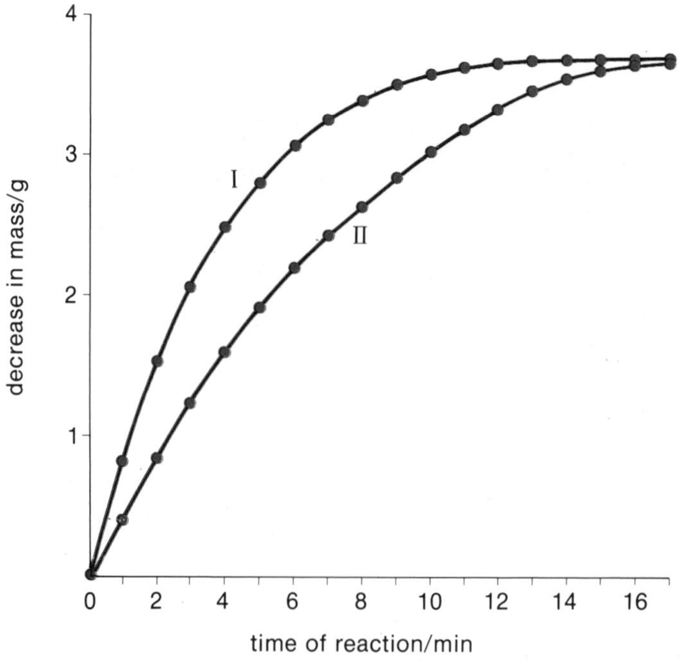

Figure 1

Concentration and reaction rate

Charcoal in a barbecue normally burns very slowly with a red glow. But if you blow onto it, so that it gets more air and more oxygen, it glows much brighter and may burst into flames. Substances that burn in air burn much more rapidly in oxygen. In oxy-acetylene burners, acetylene burns in pure oxygen. These burners produce such high temperatures that the flame will cut through sheets of metal.

$$C_2H_2 + 2\tfrac{1}{2}O_2 \rightarrow 2CO_2 + H_2O + \text{heat}$$
acetylene

> Chemical reactions occur when particles in the reacting substances collide with each other.

Collisions between acetylene molecules and oxygen molecules occur more often when oxygen is used instead of air. So the reaction happens faster and gives off more heat when the concentration of oxygen is increased by using the pure gas. Pure oxygen can also be used to speed up chemical changes in the body. This can help the recovery of hospital patients, such as those suffering from extensive burns.

> In general, reactions go faster when the concentration of reactants is increased.

In reactions between gases, the concentration of each gas can be increased by increasing its pressure. Some industrial processes use very high pressures. For example, in the Haber process (unit 5 of this section), nitrogen and hydrogen are made to react at a reasonable rate by increasing the pressure to 250 times atmospheric pressure.

Four oxyacetylene flames being used to cut through a large sheet of steel. In oxyacetylene burners, acetylene burns very rapidly in pure oxygen. The flame is hot enough to melt the metal

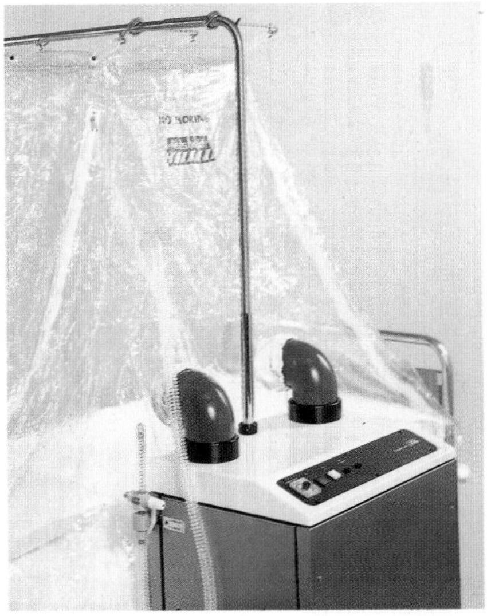

Pure oxygen is used in oxygen tents like this one, to speed up the recovery of hospital patients

QUESTIONS

1 It takes only 10 minutes to fry chips, but 20 minutes to boil potatoes. Larger potatoes take even longer to boil.
(a) Why do larger potatoes take longer to cook than small ones?
(b) Why can chips be cooked faster than boiled potatoes?
(c) Why can boiled potatoes be cooked faster in a pressure cooker?
2 Why do gaseous reactions go faster if the pressure of the reacting gases is increased?
3 Which of the following will affect the rate at which a candle burns?
the temperature of the air, the shape of the candle, the air pressure, the length of the wick
Explain your answer.
State two other factors that will affect the rate at which a candle burns.
4 Design an experiment to investigate the effect of acid concentration on the weathering of limestone.

3 TEMPERATURE AND CATALYSTS

Fish, meat and soft fruit can be kept for long periods in a deep freezer where the temperature is about −18°C. (The temperature in a house refrigerator is about 5°C)

Milk will keep for days in a cool refrigerator, but it turns sour very quickly if it is left in the sun. Other perishable foods, like fruit and cream, also go bad more quickly at higher temperatures. This is why we often keep them in a refrigerator. Lots of other processes can be speeded up by increasing the temperature, or slowed down by reducing the temperature. The chemical reactions involved in baking and cooking would never happen unless the food was heated to a high temperature.

Why do chemical reactions go faster at higher temperatures?

At higher temperatures, particles move about faster. So they collide more often and this causes the reactions to go faster. But particles do not always react when they collide. Sometimes they collide too gently. The particles do not collide with enough energy for bonds to stretch and break. In some reactions, only the molecules with high energies can react. The same sort of thing happens in a car crash (Figure 1).

Figure 1
If cars collide at very slow speeds (with low energy), they hardly dent each other. But if cars collide at high speeds (with high energy), they get smashed to pieces

So there are two reasons why reactions go faster as the temperature rises:

- The particles collide more often.
- The particles collide with more energy, so more collisions result in a reaction.

Catalysts and reaction rates

Hydrogen peroxide solution ($H_2O_2(aq)$) decomposes very slowly into water and oxygen at room temperature. When manganese(IV) oxide is added, it decomposes very rapidly.

$$2H_2O_2(aq) \rightarrow 2H_2O(l) + O_2(g)$$

The manganese(IV) oxide helps the hydrogen peroxide to decompose, but it is not used up during the reaction. The manganese(IV) oxide left at the end weighs exactly the same as that at the start of the reaction. The manganese(IV) oxide has acted as a **catalyst**.

This reaction shows the two important properties of catalysts:

- They change the rate of chemical reactions.
- They are not used up during the reaction.

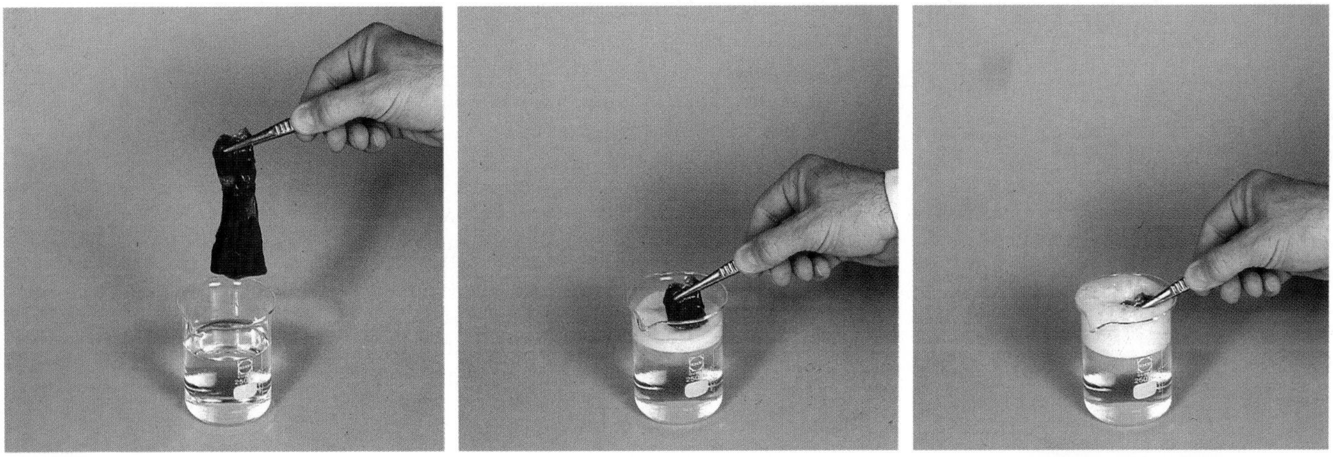

Liver and some plant tissues contain an enzyme called catalase which decomposes hydrogen peroxide in living things. As the photos show, liver can decompose hydrogen peroxide very rapidly

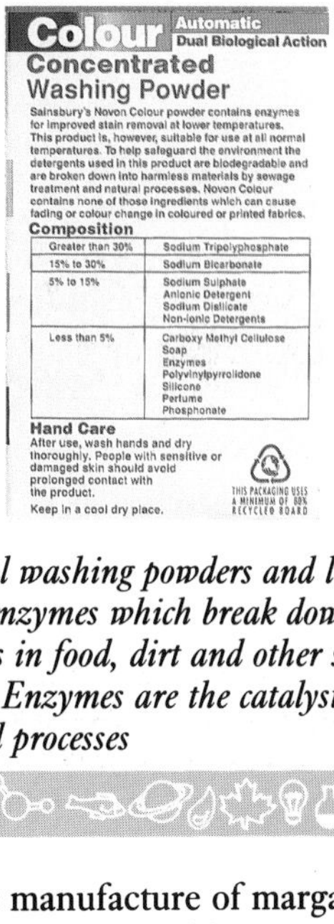

Biological washing powders and liquids contain enzymes which break down the chemicals in food, dirt and other stains on clothing. Enzymes are the catalysts in biological processes

Most catalysts are used to speed up reactions, but a few can be used to slow reactions down. These substances are called negative catalysts or **inhibitors**. For example, glycerine is sometimes added to hydrogen peroxide as an inhibitor to slow down its rate of decomposition during storage. Hydrogen peroxide is used in industry to bleach textiles, paper and pulp.

Catalysts play an important part in the chemical industry. Sulphuric acid (section G, unit 1), petrol (section I, unit 7) and ammonia (section J, unit 5) are all produced by processes involving catalysts.

Catalysts allow substances to react more easily. They do this by helping bonds to break more easily. So the particles need less energy to react and the reaction is faster.

QUESTIONS

1 Design an experiment to study the rate of reaction between magnesium ribbon and dilute sulphuric acid at different temperatures. Draw a diagram of your apparatus, say what you would do and the measurements you would make.

2 The catalysts in many reactions are transition metals or their compounds. Make a list of these reactions. (The references on this page to other catalysed reactions will help you with this question.)

3 Discuss the following questions in groups of 2 or 3.
(a) Why do plants grow faster in warm, wet weather than in cold, dry weather?
(b) Why do gardeners add fertiliser to the soil?
(c) Are fertilisers catalysts?

4 Describe an experiment that you could carry out to show that manganese(IV) oxide is not used up when it catalyses the decomposition of hydrogen peroxide.

5 In the manufacture of margarine, nickel catalyses an addition reaction involving hydrogen and vegetable oils.
(a) What is an addition reaction?
(b) What kind of bond must vegetable oils have for this to happen?
(c) How do you think the vegetable oil can be converted into different margarines that melt at different temperatures?

4 STUDYING REACTION RATES

Checking the pH level in a swimming pool

> A chemical reaction cannot happen unless particles in the reacting substances collide with each other.

This explains why reactions between gases and liquids usually happen faster than reactions involving solids. Particles in gases and liquids can mix and collide much more easily than particles in solids. In a solid, only the particles on the surface can react.

During a reaction, reactants are being used up and products are forming. The amount and the concentration of the reactants fall as the amount and the concentration of the products rise. So we can measure reaction rates by measuring how much of a reactant is used up or how much of a product forms in a given time.

$$\therefore \text{Reaction rate} = \frac{\text{change in mass (or concentration) of a substance}}{\text{time taken}}$$

For example, when 0.1 g of magnesium was added to dilute hydrochloric acid, the magnesium reacted and disappeared in 10 seconds.

$$\therefore \text{Reaction rate} = \frac{\text{change in mass of magnesium}}{\text{time taken}}$$

$$= \frac{0.1}{10} \text{ g magnesium used up per second}$$

$$= 0.01 \text{ gs}^{-1}$$

Strictly speaking, this is the *average* reaction rate over the 10 seconds for all the magnesium to react. Although reaction rates are usually measured as changes in mass (or concentration) with time, we can also use changes in volume, pressure, colour and conductivity with time.

This mechanic is tuning the engine of the car to adjust the rate at which petrol will burn in the cylinders. The performance of the car will depend on the rate of this reaction

Time (min)	**Mass of flask and contents /g**	**Decrease in mass /g**	**Decrease in mass for each minute interval /g**
0	78.00	0	
1	76.50	1.50	1.50
2	75.50	2.50	1.00
3	74.95	3.05	0.55
4	74.60	3.40	0.35
5	74.41	3.59	0.19
6	74.33	3.67	0.08
7	74.30	3.70	0.03
8	74.30	3.70	0

Table 1 The results of one experiment to measure the rate of reaction between marble chips and dilute hydrochloric acid

Calculating reaction rates

The rate of reaction between small marble chips (calcium carbon
hydrochloric acid was studied using the apparatus in Figure 1.
occurs, carbon dioxide escapes from the flask and so the mass
its contents decrease.

$CaCO_3(s) + 2HCl(aq) \rightarrow CaCl_2(aq) + H_2O(l)$
marble chips

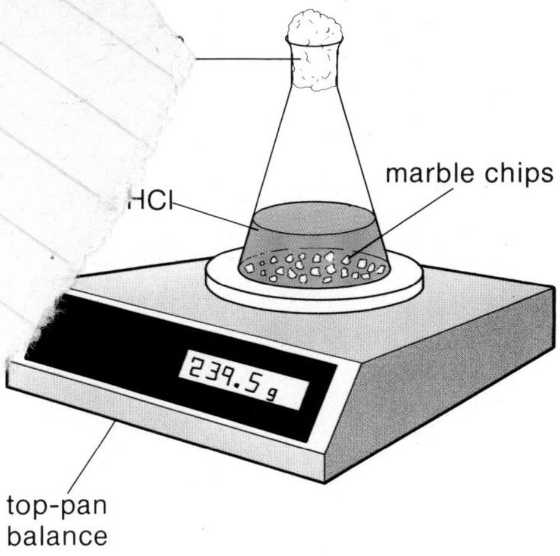

Figure 1

The cotton wool in the mouth of the flask stops liqui
as the mixture fizzes. The results of one experiment a
These results have been plotted on a graph in Figur
there is a decrease in mass of 1.5 g as carbon dioxi

∴ Average rate of reaction in the first minute

$= \frac{\text{change in mass}}{\text{time taken}} = \frac{1.5}{1} = \frac{\text{1.5 g of car}}{\text{lost per minute}}$

(Notice that the units for reaction rate are grams per minute this time.) During the second minute (from time = 1 minute to time = 2 minutes), 1.0 g of carbon dioxide escapes.

∴ Average rate of reaction in the second minute

$= \frac{1.0}{1} = 1.0$ g of CO_2 lost per minute.

Notice that the reaction is fastest at the start of the reaction when the slope of the graph is steepest. During the reaction the rate falls and the slope levels off. Eventually the reaction rate becomes zero and the graph becomes flat with a slope (gradient) of zero.

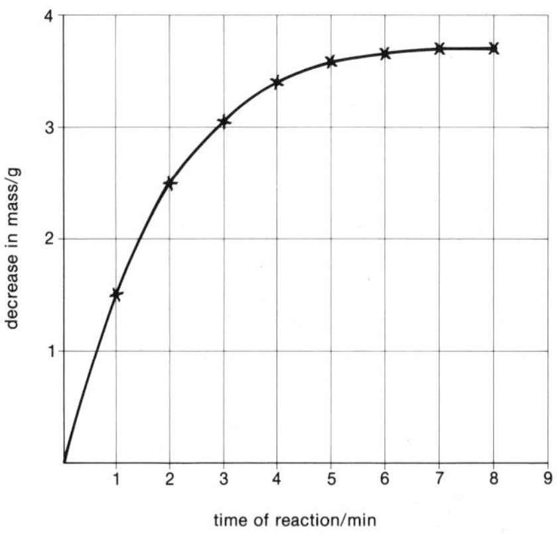

Figure 2

QUESTIONS

1 Discuss the following questions in small groups. Then write your own answers.
(a) Does magnesium react with dilute hydrochloric acid faster at the start of the reaction or at the finish? Explain your answer.
(b) Selective weedkillers can be added to a lawn either as solid pellets or as aqueous solutions. Which method will affect the weeds faster? Explain your answer.
(c) The selective weedkiller *2,4D* kills dandelions in a lawn, but not the grass. How do you think it works?

2 Look at the graph in Figure 2 and the results in Table 1.
(a) What mass of carbon dioxide is lost from the flask in
(i) the third minute (from time = 2 minutes to time = 3 minutes),
(ii) the fourth minute (from time = 3 minutes to time = 4 minutes),
(iii) the fifth minute (from time = 4 minutes to time = 5 minutes)?
(b) What is the average reaction rate in
(i) the third minute,
(ii) the fourth minute,
(iii) the fifth minute?
(c) What is happening to the reaction rate as time passes?
(d) Explain the change in reaction rate with time.
(e) Why does the graph become horizontal after a while?
(f) How would the shape of the graph in Figure 2 change if the same mass of more concentrated acid was used?

5 REACTION RATES AND INDUSTRIAL PROCESSES

Fritz Haber (1868–1934). In 1904, Haber began to study the reaction between nitrogen and hydrogen. By 1908, he had found the conditions needed to make ammonia (NH_3). The ammonia could then be used to make fertilisers

Industrial chemists try to produce materials as fast as possible and as cheaply as possible. In order to do this, they choose conditions which

1 increase the reaction rate and
2 use the most economic materials and methods.

One way to speed up a reaction is to increase the concentration of the reactants. In many industrial processes, this means increasing the pressure of reacting gases.

Catalysts are also important in industrial processes. By using a suitable catalyst it is possible to carry out some processes that would otherwise be impossible. Other processes can be carried out at lower temperatures and lower pressures when catalysts are used. This saves energy and makes the processes more economical. Temperature and pressure are also chosen carefully in the manufacture of most chemicals.

The importance of these factors in industrial processes is well illustrated by the Haber process.

The Haber process

During the last century, the populations of Europe and America rose very rapidly. More crops and more food were needed to feed more and more people. So farmers began to use nitrogen compounds as fertilisers (unit 6 of this section). The main source of nitrogen compounds for fertilisers was sodium nitrate from Chile, but by 1900 supplies of this were running out.

Another supply of nitrogen had to be found or many people would starve. The obvious source of nitrogen was the air. But how could this unreactive gas be converted into ammonium salts and nitrates for use as fertilisers? The German chemist, Fritz Haber, solved the problem. Eventually, the Haber process became the most important method of manufacturing ammonia.

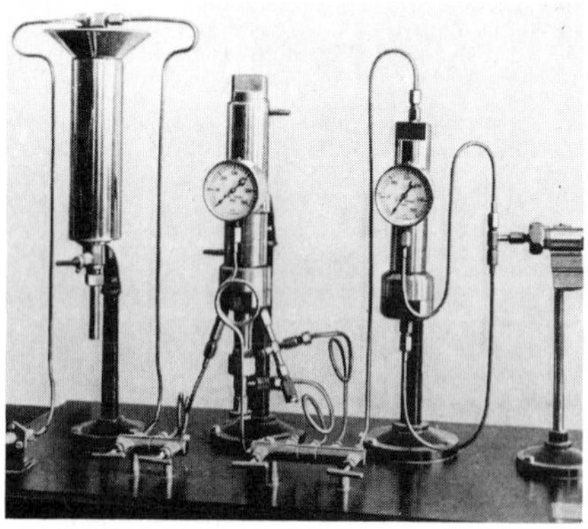

This picture shows the original apparatus used by Fritz Haber to make ammonia

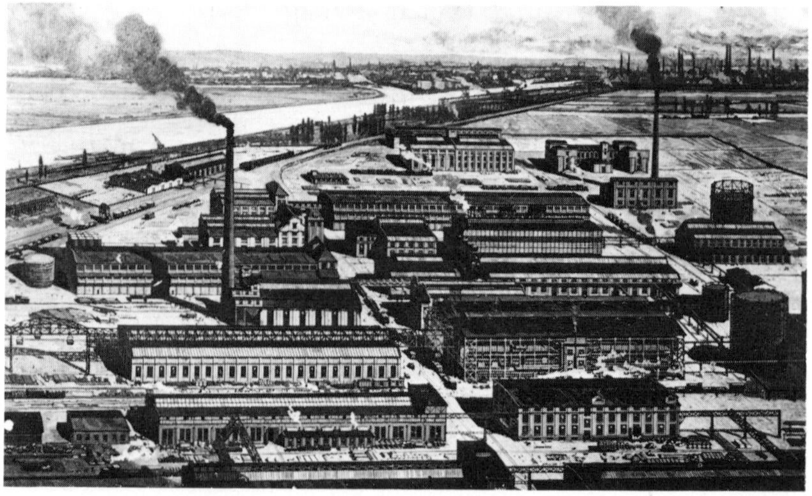

The vast ammonia manufacturing plant at Ludwigshafen, Germany. The plant was opened in 1913, only five years after Haber had found a cheap way of making ammonia

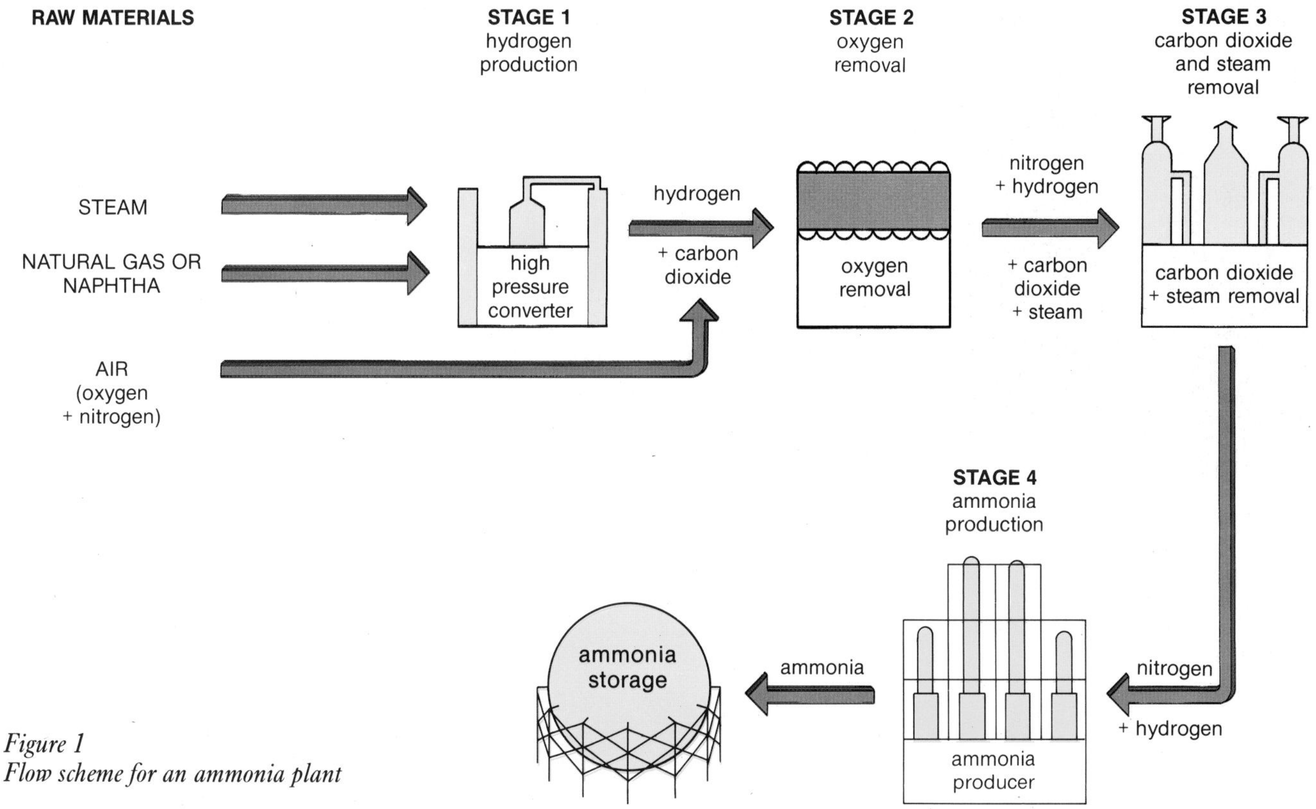

Figure 1
Flow scheme for an ammonia plant

A flow scheme for the Haber process is shown in Figure 1.

Stage 1. Steam reacts with natural gas (methane) or with naphtha from crude oil. This produces a mixture of hydrogen and carbon dioxide.

$$CH_4(g) + 2H_2O(g) \xrightarrow[\text{pressure + catalyst}]{\text{high temperature and}} CO_2(g) + 4H_2(g)$$

Stage 2. Air is added to the mixture. Oxygen in the air reacts with some of the hydrogen, forming steam

$$2H_2(g) + O_2(g) \rightarrow 2H_2O(g)$$

The product gases now contain nitrogen, hydrogen, carbon dioxide and steam.

Stage 3. The carbon dioxide and steam are removed by passing the gases through concentrated potassium carbonate solution. This leaves a mixture of nitrogen and hydrogen.

Stage 4. This is the key reaction in the Haber process. Nitrogen and hydrogen pass over

- a catalyst of iron
- at a pressure of 150–250 atmospheres and
- a temperature of 400°C

This converts 15–35% of the reactants to ammonia.

$$N_2(g) + 3H_2(g) \xrightarrow[\text{400°C, + iron catalyst}]{\text{150–250 atmospheres,}} 2NH_3(g)$$

The hot gases from the converter are cooled to liquefy the ammonia. The unreacted nitrogen and hydrogen are recycled.

QUESTIONS

1 Make a list of the main factors that affect the rate of a chemical reaction.

2 (a) Name three industrial processes that use a catalyst. Say what the catalyst is in each case.
(b) Why are catalysts important in industry?

3 At the beginning of this century Haber synthesised ammonia from nitrogen and hydrogen. Ammonia is used to make fertilisers and explosives.
(a) Why was Haber's synthesis of ammonia so important?
(b) Haber found the conditions needed to make ammonia in 1908. By 1913, a vast ammonia manufacturing plant had been opened in Germany. Why do you think the plant was working so soon after Haber's original experiments?

4 What conditions in the Haber process increase the rate of reaction between nitrogen and hydrogen?

6 FERTILISERS FROM AMMONIA

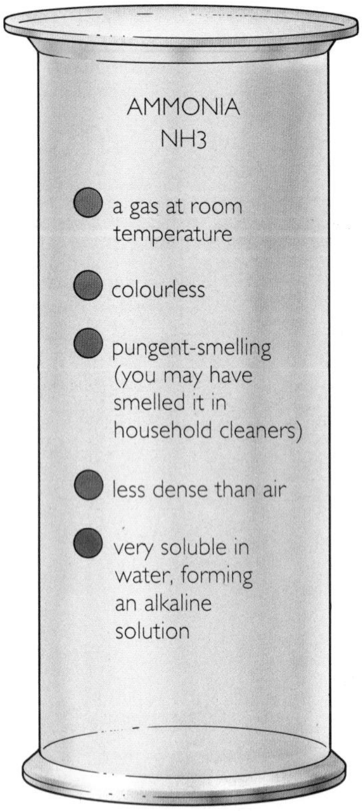

Figure 1
Properties of ammonia

A bag of NPK fertiliser. The percentages of N, P and K are indicated at the top of the bag – 17%N, 17%P and 17%K

Ammonia is an important chemical in industry and agriculture. About 2 million tonnes are manufactured in the UK each year. Three quarters of this is used to make fertilisers. The other quarter is used to make a variety of materials including nitric acid, nylon, dyes and explosives. The properties of ammonia are shown in Figure 1.

Ammonia as a base

Ammonia acts as a base in many reactions. Ammonia molecules pick up H^+ ions to form ammonium ions. So ammonia reacts with acids to form ammonium salts. This is how fertilisers, such as ammonium nitrate (*Nitram*) and ammonium sulphate, are made from ammonia.

$$\underset{\text{ammonia}}{NH_3(g)} + \underset{\text{nitric acid}}{HNO_3(aq)} \rightarrow \underset{\text{ammonium nitrate}}{NH_4NO_3(aq)}$$

Ammonia also reacts with hydrogen chloride to form a white smoke. The white smoke is tiny particles of solid ammonium chloride suspended in the air. This reaction is sometimes used as a test for ammonia.

$$NH_3(g) + HCl(g) \rightarrow NH_4Cl(s)$$

Fertilisers

Plants need essential elements (nutrients) to grow well. If crops are grown every year on the same land, these nutrients get used up. The soil becomes infertile and the plants are stunted with poor seeds and small fruit.

Three essential elements which plants obtain from soluble compounds in the soil are nitrogen, phosphorus and potassium. Plants need more of these three elements than any other (except carbon, hydrogen and oxygen). This means that shortages of nitrogen, phosphorus and potassium are soon noticed (Table 1). So these three elements must be replaced by adding fertilisers to the soil.

Fertilisers can be used as single compounds, such as ammonium nitrate, or as mixtures of compounds containing nitrogen, phosphorus and potassium ('NPK' fertilisers). The proportions of nitrogen, phosphorus and potassium in NPK fertilisers are usually shown as % nitrogen (N), % phosphorus(V) oxide (P_2O_5) and % potassium oxide (K_2O).

Nutrient	Role in plant growth	Effect of shortage
Nitrogen	Essential for synthesis of proteins and chlorophyll	Plants are short and stunted.
Phosphorus	Essential for synthesis of nucleic acids (DNA)	Plants grow slowly. Seeds and fruit are small
Potassium	Plays a part in synthesis of carbohydrates and proteins	Leaves become yellow and curl inwards

Table 1 The role of nitrogen, phosphorus and potassium in plant growth and the effects of shortages of these elements

Nitrogen fertilisers are usually nitrates or ammonium salts. Ammonium nitrate (*Nitram*), NH_4NO_3, is the most widely used fertiliser because it is soluble, it can be stored and transported as a solid and it has a high percentage of nitrogen (Table 2). The higher the percentage of nitrogen the better, because less useless material needs to be stored and transported. Other nitrogen fertilisers are ammonium sulphate, urea and nitrochalk. Nitrochalk is a mixture of ammonium nitrate and calcium carbonate. This provides calcium for the soil as well as nitrogen and it also corrects soil acidity.

Liquid ammonia being injected into the soil as a fertiliser. What are the advantages and disadvantages of using liquid ammonia?

Fertiliser	Formula	Mass of one mole	Mass of nitrogen in one mole	% of nitrogen
Ammonium nitrate	NH_4NO_3	80 g	28 g	$\frac{28}{80} \times 100 = 35$
Ammonium sulphate	$(NH_4)_2SO_4$	132 g	28 g	$\frac{28}{132} \times 100 = 21$
Urea	N_2H_4CO	60 g	28 g	$\frac{28}{60} \times 100 = 47$

Table 2 The percentage of nitrogen in different fertilisers

Fertilisers are esssential for producing high crop-yields year after year

Fertiliser being bagged ready to send out to farms all over Europe and America

QUESTIONS

1 (a) Why should a fertiliser be soluble in water?
(b) What are the problems in storing, transporting and using fertilisers?
(c) Make a list of the important properties of an ideal fertiliser.
2 (a) Use the method in Table 2 to calculate the percentage of nitrogen in ammonia.
(b) Liquid ammonia has been used as a fertiliser in some countries.
(i) What are its advantages?
(ii) What are its disadvantages?
3 (a) Describe how you would make a sample of ammonium sulphate.
(b) How does your method compare with the industrial manufacture of ammonium sulphate?
(c) Design an experiment to see if ammonium sulphate acts as a fertiliser for peas or beans.
4 (a) Why is NPK fertiliser so called?
(b) Rain water washes fertilisers into streams and rivers. What effect does this have on plants and animals that live in the water?
(c) Nitrochalk can act as a fertiliser *and* cure soil acidity. Explain why it can do both jobs.

7 THE NITROGEN CYCLE

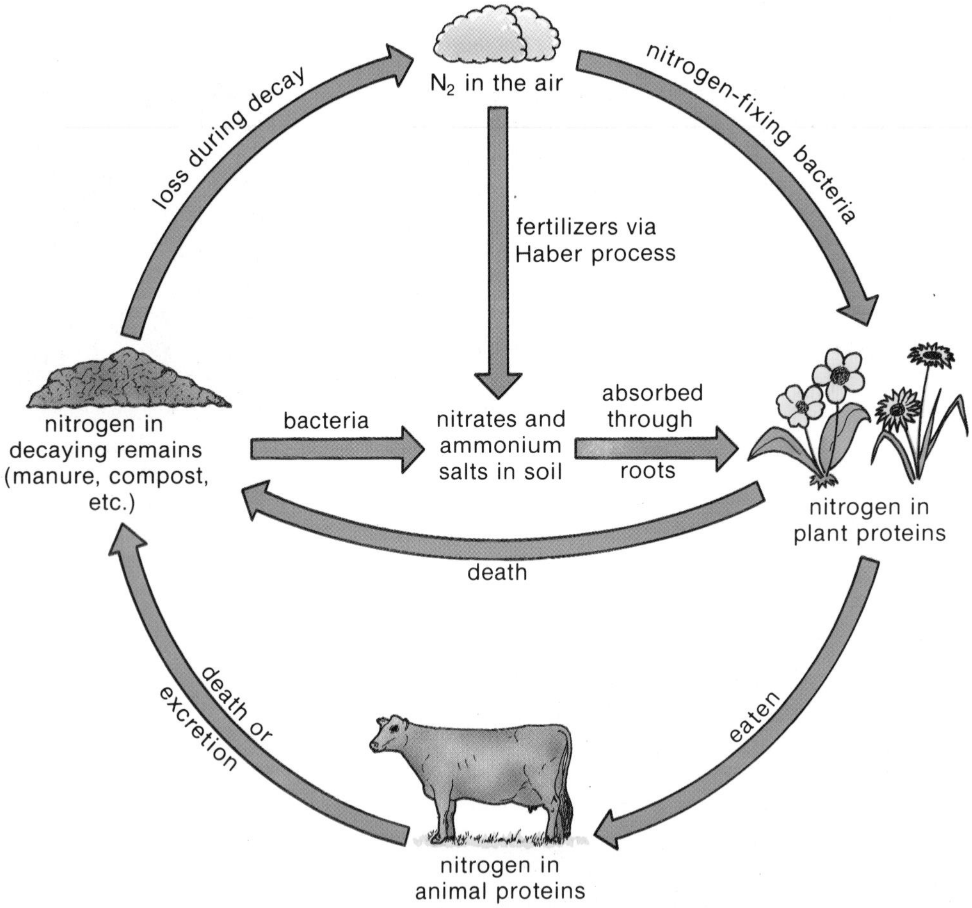

Figure 1
The nitrogen cycle

In the last unit, we looked at man-made inorganic fertilisers like ammonium nitrate. These synthetic fertilisers help to produce more food but if they are used too much they can cause serious problems.

- They may change the soil pH.
- They may harm plants and animals in the soil.
- They get washed out of the soil and lead to the pollution of rivers.

Large quantities of natural organic fertilisers are also used by farmers and gardeners. The most widely used organic fertilisers are manure and compost which contain decaying matter from animals and plants (Figure 1). The nitrogen compounds in manure and compost are decomposed by bacteria to nitrates and ammonium salts. These substances dissolve in rain water and are absorbed through the roots of plants (centre of Figure 1). The nitrates and ammonium salts are then used by plants to synthesise chlorophyll and proteins that they need for growth. Animals have to eat plants and other animals in order to get the nitrogen and the proteins which they need.

Manure and compost are excellent fertilisers, but they take time to break down and there is not enough of them for all the crops we grow today. Because of this, we need large amounts of inorganic fertilisers.

Some of the nitrogen that crops need is provided by nitrogen-fixing bacteria in the soil. These bacteria can convert nitrogen from the air into nitrogen compounds that plants can use. Some of these nitrogen-fixing bacteria live in nodules on the roots of plants such as peas, beans and clover (Figure 1, top right).

Some plants, called legumes, have nodules on their roots, like these in this photo. These nodules contain bacteria which convert nitrogen in the air into compounds the plant can use for growth

Although nitrogen is returned to the soil when living matter decays, some of the nitrogen compounds in the manure and compost are decomposed to nitrogen gas. This escapes into the air (Figure 1, top left). This is another reason why fertilisers are added to the soil in heavily cultivated areas.

Part of the European 'Grain Mountain'. Is it right that huge amounts of food are kept in store whilst thousands of people are starving?

The world food problem

There are about 4000 million people in the world. Nearly 3000 million of them are not properly fed and 400 million are starving. These figures suggest that we must increase food production, but the problem is not quite so simple.

- In Europe and North America, there are 'mountains' of surplus food but many poor countries cannot afford to buy it.
- In some poor countries, the rich have plenty to eat.
- Some countries have to export food to earn foreign currency even though some of their people are starving.

Three possible ways of reducing the food problem are as follows.

1 **Increasing birth control.** It is estimated that the world population will be 6000 million by the year 2000. Some people think that the population problem is more serious than the food problem.
2 **Improving farming methods.**
3 **Finding new food supplies.** We should try to develop new sources of food, such as farming the sea and growing bacteria on vegetable oils and cellulose to produce food.

During a lightning flash, nitrogen reacts with oxygen in the air to form nitrogen oxides. These react with rain water to produce nitric acid which increases the nitrogen content of the soil

QUESTIONS

1 Discuss the following questions with 2 or 3 others.
(a) What could the more developed nations do to solve the world food problem?
(b) Which problem is the more serious – the population explosion or the world food problem?
(c) What improvements in farming methods might reduce the world food problem?
2 (a) What is the difference between man-made and natural fertilisers?
(b) What happens to the nitrogen compounds in compost as it decays?
(c) Why do plants require nitrogen?
(d) Why do farmers and gardeners use man-made fertilisers as well as natural fertilisers?
3 Fish farming could provide new food supplies. A number of fish farms have already started working on the west coasts of Ireland and Scotland. Some people support their development and others are opposed to them. What do you think the following people would have to say about the development of a fish farm in their area?
(a) The Secretary of the local Residents' Association.
(b) The Member of Parliament.
(c) A young person unable to find work.
(d) The Secretary of the local Anglers' Club.
4 (a) What problems are caused by the over-use of fertilisers?
(b) Explain why these problems arise.

SECTION J: THINGS TO DO

Activities

1 The Nitrogen Cycle Game

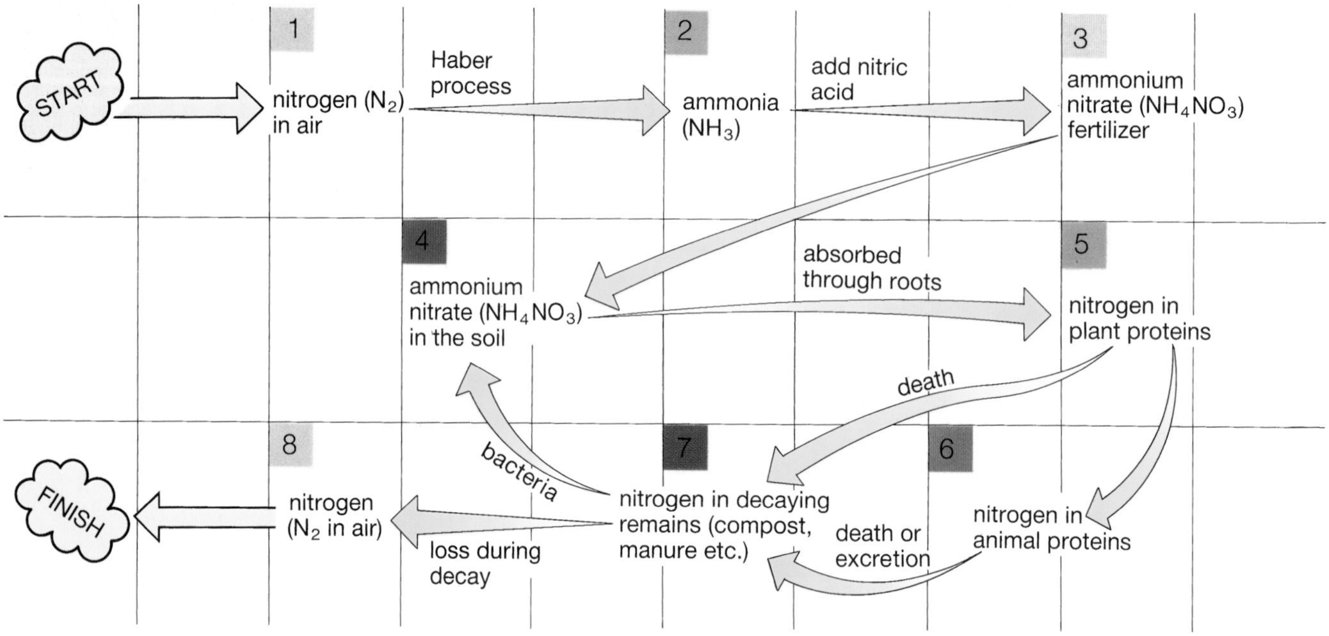

The Nitrogen Cycle Game is a mixture of 'Monopoly' and 'Trivial Pursuit'. It is based on fertilisers and the nitrogen cycle below.

Chance cards

First, make 10 CHANCE cards by writing the letters **A** to **J** on 10 pieces of card. Shuffle these cards and place the pile face downwards on the table.

Rules

(i) Any number of people can play the game.
(ii) You can play as individuals or in teams.
(iii) Each person (or team) starts at number 1 in the diagram and moves to the next number if they give a correct answer to a question from the other team.
(iv) If you answer a question wrongly, you do *not* move.
(v) When a team are asking questions, they should make up questions using pages 220–223.
(vi) Questions are put to each person (team) in turn.
(vii) At number 4, each person (team) takes a CHANCE card. The letter on the CHANCE card tells you how you must move according to the instructions on the right.
(viii) The winning person (team) is the first to reach number 8.

Instructions

A Fertiliser washed out of soil. Go back to 1.
B Fertiliser gets mixed with compost. Go on to 7.
C Fertiliser taken up by plants. Move to 5.
D Fertiliser returned to factory by farmer. Go back to 4.
E Fertiliser decomposes to ammonia Go back to 2.
F Fertiliser accidentally eaten by cattle. Go on to 6.
G Soil is frozen by cold weather. Stay at 4.
H Important nutrients missing from soil. Stay at 4.
I Fertiliser changes the soil pH and plants cannot grow. Stay on 4 and miss a turn.
J Ideal conditions of pH and temperature for plant growth. Move to 7.

2 AMMChem plant proposed for Oakbridge

AMMChem plant for Oakbridge

Yesterday, Sandra Bigg, Mayor of Oakbridge, admitted that Council officials had met representatives of AMMChem. AMMChem are the world's largest manufacturer of fertilisers. Mayor Bigg said. "The firm are seeking permission to build a chemical plant on the North side of Oakbridge."

AMMChem produces ammonia and nitric acid and uses them for the manufacture of ammonium nitrate (Nitram). It also sells large quantities of ammonia and nitric acid to other companies.

Mrs. Bigg welcomed the plant. She said, "It will help the town's economy. The plant will employ 150 local people. It is also estimated that each person employed by AMMChem will create jobs for another 3 or 4 people in local businesses. This will reduce unemployment in the town." At present, about 2000 people in Oakbridge are unemployed. This is 10% of the workforce.

Mayor Bigg also pointed to two other benefits from the plant. AMMChem would pay business rates to Oakbridge Council at the commercial level. This would make a large increase in revenue and hold down the increases in council tax to home owners. Farmers near Oakbridge would save about £10 per tonne on Nitram fertiliser because they would be paying lower transport costs. A representative for the National Farmer's Union has estimated that about 500 tonnes of Nitram fertiliser are used each year on farms near Oakbridge.

Herb Green, Secretary of the Oakbridge Environmental Group, was asked for his views. He said, 'We are concerned about emissions of ammonia and nitrogen oxides from the plant and possible accidents when chemicals are being transported. How will these affect the beautiful countryside around Oakbridge? Just imagine what would happen if the plant was not working properly and excessive amounts of ammonia or nitric acid got into the River Soak? Why can't the Mayor persuade cleaner industries to move into the area? Anyway, farmers are using less artificial fertilisers these days and growing more organic food."

Earlier today, Dr. Ken Bond, Managing Director of AMMChem issued the following statement:

"Some chemical companies have caused environmental problems but AMMChem have an excellent record. Our Safety Division checks processes and equipment continuously and ensures that all our operations are safe and pollution free.

No industry is without risks. Ammonia is very toxic and it is manufactured at high pressure and high temperature. An accident at the plant or during transportation would create a serious health hazard. But, accidents are very rare. AMMChem have been involved in only one death from such accidents in the last ten years. In recent years, industry and society have begun to realize their joint responsibility for the production of chemicals which will benefit society."

Suppose you live in Oakbridge.

1 Make a list of the possible problems from the AMMChem plant.

2 Make a list of the possible benefits from the AMMChem plant.

3 Write a letter to the Oakbridge Chronicle expressing your views about the proposed plant. (You may want to support the proposal or to be very critical of it.)

4 Suppose you are a town councillor in Oakbridge. Make a list of other points you would want to consider or other information which you would need before deciding whether AMMChem should build the proposed plant.

Study Questions

Percentage of NH_3 in the mixture at pressure of:			
Temp/°C	1 atm	100 atm	1000 atm
200	15.3	80.6	98.3
300	2.2	52.1	92.6
400	0.44	25.1	79.8
500	0.15	10.4	57.5
600	0.05	4.5	31.4
700	0.02	2.1	12.9
800	0.01	1.2	–

1 Ammonia is an important chemical in the manufacture of fertilisers. The Haber process is the industrial method by which ammonia is manufactured from nitrogen and hydrogen.

The product from the Haber process consists of ammonia together with unreacted nitrogen and hydrogen. The percentage of ammonia in this mixture varies with temperature and pressure as shown in the table.

(a) Write a balanced equation for the reaction involved in the Haber process.
(b) Use the table to comment on the effects of changing the temperature and the pressure on the percentage of ammonia produced.
(c) (i) From the table, state the conditions that produce the largest percentage of ammonia.
(ii) What conditions are normally used in the Haber process?
(iii) Why do you think that the conditions normally used in the Haber process are not the same as those conditions which give the largest percentage of ammonia.
(d) (i) What is the effect of an iron catalyst on the Haber process?
(ii) Explain the effect of the catalyst.
(e) The boiling points of the chemicals in the mixture produced by the Haber process are:

Ammonia −34°C
Nitrogen −196°C
Hydrogen −253°C

Using this information, describe how ammonia can be separated from the mixture.

2 Greenhouse trials were carried out to investigate the effect of temperature on the yield of tomatoes. Three identical greenhouses were used at 20°C, 25°C and 30°C. The mass of tomatoes growing on the 50 plants in each greenhouse was estimated weekly. The results are shown in the graph.

(a) How do you think the mass of the tomatoes growing on the 50 plants in each greenhouse could be estimated?
(b) State three environmental factors which must be controlled in the greenhouses to make a fair comparison of results.

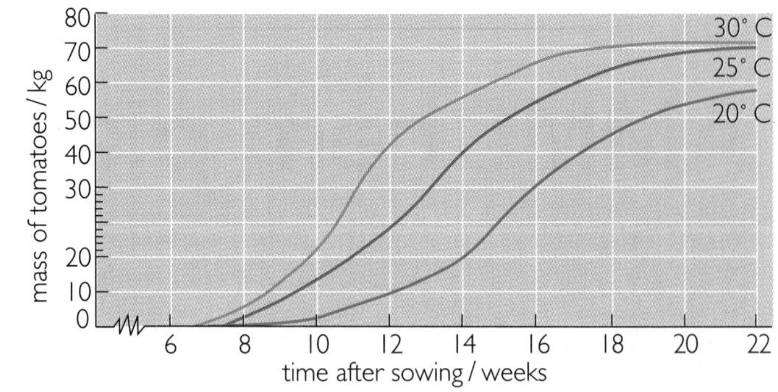

(c) What was the total mass of tomatoes in all three greenhouses at the end of week 14?
(d) What is the average weekly increase in yield between week 14 and week 18 in the greenhouse at 25°C?
(e) During which week is there the greatest difference in mass of tomatoes growing at 20°C and 30°C?
(f) What do you think will happen to the tomatoes if they are not picked during week 22 or 23?
(g) At which temperature would you grow tomatoes? Explain your choice.

3 When manganese(IV) oxide is added to hydrogen peroxide, oxygen and water are produced. The oxygen can be collected and measured at timed intervals. The equation for the reaction is

$$2H_2O_2 \rightarrow 2H_2O + O_2$$

The results for one experiment are shown below. 1.0 g of manganese(IV) oxide was added to 100 cm^3 of a solution of hydrogen peroxide.

Time/minutes	0	1	2	3	4	5	6	7	8
Volume/cm^3	1	20	33	44	52	58	59	60	60

(a) (i) Plot a graph of volume of oxygen (vertically) against time. Label this curve X.
(ii) Mark on curve X the time at which the rate of reaction is fastest. Label this point Y. Why have you chosen this point?
(iii) The experiment is repeated using 1.0 g of the same catalyst, 50 cm^3 of the hydrogen peroxide solution and 50 cm^3 of water. Sketch the curve expected from this experiment between the same axes. Label this curve Z.
(b) The speed of a chemical reaction depends on:

- the rate at which the particles collide,
- the energy possessed by these particles.

Use these factors to explain each of the following:
(i) heating the solution of hydrogen peroxide increases the speed of the reaction;
(ii) 1.0 g of a finely powdered catalyst gives a faster reaction than 1.0 g of small lumps of the same catalyst.

SECTION K

Atomic structure and radioactivity

Scientists working in radiation laboratories must handle radioactive isotopes by remote control from the other side of very thick glass windows

1 INSIDE ATOMS

J J Thomson, Professor of Experimental Physics at Cambridge University and winner of the Nobel Prize for Physics in 1906

Less than ninety years ago scientists believed that atoms were solid particles like tiny snooker balls. Since then, experiments have shown that all atoms are made of three types of particles – protons, neutrons and electrons. In this unit we shall study some of the evidence for these particles.

1897 Thomson discovers electrons

In 1897, J. J. Thomson was investigating the way that gases conduct electricity. When he applied 15 000 volts across the electrodes of a tube containing a gas, the glass walls glowed a bright green colour. Rays travelling in straight lines from the cathode hit the glass and made it glow. Thomson called these rays **cathode rays** because they come from the cathode. Experiments with a narrow beam of cathode rays (Figure 1) showed that they could be deflected by an electric field. When cathode rays passed between charged plates, they always bent towards the positive plate. This showed that the rays were negatively charged.

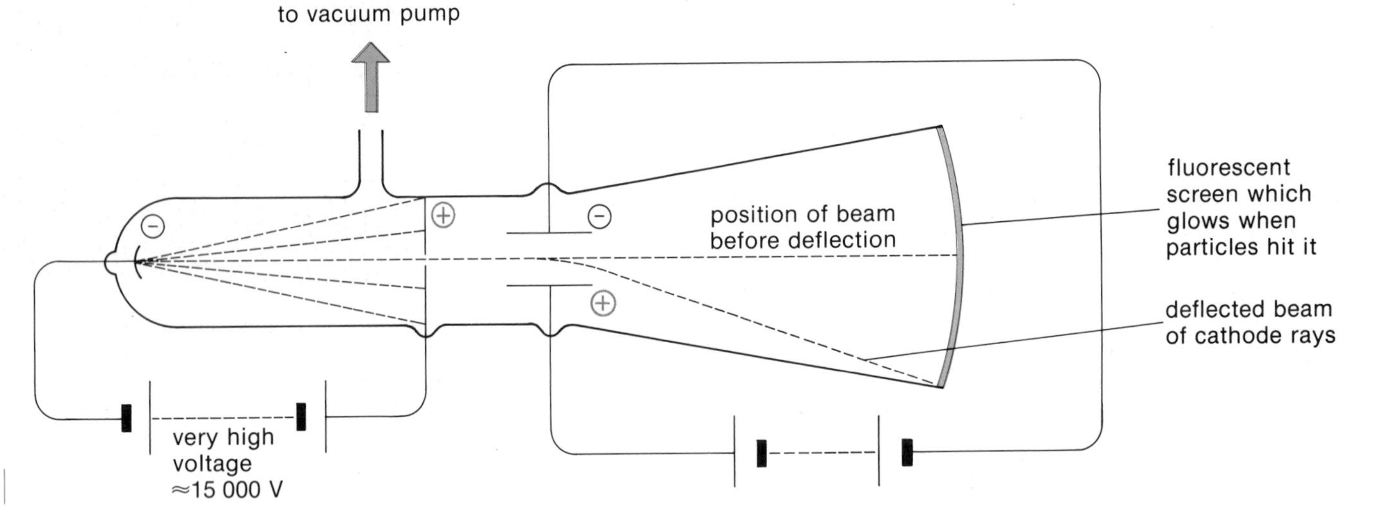

Figure 1
Deflection of cathode rays by an electric field

Further study of the deflection showed that cathode rays consisted of negative particles 1840 times lighter than hydrogen atoms. Thomson called these tiny negative particles **electrons**. The cathode rays were always the same, no matter what gas was present in the tube or what the electrodes were made of. This suggested that the electrons in all substances were identical.

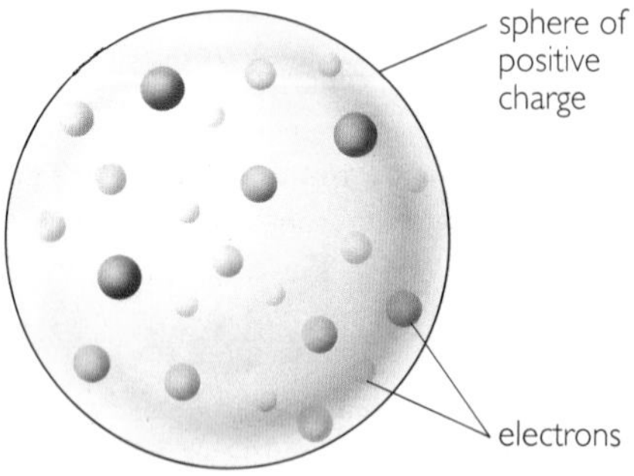

Figure 3
Thomson's 'currant bun' theory suggested that atoms were made up of a ball of positive charge dotted with negative electrons

1909 Geiger and Marsden explore the nucleus

Since atoms are neutral, they must contain positive charge to balance the negative charge on their electrons. Geiger and Marsden found a method of probing inside atoms using alpha particles from radioactive substances as 'bullets'. Alpha particles are helium ions, He^{2+}. When alpha particles from radium were fired at thin sheets of metal foil, most of the alpha particles passed straight through the foil. But some of the alpha particles were deflected by the foil and a few of them even appeared to bounce back from it (Figure 2).

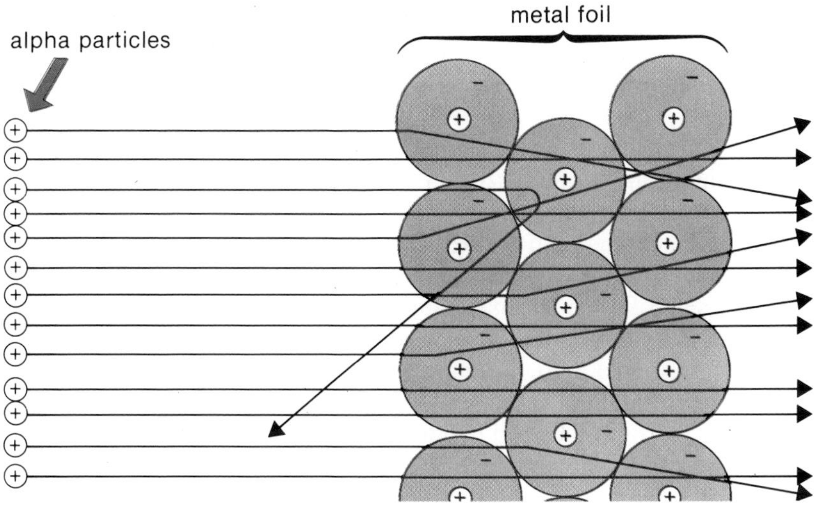

Figure 2
Most alpha particles pass straight through the foil, some are deflected, but a few rebound from the foil

Ernest Rutherford who succeeded J J Thomson as Professor of Experimental Physics at Cambridge University in 1919. Rutherford received the Nobel Prize for Chemistry in 1908 while he was working at Manchester University

1911 Rutherford explains the structure of atoms

Rutherford explained Geiger and Marsden's results by suggesting that atoms in the foil must have a very small positive **nucleus**. Surrounding this is a much larger region of empty space in which the electrons move. Most of the positive alpha particles pass straight through the large empty space where the electrons are moving. A few alpha particles pass close to the positive nucleus and get deflected. Occasionally an alpha particle approaches a nucleus head on. When this happens, the positive alpha particle is repelled by the positive nucleus and bounces back. Rutherford suggested that the structure of an atom could be compared to a miniature solar system. Each atom has a positive nucleus, orbited by tiny negative electrons like planets orbiting the Sun. He suggested that the positive charge of the nucleus was provided by positive particles which he called **protons**. Hydrogen, the smallest atom, has one proton in the nucleus, balanced by one orbiting electron. Atoms of helium, the next smallest, contain two protons and two electrons; lithium atoms have three protons and three electrons, and so on.

QUESTIONS

1 Get into a group with 2 or 3 others. Read this unit and then discuss the following questions.
(a) What evidence is there that electrons are (i) negatively charged, (ii) the same in all substances?
(b) What sort of experiments might be carried out to show that electrons are about 2000 times lighter than hydrogen atoms?

2 After discovering electrons, Thomson suggested that atoms were composed of positively charged material dotted with electrons, like currants in a currant bun (Figure 3).
(a) Why did Thomson's discovery of electrons prompt Geiger and Marsden into experiments to find positive particles?
(b) Why did Geiger and Marsden's experiment lead to
(i) the rejection of Thomson's 'currant bun' theory,
(ii) the suggestion by Rutherford that the positive charge in atoms must be concentrated in a small nucleus?

3 Have a brainstorm on atomic structure. Write 'Atomic Structure' in the centre of a sheet of paper. Then, at the end of lines fanning out from this heading, write down all you know about atomic structure. Try to group related ideas and facts together.

When you have finished the brainstorm, reorganise your notes to give a neat, well-ordered summary of what you have learned.

2 THE STRUCTURE OF ATOMS

James Chadwick discovered the neutron in 1932 when he was working with Rutherford in Cambridge. In 1935 he won the Nobel Prize for this achievement, and soon after became Professor of Physics at Liverpool University

In spite of Rutherford's success in explaining atomic structure, one big problem remained. Hydrogen atoms contain one proton and helium atoms contain two protons. So, if the relative atomic mass of hydrogen is one, that of helium should be two. Unfortunately, the relative atomic mass of helium is four and *not* two.

James Chadwick, one of Rutherford's colleagues, showed where the extra mass in helium came from. Chadwick discovered that the nuclei of atoms contained *uncharged* particles as well as positively charged protons. Chadwick called these uncharged particles **neutrons**. Further experiments showed that neutrons have the same mass as protons. This enabled Chadwick to explain the relative atomic masses of hydrogen and helium (Table 1). Hydrogen atoms have one proton, no neutrons and one electron. Since the mass of the electron is almost zero compared to the proton and neutron, a hydrogen atom has a relative mass of one unit. Helium atoms have two protons, two neutrons and two electrons. The two protons and two neutrons give a helium atom a relative mass of four units. Thus, a helium atom is four times as heavy as a hydrogen atom and the relative atomic mass of helium is four.

	Hydrogen atom	Helium atom
Number of protons	1	2
Number of neutrons	0	2
Relative mass	1	4
Relative atomic mass	1	4

Table 1 The relative atomic masses of hydrogen and helium

Protons, neutrons and electrons

We now know that all atoms are made up from three basic particles – protons, neutrons and electrons. The nuclei of atoms contain protons and neutrons. Both of these particles have a mass about the same as a hydrogen atom. Neutrons have no charge, but protons have a positive charge. Moving around the nucleus are electrons that are negatively charged. The electrons are arranged in layers or **shells** at different distances from the nucleus. The mass of the electron is so small that it can be ignored when working out the total mass of the atom. The positions, masses and charges of these three sub-atomic particles are shown in Table 2.

Particle	Position	Mass (relative to a proton)	Charge (relative to that on a proton)
Proton	nucleus	1	+1
Neutron	nucleus	1	0
Electron	shells	$\frac{1}{1840}$	−1

Table 2 Properties of the three sub-atomic particles

Different atoms have different numbers of protons, neutrons and electrons. The hydrogen atom is the simplest of all atoms. It has one proton in the nucleus, no neutrons and one electron (Figure 1). The next simplest atom is that of helium, with two protons, two neutrons and two electrons. The next, lithium, has three protons, four neutrons and three electrons. Some of the heavier atoms can have large numbers of protons, neutrons and electrons. For example, atoms of uranium have 92 protons, 92 electrons and 143 neutrons.

Notice that hydrogen, the simplest atom and the first element in the periodic table, has one proton. Helium, the second element in the periodic table, has two protons. Lithium, the third element in the periodic table, has three protons and so on. Thus, the position of an element in the periodic table tells you how many protons it will have. Furthermore, an atom must always have an equal number of protons and electrons, so that the positive charges (on the protons) balance the negative charges (on the electrons)

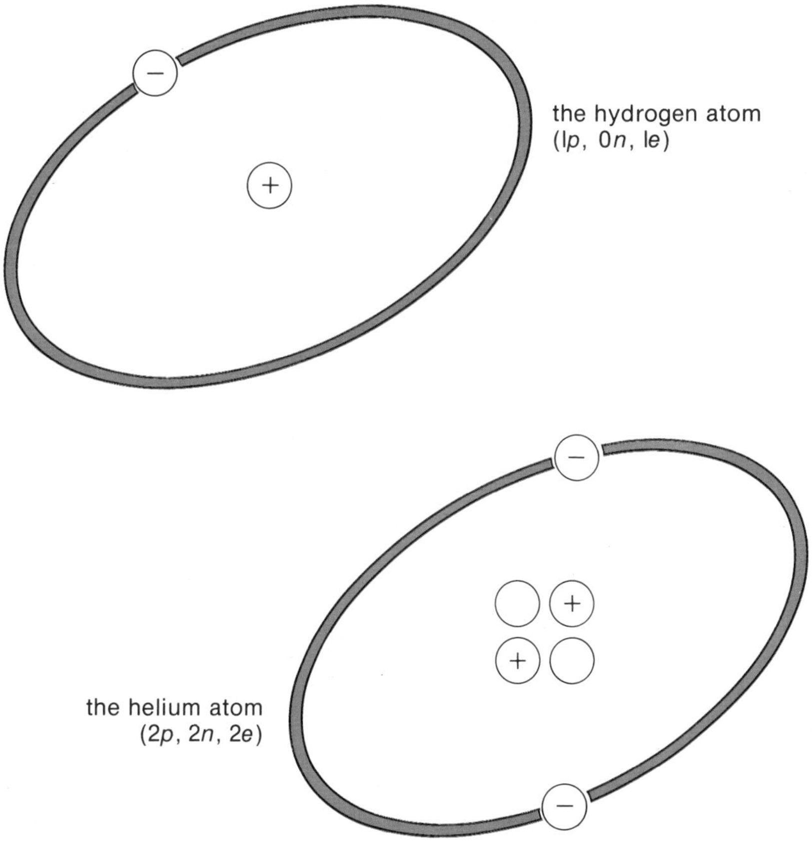

Figure 1

If the nucleus of an atom was enlarged to the size of a pea and placed on top of Nelson's Column, the electrons furthest away would be on the pavement

QUESTIONS

1 How many protons, neutrons and electrons are there in one (i) H atom, (ii) H^+ ion, (iii) Li atom, (iv) Li^+ ion?

2 Oxygen is the eighth element in the periodic table. How many protons and electrons are there in one (i) O atom, (ii) O^{2-} ion, (iii) O_2 molecule, (iv) H_2O molecule?

3 Lithium atoms have three protons, four neutrons and three electrons. Why is the relative atomic mass of lithium about seven?

4 Read this unit and the last one carefully.

(a) Make a poster showing the important dates, scientists and facts in the development of our ideas about atomic structure.

(b) How did our theory about atomic structure change between 1895 and 1935?

(c) How have (i) experiments, and (ii) predictions helped to develop our theories of atomic structure?

3 ATOMIC NUMBER AND MASS NUMBER

Only hydrogen atoms have one proton. Only helium atoms have two protons. Only lithium atoms have three protons, and so on. This shows that the number of protons in an atom decides which element it is. Because of this, scientists have a special name for the number of protons in the nucleus of an atom. They call it the **atomic number** (symbol Z). Thus, hydrogen has an atomic number of one (Z = 1), helium has an atomic number of two (Z = 2), lithium has an atomic number of three (Z = 3) and so on. Aluminium, the thirteenth element in the periodic table with 13 protons and 13 electrons, has an atomic number of 13.

Protons alone do not account for all the mass of an atom. Neutrons in the nucleus also contribute to the mass. Therefore, the *mass* of an atom depends on the number of protons and the number of neutrons added together. This number is called the **mass number** of the atom (symbol A). (See left).

Atomic number = number of protons.
Mass number = number of protons + number of neutrons.

Thus, hydrogen atoms (with one proton and no neutrons) have a mass number of one (A = 1). Helium atoms (two protons and two neutrons) have a mass number of four (A = 4) and lithium atoms (three protons and four neutrons) have a mass number of seven (A = 7). We can write the symbol ${}^{7}_{3}Li$ (Figure 1) to show the mass number and the atomic number of a lithium atom. The mass number is written at the *top* and to the left of the symbol. The atomic number is written at the *bottom* and to the left. A lithium ion is written as ${}^{7}_{3}Li^{+}$. A sodium atom (11 protons and 12 neutrons) is written as ${}^{23}_{11}Na$.

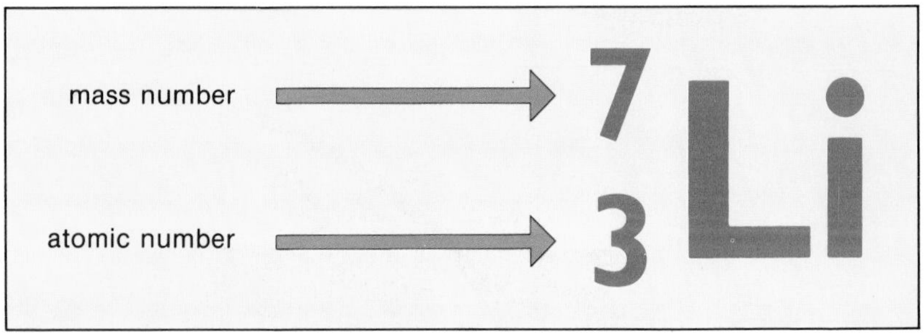

Figure 1

Isotopes

Many elements have relative atomic masses which are nearly whole numbers. For example, the relative atomic mass of nitrogen is 14.007 and that of sodium is 22.99. This is not surprising, since the mass of an atom depends on the mass of protons and neutrons in its nucleus and the relative mass of both particles is 1.00.

However, some elements have relative atomic masses that are nowhere near whole numbers. For example, the relative atomic mass of chlorine is 35.5 and that of copper is 63.5. At one time, chemists could not understand why the relative atomic masses of these elements were not close to whole numbers. F. W. Aston discovered the answer in 1919 when he built the first mass spectrometer. This instrument can compare the relative masses of atoms. Aston discovered that some elements contained atoms with different masses. When atoms of these elements were ionised and passed through a mass spectrometer, the beam of ions separated into two or more paths (Figure 2). This suggested that one element could have atoms with different masses. These *atoms of the same element with different masses are called* **isotopes**. Each isotope has a relative mass close to a whole number, but the average atomic mass for the mixture of isotopes is not always close to a whole number. This is studied further in the next unit.

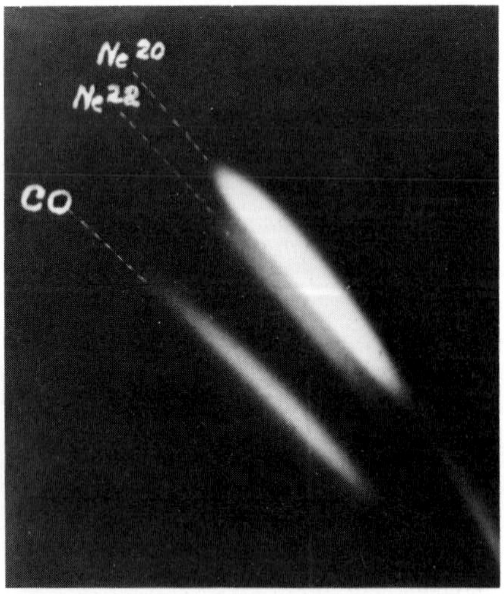

This photograph shows evidence for the two isotopes in neon, neon-20 and neon-22. Notice that the trace from neon-20 is much more prominent than that from neon-22. What does this tell you about the two isotopes?

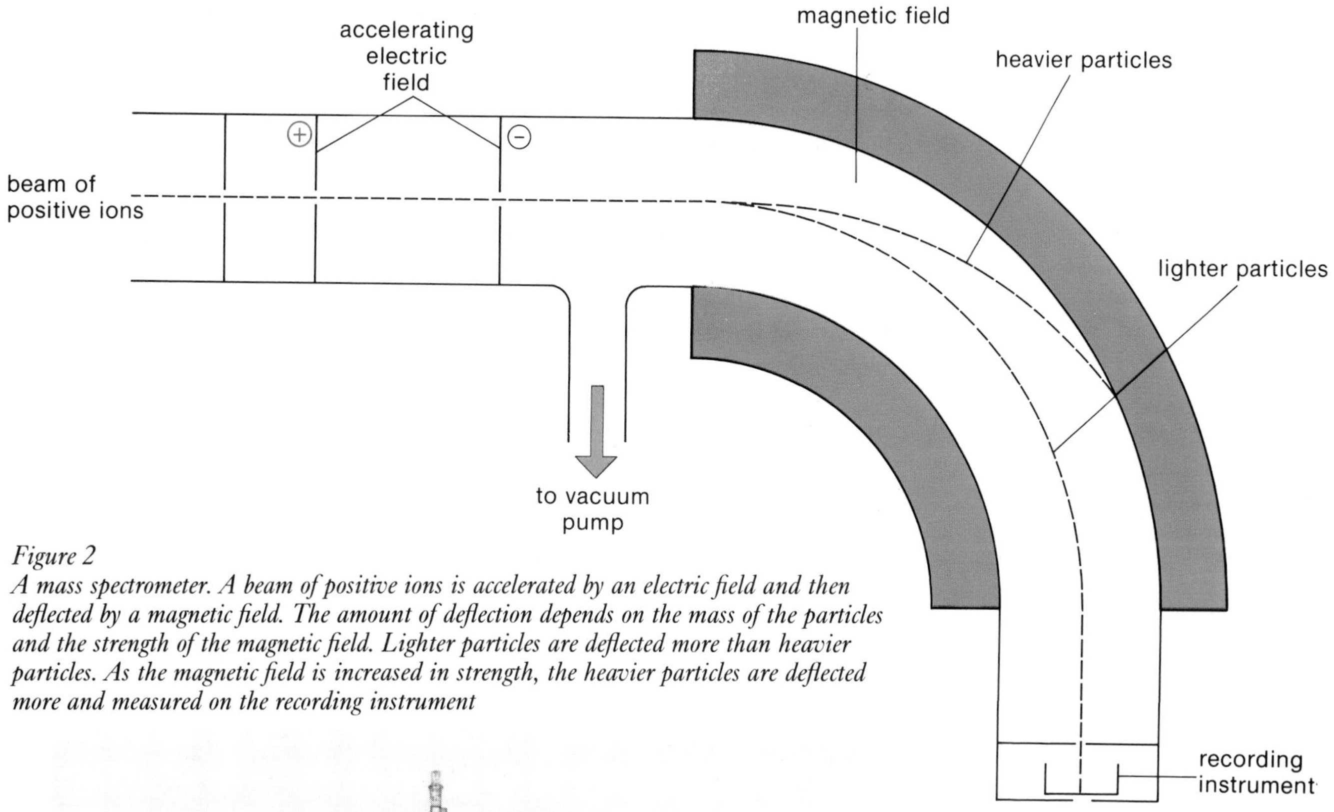

Figure 2
A mass spectrometer. A beam of positive ions is accelerated by an electric field and then deflected by a magnetic field. The amount of deflection depends on the mass of the particles and the strength of the magnetic field. Lighter particles are deflected more than heavier particles. As the magnetic field is increased in strength, the heavier particles are deflected more and measured on the recording instrument

Aston's original mass spectrometer. Positive ions were accelerated along the cylindrical metal tube at the top of the instrument. The coils of the electromagnet produced a magnetic field which deflected the particles

QUESTIONS

1 (a) How would you explain to a friend what we mean by (i) atomic number, (ii) mass number, (iii) isotope?
(b) How would you explain the way in which our understanding of atomic structure has changed since 1850?
2 (a) What is the atomic number of fluorine?
(b) How many protons, neutrons and electrons are there in one fluorine atom of mass number 19?
3 (a) What do 16,8,2– and O mean, in the symbol $^{16}_{8}O^{2-}$?
(b) How many protons, neutrons and electrons are there in one $^{23}_{11}Na^{+}$ ion?
4 Why do some elements have relative atomic masses which are *not* close to whole numbers?
5 Look at the photo on the left-hand page.
(a) What does the photo tell us about the two isotopes of neon?
(b) What does CO represent?
(c) How do you think the photo could be used to estimate the relative amounts of the two isotopes?

4 ISOTOPES

Isotopes have the same
Number of protons
Number of electrons
Atomic number
Chemical properties
Isotopes have different
Numbers of neutrons
Mass numbers
Physical properties

Table 1 The similarities and differences between isotopes of the same element

Isotopes are atoms of the same element with different masses. All the isotopes of one element have the same number of protons. Therefore, they have *the same atomic number.* As isotopes have the same number of protons, they must also have the same number of electrons. This gives them the same chemical properties because chemical properties depend upon the number of electrons in an atom and the way in which these electrons are transferred and shared during reactions.

Isotopes do, however, contain different numbers of neutrons. This means *they have different mass numbers*. For example, neon has two isotopes. Each isotope has 10 protons and 10 electrons and therefore an atomic number of 10. But one of these isotopes has 10 neutrons and the other has 12 neutrons. Their mass numbers are therefore 20 and 22 (Figure 1). They are sometimes called neon-20 and neon-22. These 2 isotopes of neon have the same chemical properties, but they have different physical properties because they have different masses. Samples of $^{20}_{10}Ne$ and $^{22}_{10}Ne$ therefore have different densities, different melting points and different boiling points. The similarities and differences between isotopes of the same element are summarised in Table 1.

	neon-20	neon-22
	$^{20}_{10}Ne$	$^{22}_{10}Ne$
number of protons	10	10
number of electrons	10	10
atomic number	10	10
number of neutrons	10	12
mass number	20	22

Figure 1
The two isotopes of neon

Chemists can obtain samples of uranium with a higher percentage of uranium-235 because of the different physical properties of uranium-235 and uranium-238. Uranium-235 is needed for use in nuclear reactors. Natural uranium contains only about 0.7% of uranium-235 and 99.3% of uranium-238. Nuclear reactors use uranium with 3 or 4% of uranium-235. The natural uranium is converted to uranium hexafluoride (UF_6) which is very volatile. The uranium hexafluoride is vaporised and allowed to diffuse through a porous solid. Particles of $^{235}UF_6$ are slightly lighter than those of $^{238}UF_6$. So they can move faster and diffuse faster than those of $^{238}UF_6$. After repeated diffusion through the porous solid, the uranium hexafluoride contains 3 or 4% $^{235}UF_6$. Finally, this is converted back to uranium and used as nuclear fuel.

Uranium ore being treated in giant tanks at the Rossing Mine in Namibia

Relative atomic mass

Most elements contain a mixture of isotopes. This explains why their relative atomic masses are *not* whole numbers. The relative atomic mass of an element is the average mass of one atom, taking account of its isotopes and their relative proportions. For example, the mass spectrometer trace in Figure 2 shows that chlorine consists of two isotopes with mass numbers of 35 and 37. These isotopes can be written as ${}^{35}_{17}Cl$ and ${}^{37}_{17}Cl$.

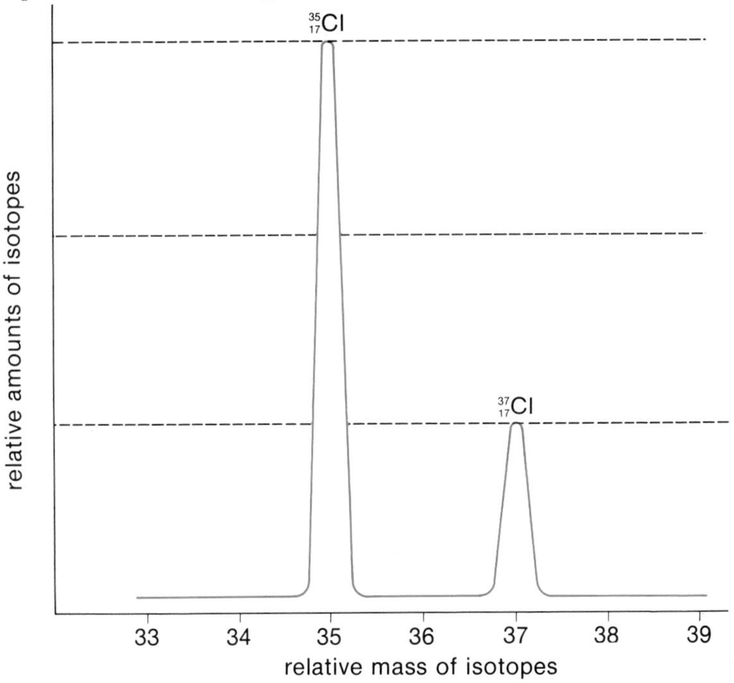

If chlorine contained 100% ${}^{35}_{17}Cl$, then its relative atomic mass would be 35. If it contained 100% ${}^{37}_{17}Cl$, then its relative atomic mass would be 37. A 50:50 mixture of ${}^{35}_{17}Cl$ and ${}^{37}_{17}Cl$ would have a relative atomic mass of 36. Figure 2 shows that naturally occurring chlorine contains three times as much ${}^{35}_{17}Cl$ as ${}^{37}_{17}Cl$, i.e. 75% to 25%. This gives a relative atomic mass of 35.5 as shown in Table 2.

Single fuel pins being inspected in a bundle of them, at Dounreay near Thurso in Scotland

Figure 2
A mass spectrometer trace for chlorine. What are the relative amounts of ${}^{35}_{17}Cl$,

Percentage of ${}^{35}_{17}Cl$	Percentage of ${}^{37}_{17}Cl$	Relative atomic mass
100	0	35
75	25	35.5
50	50	36
25	75	36.5
0	100	37

Table 2 The relative atomic mass of different mixtures of the isotopes of chlorine

QUESTIONS

1 Read the paragraphs describing the production of nuclear fuel from natural uranium, on the opposite page.
Draw a flow diagram to show the production of nuclear fuel containing 3 or 4% uranium-235 from natural uranium.

2 There are three isotopes of hydrogen with mass numbers of one, two and three. (Naturally occurring hydrogen is almost 100% ${}^{1}_{1}H$). How many protons, neutrons and electrons do each of the three hydrogen isotopes have?

3 Neon has two isotopes, with mass numbers of 20 and 22.
(a) How would you expect the boiling point of ${}^{20}_{10}Ne$ to compare with that of ${}^{22}_{10}Ne$? Explain your answer.
(b) Suppose a sample of neon contains equal numbers of the two isotopes. What is the relative atomic mass of neon in this sample?
(c) Neon in the air contains 90% of ${}^{20}_{10}Ne$ and 10% of ${}^{22}_{10}Ne$. What is the relative atomic mass of neon in the air?

4 Discuss the following questions with two or three others.
(a) Why do isotopes have the same chemical properties, but different physical properties?
(b) Why do samples of natural uranium from different parts of the world have slightly different relative atomic masses?
(c) Why can chlorine form molecules with three different relative molecular masses?

5 RADIOACTIVITY

Henri Becquerel – the discoverer of radioactivity

1896 Becquerel discovers radioactivity

In 1896, the Frenchman Henri Becquerel was investigating the reactions of uranium salts. Most of his experiments were carried out in bright sunlight. By chance, Becquerel left some uranium salt *in the dark* and on a photographic plate wrapped in black paper. When he developed the photographic plate, Becquerel was very surprised to see that it had been darkened **(fogged)** in the area near the uranium salt. Further experiments showed that the uranium salt was giving off some kind of radiation. Becquerel called the process **radioactivity** and described the uranium salts as **radioactive**. He also suggested that the radiation was a form of energy like light. This would explain why it caused a reaction on the photographic plate.

1898 Marie Curie discovers radium

Two of Becquerel's colleagues, Marie and Pierre Curie, decided to examine the radioactivity of uranium salts in more detail. They found that all uranium salts showed radioactivity and affected photographic plates. They also found that radioactive substances would either increase or decrease the charge on an electroscope. This meant that the radiation itself must carry a charge. During their investigations the Curies noticed that pitchblende (impure uranium sulphide) was much more radioactive than they had expected from its uranium content. This led them to think that the pitchblende contained an element more radioactive than uranium.

Marie Curie (1867–1934) and her husband Pierre (1859–1906). The Curies spent nearly four years isolating the radioactive elements radium and polonium from pitchblende. In 1903 they shared the Nobel Physics Prize with Becquerel. Then in 1911, Marie was awarded the Chemistry Prize – the first person to win two Nobel Prizes. In 1934, Marie died of cancer, probably caused by the radioactive materials she had spent her life studying

In 1898, after months of hard work, Marie Curie isolated two other radioactive elements from pitchblende. She named these elements radium (after the term 'radioactivity') and polonium (after Poland where she was born). Radium was found to be two million times more radioactive than uranium.

The fogging of a photographic plate by radioactive materials has practical applications today. For example, we can follow the way in which a plant leaf takes in carbon dioxide by exposing the leaf to carbon dioxide containing radioactive carbon-14. The leaf is then held against a photographic plate. When the plate is developed, the darkest parts show where the carbon-14 has gone.

Radioactivity can be used to study the uptake and distribution of metal ions in plants. In the picture, the plant has been grown in a solution of a radioactive mercury compound for 48 hours. The radioactive mercury can then be detected in the stem and leaves by laying the plant on photographic film and allowing sufficient exposure time

Before very long, another effect of radioactive substances was discovered. When a screen coated with zinc sulphide is placed near a radium salt and examined with a magnifying lens, tiny flashes of light appear on the surface of the zinc sulphide. The flashes of light are called **scintillations**. When particles hit the zinc sulphide their kinetic energy is converted to light which causes the scintillations. The particles are produced as the radioactive radium atoms break up.

The radiations and particles emitted by radioactive substances were originally detected in one of three ways:

1 the fogging of photographic plates,
2 the discharge of an electroscope, or
3 scintillation methods.

Nowadays, the radiations are detected using a Geiger-Müller tube (GM tube) connected to a counter or ratemeter.

QUESTIONS

1 (a) Why are photographic plates darkened ('fogged') when they are left in the dark near uranium salts?
(b) Why did the Curies think that pitchblende contained an element more radioactive than uranium?
(c) Why do radioactive materials either increase or decrease the charge on an electroscope?
2 (a) What are scintillations?
(b) What causes the scintillations when a radium salt is placed near zinc sulphide?
3 (a) Find out what you can about the personal life of Marie Curie.
(b) Find out what you can about the scientific work of Marie Curie.
(c) Why do you think that Marie Curie is one of the most admired scientists of the twentieth century?
4 How did unexpected results in their work stimulate Becquerel and the Curies into further experiments?

6 NUCLEAR REACTIONS

Alchemists dreamed of changing base metals to gold

Mixtures of zinc sulphide and a small amount of any radium salt glow in the dark. These mixtures are used in luminous paints on the hands and numbers of clocks and watches. We now know that the glow is caused by **alpha particles** hitting the zinc sulphide in the paint. This stops the alpha particles and their kinetic energy is converted to light energy. The alpha particles are produced as radium atoms split up (disintegrate) of their own accord. All the time, radium atoms are breaking up and losing alpha particles. This spontaneous break-up of atoms results in **radioactivity (radioactive decay)**. Alpha particles are helium ions, ${}^{4}_{2}He^{2+}$ – helium atoms that have lost both of their electrons. The mass number of an alpha particle is four and its atomic number is two. Thus, when an atom of ${}^{226}_{88}Ra$ loses an alpha particle, the fragment left behind will have a mass number which is four less than ${}^{226}_{88}Ra$ and an atomic number which is two less than ${}^{226}_{88}Ra$. So the fragment will have a mass number of 222 and an atomic number of 86. All atoms of atomic number 86 are those of radon, Rn. Thus, the final products are ${}^{222}_{86}Rn$ and ${}^{4}_{2}He$. The nuclear decay for radium-226 can be summarised in a nuclear equation as

$${}^{226}_{88}Ra \rightarrow {}^{222}_{86}Rn + {}^{4}_{2}He$$

Radioactive radium compounds are used in the luminous paint on the hands and numbers of clocks and watches

Radioactive decay with the loss of an alpha particle is common for large isotopes with an atomic number over 83. These include uranium-238, radium-226 and plutonium-238. These isotopes decay because they are simply too heavy. They lose mass and try to become stable by losing an alpha particle.

Radioactive isotopes with atomic numbers below 83 usually lose **beta particles** when they decay, rather than alpha particles. Experiments show that beta particles are electrons ${}^{0}_{-1}e$. Carbon-14 (${}^{14}_{6}C$) decays by losing beta particles to form ${}^{14}_{7}N$. The decay process can be represented as

Mass →
$${}^{14}_{6}C \rightarrow {}^{14}_{7}N + {}^{0}_{-1}e$$
Charge →

Notice that the *total* mass and the *total* charge are the same after beta decay as they were before. During beta decay, a neutron in the nucleus of the radioactive atom splits up into a proton and an electron. The proton stays in the nucleus, but the electron is ejected as a beta particle. Thus, the mass number of the remaining fragment stays the same, but its atomic number increases by one (Table 1).

	Isotopes involved	Particle lost	Change in mass number	Change in atomic number
Alpha decay	Atomic number usually >83	${}^{4}_{2}He^{2+}$	−4	−2
Beta decay	Atomic number usually ≤83	${}^{0}_{-1}e$	0	+1

Table 1 Comparing alpha decay and beta decay

Chemical reactions and nuclear reactions

Chemical reactions involve changes in the outer parts of atoms, the electrons. During chemical reactions electrons are either

1 *transferred* from one atom to another, or
2 *shared* between two atoms.

Nuclear reactions, on the other hand, involve changes in the central parts of atoms, the nucleus. During nuclear reactions, one element may be converted to another element by

1 radioactive decay,
2 atomic fission (unit 9), or
3 atomic fusion (unit 12).

The first scientist to show that one element could be converted to another was Rutherford. In the 1920s, Rutherford and his colleagues turned sodium into magnesium and aluminium into silicon. At first, the changes from one element to another involved only elements with low mass numbers. However, in 1940, American chemists managed to build up heavier elements from uranium, the heaviest natural element. For example, neptunium-239 and plutonium-239 were obtained by bombarding uranium-238 with neutrons. By 1972, thirteen new elements had been synthesised, all of which come after uranium in the periodic table. All these new elements are radioactive and most of them disintegrate rapidly.

QUESTIONS

1 Explain the following words.
radioactivity, alpha particles, beta particles, nuclear reactions
2 Uranium-238 (${}^{238}_{92}U$) decays to give an alpha particle and a new atom.
(a) What is the mass number of the new atom?
(b) What is the atomic number of the new atom?
(c) What is the symbol of the new atom?
(d) Write an equation for the decay of uranium-238.
3 Write nuclear equations for (i) the beta decay of ${}^{43}_{19}K$, ${}^{24}_{11}Na$ and ${}^{131}_{53}I$, (ii) the alpha decay of ${}^{212}_{84}Po$ and ${}^{237}_{93}Np$.
4 Look up the names and symbols of the elements after uranium in the periodic table. Most of these elements were first synthesised in the USA, the other two or three in Russia.
(a) Name four elements that you think were first synthesised in the USA.
(b) Name two elements that you think were first synthesised in Russia.
(c) Suggest why the particular name was chosen for four of these elements.
(d) Why do you think these elements are radioactive?
5 How do nuclear reactions and chemical reactions differ?
6 What precautions should be taken in experiments and work with radioactive isotopes?
7 Why is radioactive waste not destroyed by incineration?

7 DETECTING RADIOACTIVITY

When radioactive substances decay, they emit alpha particles ($^{4}_{2}He^{2+}$ ions) and beta particles (electrons). At the same time, they usually give off **gamma rays**. Gamma rays are electromagnetic waves, like light. Their properties are very similar to those of X-rays. Gamma rays contain so much energy that they can damage cells and kill organisms. As a radioactive atom gives off gamma rays, it loses energy and becomes more stable.

When alpha particles, beta particles and gamma rays pass through air and other materials, they cause particles in the materials to lose electrons and form ions. Because of this, they are sometimes called **ionising radiations**. Although ionising radiations can be harmful, they do have beneficial uses. These uses are discussed in unit 8.

When alpha particles are emitted by radioactive substances, they travel a few centimetres in air and they can be deflected by electric and magnetic fields. Beta particles are much lighter than alpha particles. They travel several metres in air and are deflected much further by electric and magnetic fields. Gamma rays are unaffected by electric and magnetic fields, but they travel a long way in air and penetrate bricks and metal sheets. The properties of alpha particles, beta particles and gamma rays are summarised in Table 1 and Figures 1 and 2.

Table 1 The nature and properties of alpha particles, beta particles and gamma rays

Radiation	Nature	Effect of electric and magnetic fields	Distance travelled in air	Penetration of		
				paper	thin aluminium	thick lead
Alpha	helium nuclei, $^{4}_{2}He^{2+}$	small deflection	a few cms	×	×	×
Beta	electrons, $^{0}_{-1}e$	large deflection	a few metres	√	×	×
Gamma	electromagnetic waves	no deflection	many kms	√	√	×

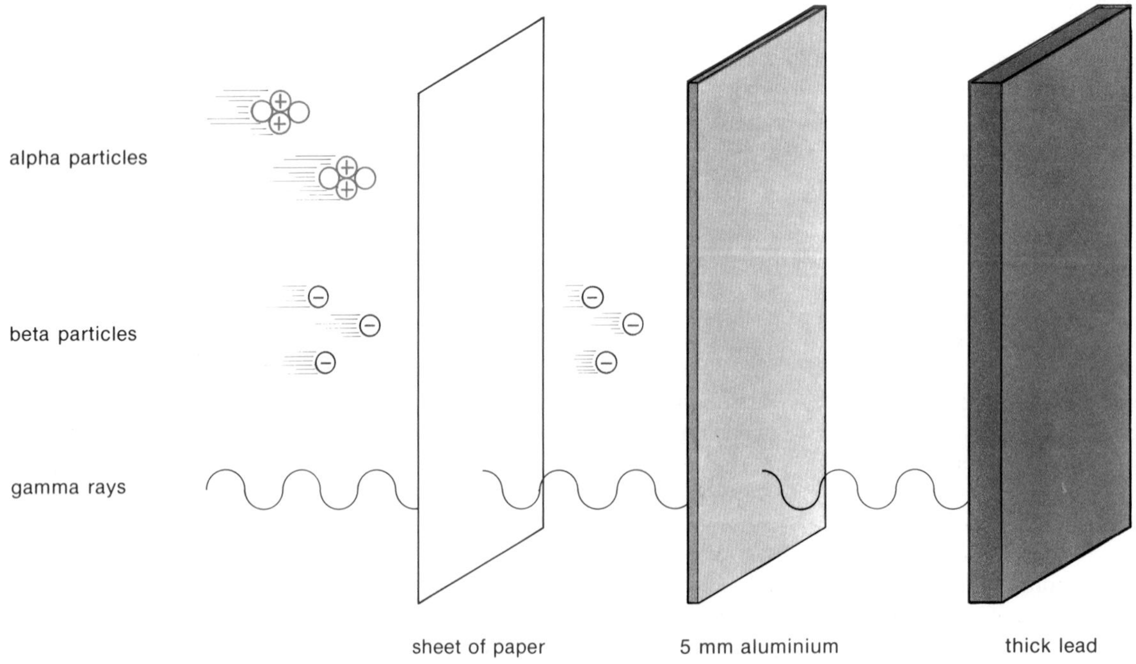

Figure 1 The relative penetrating power of alpha particles, beta particles and gamma rays

Background radiation

Various rocks in the Earth, including granite, contain small amounts of radioactive uranium, thorium and potassium compounds. Our bodies also contain traces of radioactive materials as do the bricks which are used to build our homes, schools and factories. In addition to these sources, we are also exposed to radiation from the Sun. These four *natural* sources of radiation are called **background radiation**. We are exposed to this background radiation all our lives. Normal background radiation is very low. It causes no health risk.

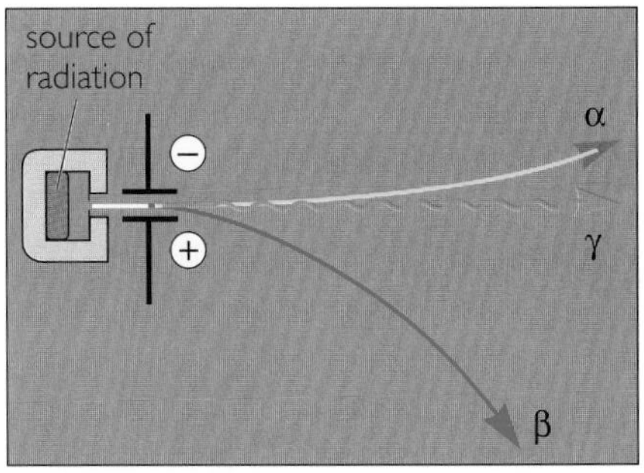

Figure 2
The effect of an electric field on different ionising radiations

Half life

The rate of decay of a radioactive isotope is shown by its **half life**. *This is the time it takes for half of the atoms of the isotope to decay*. Half lives can vary from a few milliseconds to several million years. The shorter the half life, the faster the isotope decays and the more unstable it is. The longer the half life, the slower the decay process and the more stable the isotope. Uranium-238 with a half life of 4500 million years is 'almost stable', but polonium-234, with a half life of only 0.15 milliseconds is quite the reverse.

The radioactive isotope iodine-131 has been used to measure the uptake of iodine by the thyroid, an important gland in the human body.
Iodine-131 has a half life of eight days. This means that, starting with 1 g of iodine-131, only ½ g remains eight days later. After another eight days, the half gram will have decayed to ¼ g and eight days after that only ⅛ would remain (Figure 3). In fact, only tiny quantities of iodine-131 are used in this procedure. By measuring the amount of radioactivity in the thyroid gland, doctors can tell how much of the iodine-131 has been taken up by the gland.

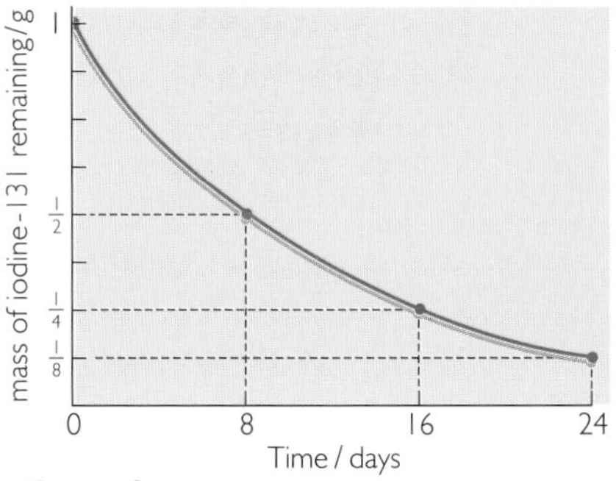

Figure 3
A radioactive decay curve for I-131. The graph shows the mass of I-131 remaining when 1 g of the isotope decays for 24 days

QUESTIONS

1 In the UK, an average of 78% of the total radiation received by a person comes from background radiation. This is composed of 16% from rocks and soil, 16% from natural radioactive substances in our bodies, 33% from bricks and other building materials and 13% radiation from the Sun.
(a) Use these figures to draw a pie chart showing where the total radiation comes from.
(b) What percentage of the total radiation to which we are exposed comes from artificial sources?
(c) What do you think are the main sources of this artificial radiation?

2 The figure shows a gamma ray unit used in a hospital. **A** is a container of radioactive material. The unit is made 'safe' by moving the container to position **B**. The radioactive material is cobalt-60 ($^{60}_{27}Co$). Cobalt-60 has a half life of 5 years.

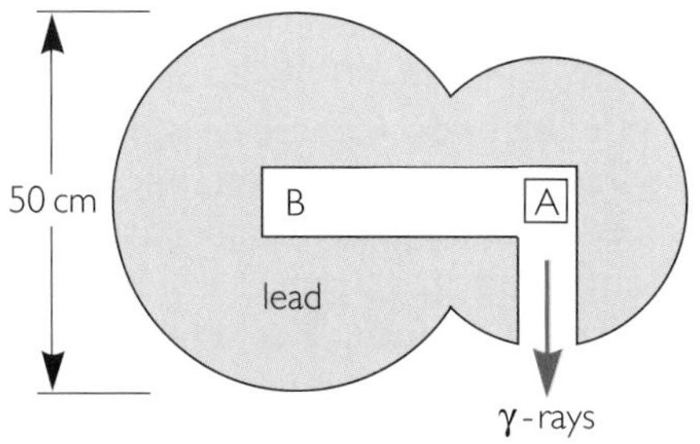

(a) Why is the unit made mainly of lead?
(b) What is (i) the atomic number, (ii) the mass number of $^{60}_{27}Co$?
(c) How many protons, neutrons and electrons are there in one cobalt-60 atom?
(d) Why is the gamma ray unit 'safe' when the container is in position **B**?
(e) Suggest *three* safety precautions which someone should take when removing a container of cobalt-60 from the unit?

3 A medical gamma ray unit was supplied with 10 g of caesium-137 which has a half life of 35 years.
(a) How much caesium-137 would be left after (i) 35 years, (ii) 70 years, (iii) 140 years?
(b) When will its rate of decay be half as great as it was at the start? Explain your answer.
(c) Suppose the medical unit is closed after 20 years. How much caesium-137 is left at this time? (Hint: it may help you to draw a graph.)

4 (a) What do you understand by the term 'radioactive element'?
(b) Name two radioactive elements which occur naturally.
(c) Which part of an atom is responsible for radioactivity?
(d) Which of the particles or rays emitted by radioactive substances (i) is most penetrating, (ii) contains positive particles, (iii) is not deflected by a magnetic field?

8 USING RADIOACTIVE ISOTOPES

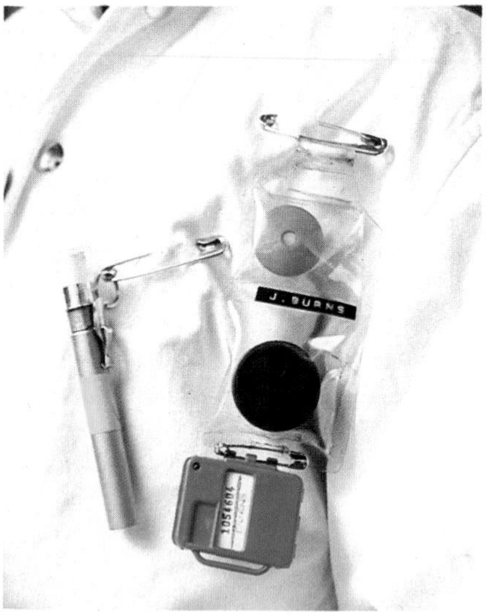

A technician wearing a blue radiation-level badge

Radioactive isotopes are widely used in industry and medicine. People who work with radioactive materials must wear a radiation badge which measures the radiation to which they have been exposed. Gamma rays are the most penetrating form of radiation. They can cause changes in the structure of chemicals in our bodies and kill living cells. Scientists and technicians who use dangerous isotopes must be protected by lead or concrete shields and handle radioactive materials by remote control.

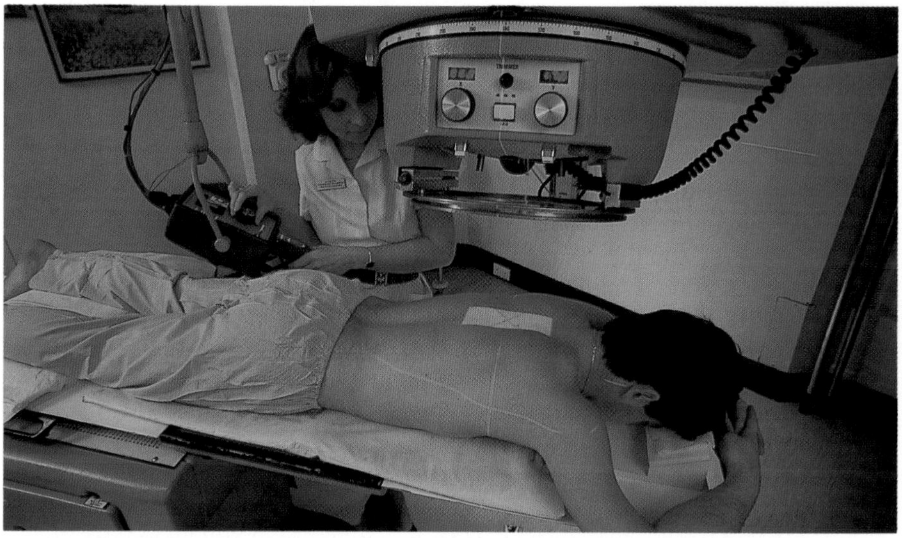

This patient has cancerous cells inside his body. He is being treated using gamma radiation from cobalt-60, which penetrates the body and kills the cancer cells.

Medical uses

When a living cell is exposed to radiation, the structure of chemicals (genes) in the nucleus may be changed. This can cause the cell to die. Penetrating gamma rays from cobalt-60 ($^{60}_{27}Co$) are used to kill cancer cells and to treat growths inside the body. Cancer cells on the surface of the body, as in skin cancer, can be treated with less penetrating beta rays. This is done by strapping a plastic sheet containing $^{32}_{15}P$ (radioactive phosphorus) or $^{90}_{38}Sr$ (radioactive strontium) on the affected area. Cancer can be fatal, but doctors are becoming more and more successful at curing less serious cases. Medical items, such as dressings and syringes, are often sealed in polythene bags and sterilised by gamma rays. This method of sterilisation is much easier than the old method which used steam. Intense doses of gamma radiation sterilise the articles by killing any bacteria on them.

Tracer studies

Radioactive isotopes are easy to detect. Therefore they are used to *trace* what happens to different substances in chemical, physical and biological processes. Radioactive isotopes are usually mixed with non-radioactive atoms of the substance under investigation. For example, the way that plants take in and use phosphates can be studied using a fertiliser containing $^{32}_{15}P$. Tracer studies using $^{14}_{6}C$ have helped in the study of photosynthesis and protein synthesis. The activity of the thyroid gland can be studied by measuring the uptake of iodine-131 (unit 7 of this section).

Archaeological uses

The common isotope of carbon is carbon-12. The carbon in living things is therefore mainly carbon-12 with a small constant percentage of *radioactive* carbon-14. This gets into living things from the carbon-14 which is part of the carbon dioxide in the air. When an animal or plant dies, the carbon-14 in it continues to decay. However, the replacement of decayed carbon-14 from food and carbon dioxide stops. The amount of carbon-14 left in the remains of the animal or plant can be measured. Then, knowing the half life of carbon-14, it is possible to work out how long it is since the animal or plant died.

'Carbon dating' has been used to check the age of ancient documents and the bones of ape men. This method was used to show that the famous Dead Sea Scrolls were in fact 2000 years old and that the skull of the Piltdown Man was a fake.

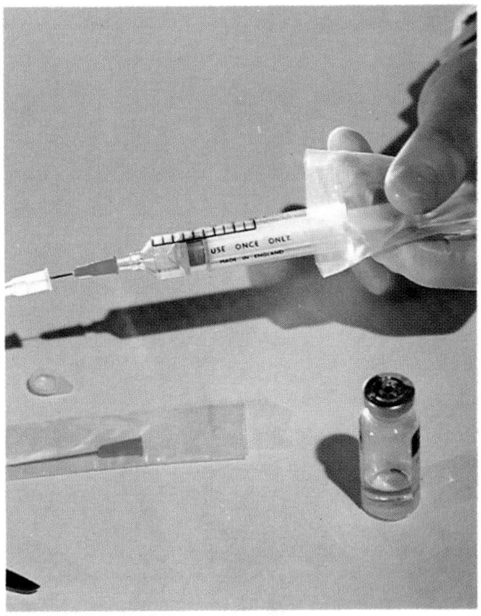

Syringes and other medical items are sterilised using gamma radiation. In some countries, gamma radiation is used to prevent food going bad

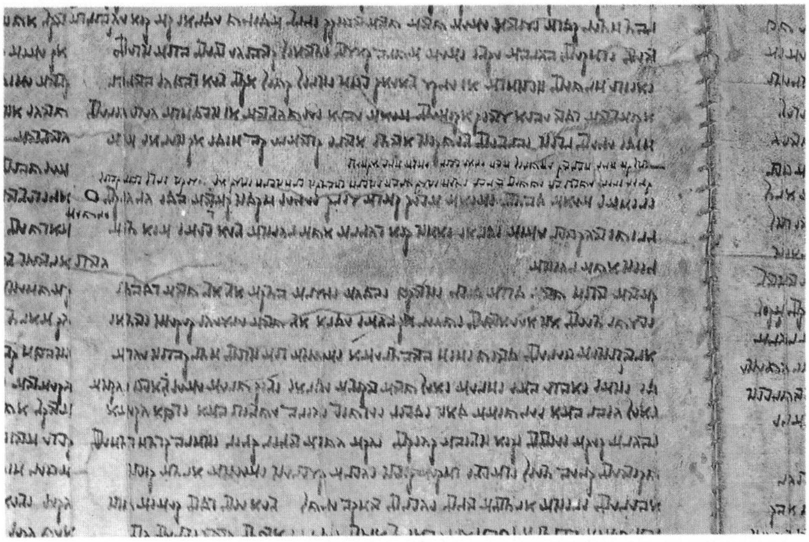

Part of the Dead Sea Scrolls. Carbon dating shows that the Dead Sea Scrolls were probably authentic

QUESTIONS

1 What precautions are taken to ensure that scientists and technicians who work with radioactive isotopes are not exposed to dangerous radiations?

2 Suppose that carbon-14 makes up $x\%$ of the carbon in living things and that the half life of carbon-14 is 5700 years.
(a) How long will it take for the percentage of carbon-14 to fall from (i) $x\%$ to $\frac{1}{2}x\%$, (ii) $\frac{1}{2}x\%$ to $\frac{1}{4}x\%$?
(b) How old is an object which has $\frac{1}{8}x\%$ of its carbon as carbon-14?

3 Why are radioactive isotopes important?

4 In some countries, gamma radiation is used to prevent food going bad.
(a) What will gamma radiation do to bacteria to prevent the food going bad?
(b) This process is not yet used in the UK. Why do you think that some people are opposed to its use?

5 The Holy Shroud of Turin is a sheet of cloth. Some people believe that the body of Jesus was wrapped in it after his crucifixion. The Shroud has been venerated through the centuries. It is carefully protected and shown to the public only on rare occasions. The last time it went on show was in 1978. At that time, Pope John Paul II was among the pilgrims who went to Turin to see it.

In 1978, a request was made to test the age of the Holy Shroud by scientific methods. In October 1986, Pope John Paul II gave permission for carbon tests to be carried out on it. The tests showed that the Shroud was only about four hundred years old. It could not be the authentic sheet which covered the body of Jesus.
(a) What is the meaning of (i) authentic, (ii) venerated, (iii) pilgrims?
(b) Describe how you think the scientists found the age of the Shroud.
(c) How would their results have been different if the Shroud had been 2000 years old?
(d) Do you think that it is right to carry out scientific tests on holy objects like the Shroud?

9 NUCLEAR ENERGY

Enormous amounts of energy can be obtained from nuclear reactions, but this discovery was made almost by chance. In 1938, the German scientists Hahn and Strassmann were trying to make a new element by bombarding uranium with neutrons. Instead of producing a new element, the neutron caused the uranium nucleus to break up violently into smaller nuclei. Two or three separate neutrons were also released together with large quantities of energy (Figure 1).

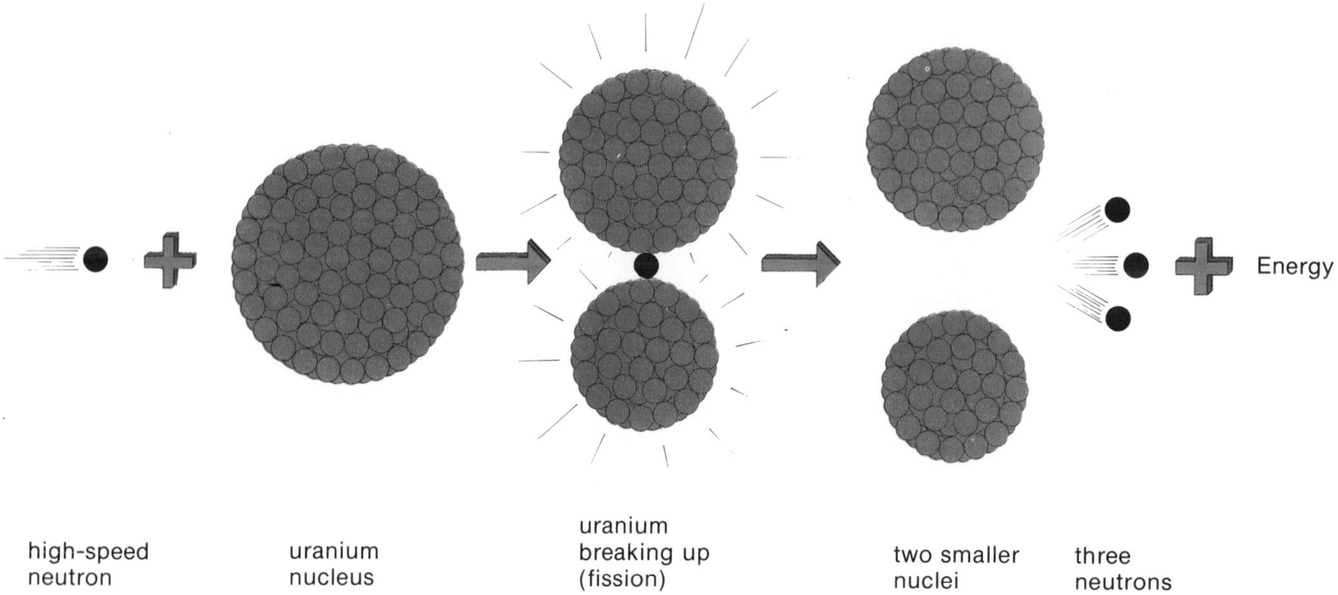

Figure 1
The fission of a nucleus of uranium-235

Notice how this nuclear reaction differs from radioactive decay. First, it does not happen of its own accord like radioactive decay. It only happens when the uranium is bombarded by neutrons. Second, it involves the break-up of one large nucleus into two fragments of roughly the same size. During radioactive decay, the products are one large fragment and one very small fragment (either an alpha particle or an electron). The special name **nuclear fission** (atomic fission) is used to describe the splitting of an atom into two fragments of roughly the same size.

Atomic bomb explosions release vast amounts of energy

Atomic bombs and atomic reactors

Natural uranium contains two isotopes, $^{235}_{92}U$ and $^{238}_{92}U$. Only 0.7% is uranium-235. Experiments showed that only uranium-235 took part in nuclear fission during neutron bombardment. This led scientists to realise that if the uranium sample contained a larger fraction of uranium-235, the neutrons released during fission would split more uranium nuclei and cause a chain reaction.

Figure 2 shows how an exploding chain reaction occurs in uranium-235. One reaction releases three neutrons, three neutrons release nine, then 27, 81 and so on. Each time more and more energy is produced as more and more uranium-235 atoms undergo fission.

In an atomic bomb, the fission of *enriched* uranium-235 happens in an *uncontrolled* manner and enormous amounts of energy and radiation are released. In an atomic reactor, a *mixture* of uranium-235 and uranium-238 is used to give a *controlled* chain reaction. In this case, the heat produced is used to generate electricity. Figure 3 shows a simplified diagram of a gas-cooled nuclear reactor. At the centre of the reactor, rods of 'enriched' uranium containing 3% of uranium-235 are stacked inside a large block of graphite. Neutrons released by the fission of the uranium are slowed down by collision with carbon atoms in the graphite.

The temperature of the reactor is controlled by moveable rods of boron or cadmium which absorb neutrons. The deeper these rods are inserted, the more neutrons are absorbed and the slower the reaction becomes. By carefully adjusting the neutron-absorbing rods, a controlled chain reaction can be obtained. Heat is produced steadily and taken away by carbon dioxide gas circulating through the reactor. The hot gas is then used to make steam which drives turbines and generates electricity. In the USA, most reactors use water as a coolant in place of carbon dioxide.

Recently, reactors have been developed that use plutonium as the fuel and liquid sodium as the coolant. Unlike uranium, which reacts best with slow neutrons, plutonium can use *fast* neutrons. This means that the reactor does not need a graphite block to slow them down. These reactors are called fast reactors. The first power station to use a fast reactor for generating electricity was the one at Dounreay in Scotland.

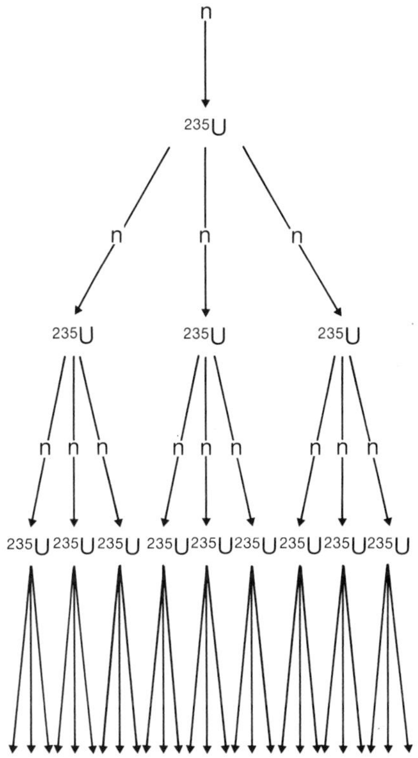

Figure 2
A chain reaction in uranium-235

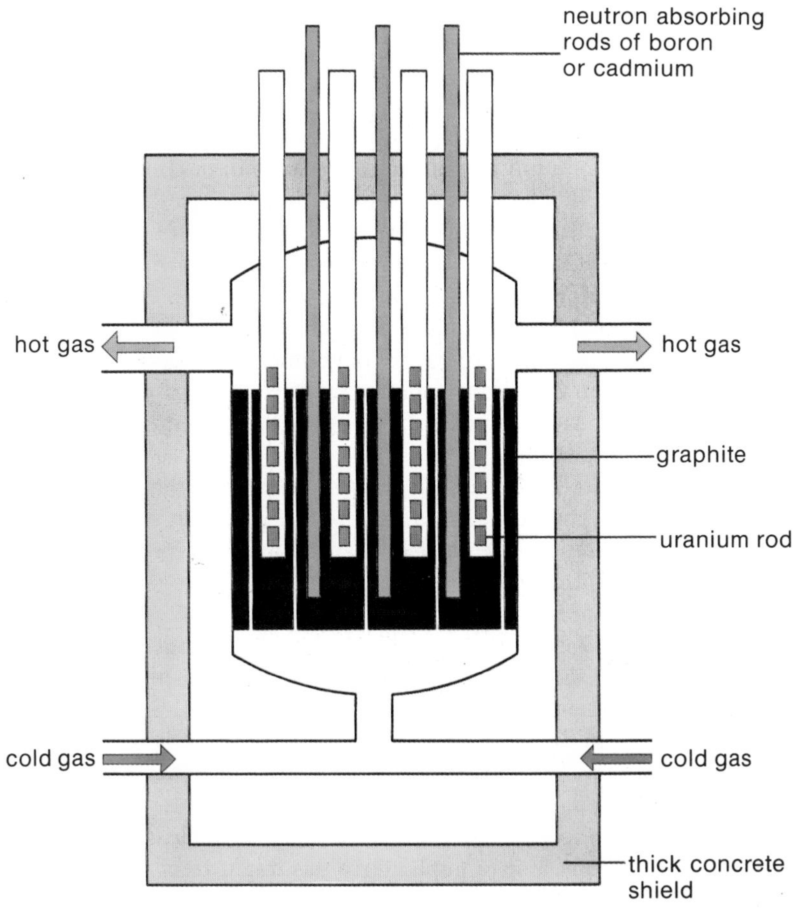

Figure 3
A simplified diagram of a gas-cooled nuclear reactor

QUESTIONS

1 Explain the following terms in your own words:
nuclear fission, nuclear energy, chain reaction, fast reactor?

2 What are the differences between nuclear fission and radioactive decay?

3 In what ways are nuclear reactors similar to nuclear bombs? In what ways are they different?

4 Read the section on nuclear reactors.
(a) Why are the uranium rods sunk in a graphite block?
(b) What is the function of the boron or cadmium rods?
(c) How is the rate of nuclear fission controlled?
(d) Why is the reactor surrounded by thick concrete?
(e) How is the energy from nuclear fission converted into electricity?
(f) Recently, some reactors have used liquid sodium as the coolant. Mention one advantage and one disadvantage of using sodium.

SECTION K: THINGS TO DO

Activities

The nuclear debate

Hiroshima, in Japan, a few weeks after the atomic bomb exploded in August 1945. The long wooden building in the foreground was, of course, erected after the blast

CND marchers carrying banners and CND slogans. During the 1960s, 1970s and 1980s, the USA and the USSR continued to develop more and more powerful nuclear weapons. This led to more support for CND and the anti-nuclear movement

Hardly a week passes without some mention of nuclear technology in the newspapers or on television. A lot of the publicity concerns the dangers involved. These include pollution by radioactive waste and, of course, the horrors of nuclear war.

The first atomic reactor was built in America in 1942 by the Italian scientist, Enrico Fermi. The first industrial nuclear power station was opened in 1956 at Calder Hall, now Sellafield, in Cumbria, UK.

The first nuclear (atomic) bomb was tested in July 1945. Robert Oppenheimer, who had played a large part in the bomb's development, was shocked by its enormous power. Afterwards, he said, 'we knew the world would not be the same'. A month later a nuclear bomb was dropped on Hiroshima in Japan, killing over 70 000 people.

Oppenheimer was opposed to the development of the hydrogen bomb, which was even more destructive than the nuclear bomb. Nevertheless, a hydrogen bomb was tested in the Pacific Ocean in 1953. Many people were worried about the possibility of such weapons being used in a war. Large numbers of people joined CND (the Campaign for Nuclear Disarmament).

As more nuclear reactors are built, there is an increasing danger of pollution by radioactive waste. After use, the spent fuel is still very radioactive. Some of the waste material will remain dangerous for hundreds of years. Several methods have been used to dispose of radioactive waste. These include dumping the waste in deep mines or solidifying it in glass or concrete and then burying it below the sea bed.

Many people are also concerned about the link between nuclear power stations and nuclear weapons. For example, terrorists might steal uranium or plutonium and use it to make bombs. There is also the risk that terrorists might capture a nuclear power station and use it as a bargaining device by threatening to destroy the reactor.

In 1986, a nuclear reactor exploded at Chernobyl in Russia. 31 people died in the explosion and 135 000 people evacuated from their homes. Clouds of radioactive materials were carried by winds to all parts of Europe. The contamination led to restrictions on the sale of meat and vegetables as far away as Scotland and Wales.

The main causes of the accident at Chernobyl were serious faults in the reactor design. This type of reactor is used only in the Soviet Republics. As a result of the accident, they have improved the design and the safety measures in these reactors. A reactor like the one at Chernobyl could not be built in the UK because it does not meet our safety requirements.

Many people feel that the problems created by nuclear technology outweigh the advantages. Other people say that we must learn to live with the risks if we are to maintain our industrial activity, preserve our standards of welfare and improve the quality of life in the Third World.

Any discussions on nuclear technology are always linked to energy supplies. The world consumption of coal, oil and gas will continue to increase. This is particularly so in developing countries. We must find another large source of energy and the only possibility seems to be nuclear power. Solar energy, tidal energy and wind energy may help in certain areas, but none of them could provide the huge amounts of energy that we need.

Unfortunately, the reserves of uranium are also limited. Some experts think that there is not enough for the next century. Eventually, we may be forced to build more fast reactors that use plutonium rather than uranium. These can extract 50 to 60 times more energy from the fuel than ordinary uranium reactors.

Some radioactive waste used to be sealed in drums and then dumped at sea. There has been a worldwide agreement to stop this. Instead, waste is to be buried deep under ground on land

When the nuclear reactor exploded at Chernobyl in Russia, radioactive material was carried on the wind all over Europe. Many people say accidents like this make nuclear power too risky, but others believe we will not be able to supply the world's future energy needs without it

1 (a) What methods are used to dispose of waste from nuclear power stations?
(b) Why are such thorough methods of disposal needed?

2 Why do some people think that we will have to rely on nuclear energy in the future?

3 Design a poster to show the main dangers associated with nuclear technology.

4 Is nuclear power the answer to our energy problems or are the risks too great? What do you think?
Plan a questionnaire and a survey to find out what other people think.
(a) Decide the questions for your survey.
(b) Carry out your survey.
(c) Prepare a report of your survey.

Study Questions

1 Question 1 concerns the five particles labelled *A* to *E* below.

A alpha particle
B beta particle
C hydrogen atom
D proton
E neutron

Which of these particles
(a) would be attracted by a positive plate?
(b) is the heaviest?
(c) causes fission of uranium-235?
(d) could be used in a thickness gauge for polythene or paper?
(e) contains the same number of protons and electrons?

2 Dalton put forward his Atomic Theory in 1807. The main points in his theory are listed below.

- All matter is made up of tiny particles called atoms.
- Atoms cannot be made or broken apart.
- All the atoms of one element are exactly alike.
- Atoms of one element are different from those of another element.
- Atoms combine in small numbers to form molecules.

Study each of these five points in turn.
(a) Say whether each point is still true or whether modern knowledge about atoms, molecules and atomic structure have shown them to be inaccurate.
(b) If a point is now inaccurate, explain why.

3 When a stream of alpha particles is directed at thin aluminium foil, most of the alpha particles pass straight through. The remaining alpha particles are either deflected or seem to bounce back off the foil. If the energy (speed) of the alpha particles is increased, some of them are absorbed by the aluminium.
(a) Why do you think most alpha particles pass straight through the foil?
(b) Why do you think some alpha particles are deflected as they pass through the foil?
(c) Why do you think some alpha particles seem to bounce back off the foil?
(d) What do you think happens to alpha particles which are absorbed by the aluminium?
(e) Why do you think the alpha particles must have high energy before they are absorbed by the foil?
(f) Write a nuclear equation for the process which occurs when high-energy alpha particles are absorbed by aluminium.

4 Radioactivity can be used in hospitals to investigate inside a patient's body without having to cut the person open. Technetium-99 (^{99}Tc) is one of the radioactive isotopes which is used. This gives out gamma rays and has a half life of six hours.
(a) Why is lead used to shield hospital workers while they are using ^{99}Tc?
(b) Why is an isotope with such a short half life used?
(c) Use a periodic table to suggest what might happen to the ^{99}Tc when it decays.
(d) Medical equipment such as disposable syringes are often sealed in polythene bags and then sterilised using gamma rays. Why is this a better method of sterilisation than heating them to high temperatures in steam and then packing them?

5 The ion X^{2-} contains 54 electrons.
(a) How many electrons are present in one atom of *X*?
(b) Use a periodic table to find the group to which element X belongs.
(c) Write the formula of the compound formed between caesium and X. (Caesium has the symbol Cs and is in Group 1 of the periodic table.)
(d) How many protons are present in the nucleus of an atom of X?

6 The following passage was written to support the building of more nuclear power stations.

'World energy requirements will rise rapidly in the next decade. By the year 2000, it is estimated that the average world citizen will consume twice as much energy as the present average citizen in the USA. This energy can only be obtained by building 4000 nuclear fission reactors using the inexhaustible supplies of nuclear fuel.

We must stop using fossil fuels to produce energy because they are irreplaceable as raw materials for the chemical industry. Their use also damages the environment. Atmospheric pollution is caused by burning coal and oil, oil drilling and transport cause marine pollution and coal mining damages the environment.'

(a) Criticise the arguments used in the passage.
(b) Suggest four alternative sources of energy not considered in the passage.
(c) Discuss three important arguments against increasing the number of nuclear power stations.

7 Natural boron usually contains 20% boron-10 and 80% boron-11. Boron is the fifth element in the periodic table.
(a) What is the atomic number of boron?
(b) What are the mass numbers of the two boron isotopes?
(c) How many protons, neutrons and electrons does one atom of boron-11 possess?
(d) What is the relative atomic mass of boron?
(e) Samples of boron obtained from naturally occurring boron compounds in different parts of the world have slightly different relative atomic masses. Why is this?

SECTION L

Earth and atmosphere

The Earth and its atmosphere, a delicate balance of physical processes

1 WEATHER AND ATMOSPHERE

The Earth is a small, rocky planet. It is the only planet in our solar system which supports life. 4600 million years ago, the Earth was molten but it slowly cooled. Gases including water vapour were given off from the hot surface, forming an amosphere. Eventually, the Earth cooled enough for water to collect on its surface.

Weather and climate

We all take an interest in the weather, particularly in the UK where the weather sometimes changes from one hour to the next. **Weather** is the state of the atmosphere at any one time or place. It includes the temperature, rainfall, windspeed and wind direction. **Climate**, on the other hand, is the average of these weather records over a long period of time.

The weather affects our daily lives. Many people try to watch the weather report on television every day. This is Ian McCaskill from the Meteorological Office giving a weather report. He is using standard symbols to explain the weather

Information about the weather is collected from weather stations on land and sea, from weather balloons and from satellites which view the Earth from space. They provide the vital basis for weather forecasts. There are various ways to describe the weather. The symbols used on television weather maps are shown in Figure 1. You can see some of these in the photograph too.

Figure 1
TV weather symbols

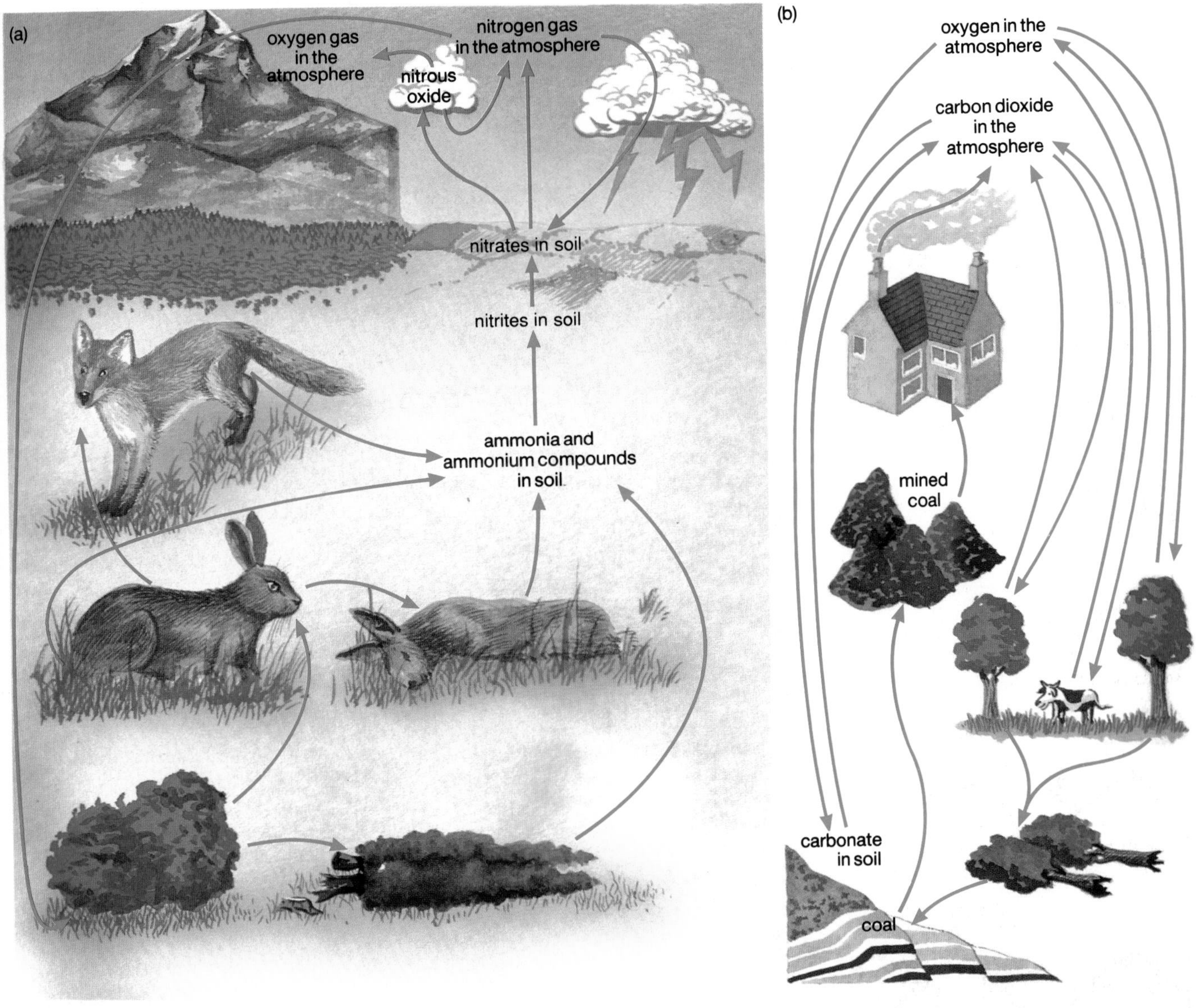

Figure 2
Gaseous cycles in the atmosphere
(a) Circulation of nitrogen and ammonia
(b) Circulation of carbon and methane

The evolution of our atmosphere

4600 million years ago, the Earth was still forming. At first, the atmosphere was rich in hydrogen and helium. As the molten, volcanically active surface cooled, other gases were added. These included methane, carbon dioxide and water vapour. When simple, green plants appeared 3500 million years ago, oxygen was formed from carbon dioxide by photosynthesis. Gradually complex plants evolved, adding more oxygen to the atmosphere. In time, animals evolved and used the oxygen for respiration. This helped to maintain a balance between the production and use of both oxygen and carbon dioxide. The composition of the atmosphere has remained more or less in balance for the last 500 million years.

Today, our industries are altering this delicate balance. Burning fossil fuels in excess of the quantity plants can cope with is removing oxygen and adding more carbon dioxide to the atmosphere. The clearing of forests has also resulted in a decrease in oxygen production by photosynthesis. Figure 2 shows how air is renewed by living creatures. Nitrogen, oxygen and carbon dioxide are used and restored to the soil and atmosphere in their cycles.

Figure 3
Global airstreams (over one quarter of the globe)

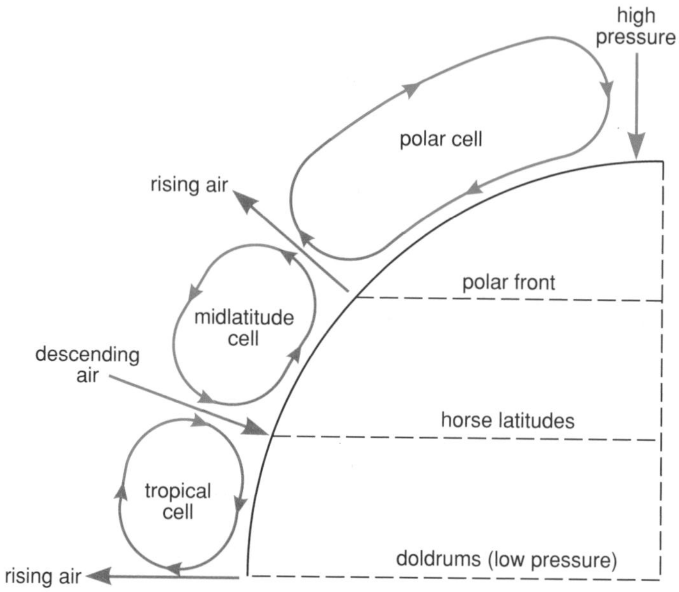

Global airstreams

The airstreams, or **global winds**, of the world are the result of heat energy from the Sun. Warm, less dense air rises, and cold, denser air flows in to replace it. Powered by the Sun, airstreams constantly circulate between the hot tropics and the freezing poles.

Figure 3 shows the global airstreams over one quarter of the globe. At the equator, where the Sun's rays are most intense, the air is heated and the hot air rises. This creates the **doldrums**, a low pressure area of light variable winds. As the air rises, it cools and spreads outwards, then becomes denser and sinks. This movement of air forms belts of high pressure and relative calm, the **horse latitudes**. This dense, high pressure air is drawn back to the low pressure of the equator. This circular flow of a large mass of air is known as a **cell**. Some air from the tropical cells flows towards the temperate latitudes, creating **midlatitude cells**. These cells end where polar cells of intensely cold air, flow outwards from the poles, displacing the warm air. The cold polar air warms in the temperate latitudes, rises and flows back to the poles where it cools and sinks again.

Water vapour, trapped under a layer of cold air, results in fog. Fog can be a serious problem. It reduces road visibility for motorists, and can be very dangerous if it freezes on the road surface

QUESTIONS

1 What measurements are always taken at a weather station?
2 (a) Using the weather symbols in Figure 1, record your own observations of the weather for today. Write a paragraph to record these same conditions in the kind of language used in weather forecasts.
(b) Now use the television symbols to plot your own weather map for the British Isles for a cold, windy winter day.

2 WINDS AND WEATHER

The Sun's rays are strongest where the Sun is directly overhead. The ground is heated up, and the air above it warms. This warm air is less dense than cold air so it rises. The rising air forms 'low pressure' belts found in equatorial regions such as the one in the photo

The Sun is the Earth's major source of energy. As well as supporting life, solar energy controls our planet's climate and weather. Figure 1 shows what happens to the energy from the Sun as it reaches the Earth.

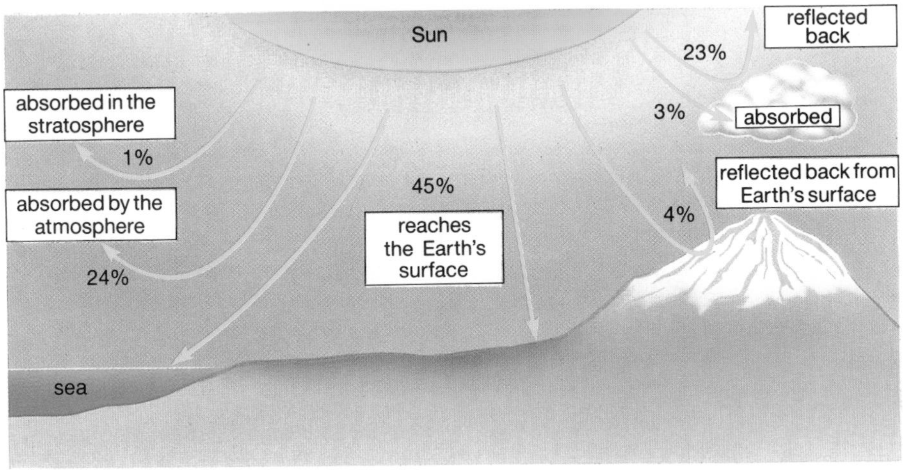

Figure 1
Distribution of the Sun's energy

As the Earth is not appreciably heating up or cooling down, there must be a balance between incoming energy (**solar radiation**) and outgoing heat loss (**terrestrial radiation**). 45% of the solar radiation reaches the Earth's surface. 55% of the Sun's energy is lost in the atmosphere, it never reaches the Earth.

Heat transfer

Heat is lost from the Earth's surface and out into the atmosphere by radiation, convection and evaporation. This is called **vertical heat transfer** and it prevents the Earth's surface from overheating. There is also **horizontal heat transfer** between different regions on the Earth. It stops the equator overheating, and prevents the poles becoming even colder. This horizontal heat transfer is brought about largely by airstreams, by winds and, to a smaller extent, by ocean currents. The movements of air and clouds in winds and convection currents dictate our weather.

Water molecules turn to ice crystals if the temperature drops below freezing. The frost in this photo is an example. Water may become trapped in cracks in solid material such as rocks or building stone. If this water freezes, its expansion can cause breakages in the rock or stone

Mist, fog, dew and frost

During the day, surface water evaporates from oceans, rivers and even puddles into the atmosphere. If the air temperature then falls, tiny droplets of water condense in the atmosphere and a **mist** forms. If the temperature drops further, more water droplets condense, the mist gets thicker and we call this **fog**. If the temperature of the Earth falls at the same time, then water condenses at ground level as **dew**.

Frost will form when the temperature of the Earth and the atmosphere falls below 0°C. This happens particularly on starry nights with no cloud cover to prevent heat loss by radiation. Air temperature falls quickly below freezing point, and moisture turns into needle-shaped ice crystals like those you can see in the photo.

Weather patterns

The movement of large masses of air gives rise to weather conditions over longer periods.

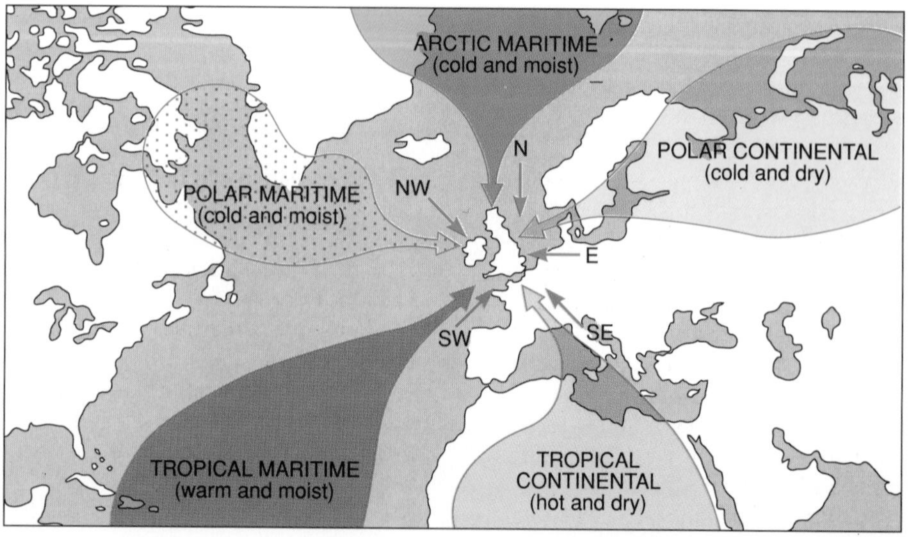

Figure 2
The air masses that affect the British Isles

Five large air masses jostle for dominance in dictating our weather (Figure 2). These air masses push against each other forming **fronts** where they meet. At a **cold front**, the cold air is denser and so it tends to wedge its way beneath warm air. Eventually, the warm and cold air mix forming a **mixed** or **occluded front**.

Frontal systems passing over Britain tend to form where **polar continental** and **tropical maritime** air masses meet over the Atlantic Ocean (Figure 2).

Figure 3
The weather conditions associated with (a) a warm front (b) a cold front

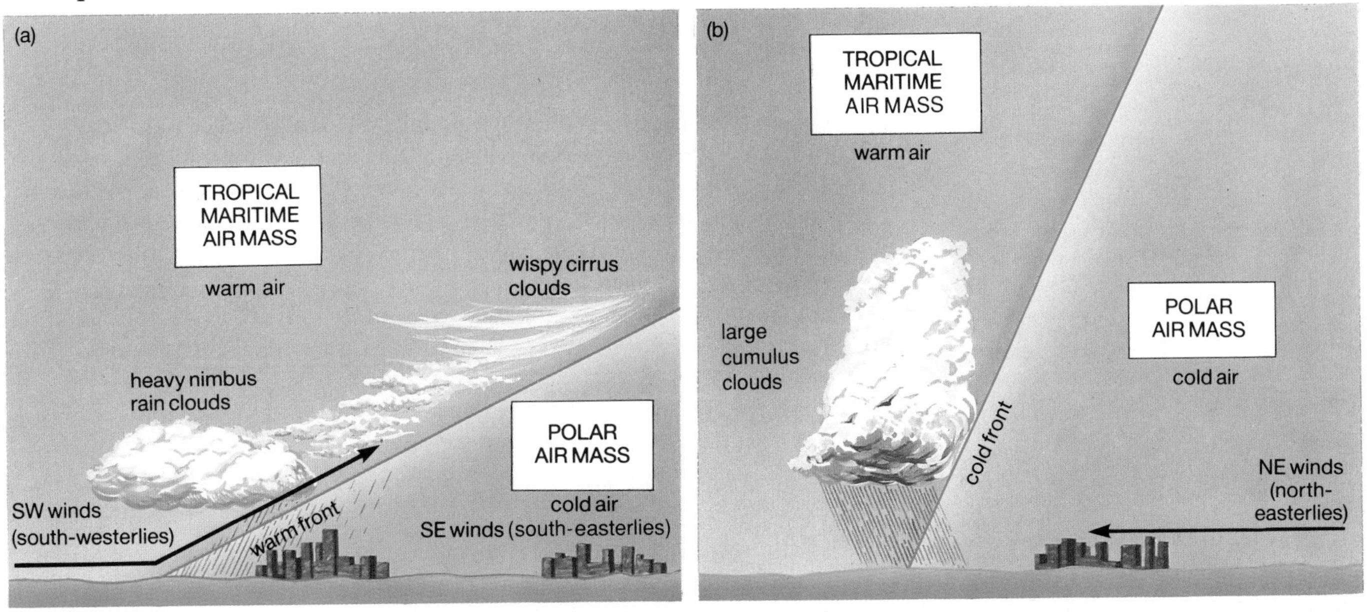

Figure 3 shows the weather conditions associated with cold and warm fronts. As the two air masses meet, they begin to curl round. The warm south-west winds (**south-westerlies**) turn northwards, forming a warm front (Figure 3(a)). The cold north-easterly airstream turns southwards, forming a cold front (Figure 3(b)). These frontal systems drift eastwards over Britain as the Earth turns on its axis. Each front may take two or three days to pass over Britain, with a sequence of pronounced changes in the weather. As the warm front approaches, first high level, then medium level and finally low level clouds come from the west. This usually brings rain because condensation occurs as the warm, moist air rises and cools. This is followed by a settled warm spell as the warm air mass moves over. Cloud cover becomes lighter, temperatures rise and air pressure increases.

When a cold front is passing over, there are usually short, heavy rain showers. At the same time, there is a drop in temperature and an increase in cloud cover.

QUESTIONS

1 Using Figure 1 draw a pie graph for the distribution of the Sun's energy. Label each of the segments. In yellow, colour the amounts of solar energy not absorbed. In red, colour the amounts of solar energy absorbed by the Earth.

2 People ski at high altitudes, on snow or ice covered ground where temperatures are low. Why do skiers have to protect their exposed skin with sun block creams?

3 Sunlight contains harmful ultra-violet rays which can damage life on Earth.
(a) Which part of the atmosphere protects the biosphere from these harmful rays?
(b) What molecules play an active part in this control?
(c) What is happening to this layer of the atmosphere, and what are the possible consequences for the human race?

4 The British Isles benefit a great deal from the horizontal transfer of energy. Although further north than much of Europe, the British Isles enjoy a milder winter climate. What brings us this warmth?

5 What changes take place in the atmosphere and the weather during the passage of (a) a warm front, (b) a cold front?

6 Use your knowledge of convection currents to explain the flow of air shown in Figure 3.

3 WEATHERING AND EROSION

Our landscape is slowly and constantly changing. Several processes act upon it, wearing down old rocks and creating new ones. These processes involve **weathering**, **erosion**, **transport** and **deposition**. They have operated constantly since the Earth was first formed. They have lowered mountains to plains, creating sediments which may later be built up into mountains.

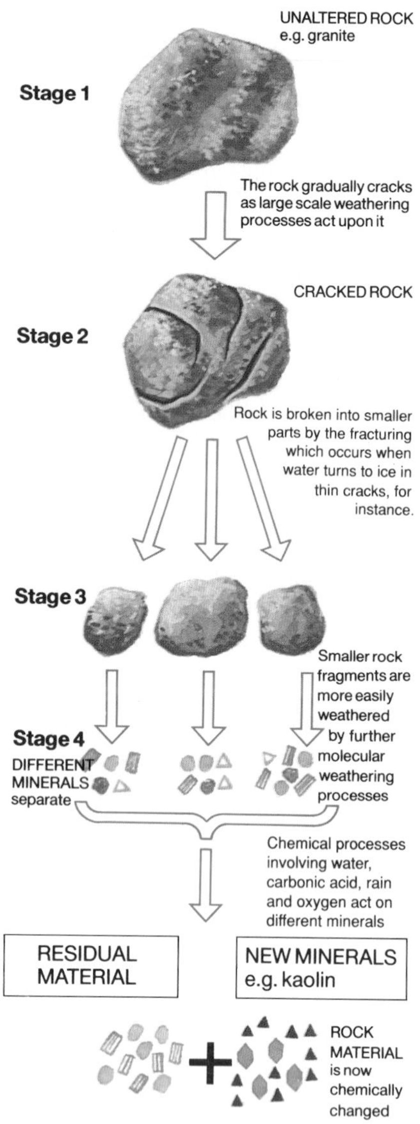

Figure 1
Sequence of weathering processes

Weathering

The first stage in the cycle which wears down the landscape is **weathering**. Weathering involves the breakdown, fracturing and decomposition of rocks on or near the surface of the Earth. There are many different kinds of weathering. Some occur on a large scale, others at molecular level. They take place in normal atmospheric conditions and are helped along by changes in temperature and the presence of water.

Hard, crystalline rocks are not necessarily more difficult to weather than the 'softer' sediments. Granite is very hard, yet one of its main components, feldspar, is particularly vulnerable to reaction with water (**hydrolysis**). In hydrolysis, the feldspar weathers to china clay (kaolin). On the other hand clay rocks are very soft, but they weather very little because clay minerals are the end product of weathering (they cannot be broken down further).

The main stages of weathering that most rocks pass through are shown in Figure 1.

Erosion and transport

Erosion is different from weathering because it involves transport.

Erosion = transport of weathered material
= wearing down of land surface

Water is an agent of erosion. Weathered material is transported by rivers to the sea where it is carried by water currents and wave action. The Scarisdale River, shown here, is able to transport fine material, as well as large boulders when it is in flood. The load it carries helps to erode its valley

An agent of erosion is something which can pick up particles and carry them along. So moving water in rivers and waves, winds (especially in deserts) and glaciers are the important agents of erosion. You can see examples of erosion in the photographs.

For wind and water, a critical speed of flow is needed before particles of a certain size can be picked up and carried along. If the speed of flow falls below this critical level, heavier particles will be deposited. The same is true for glaciers, like the one in the photograph, although the slow-moving ice of glaciers works on a totally different time scale to wind and rivers.

The rock particles carried in the agent wear away the landscape, like a scouring pad. Rivers wear away their channels, deepening and widening valleys. Coastal waves undercut cliffs and scrape out caves. Glaciers carve slowly and deeply like giant files. Desert winds sculpture and polish bare rock surfaces over wide areas. Little by little, these agents erode and alter the landscape.

Glaciers are made of ice. They transport material very slowly, some only move a few metres a year, and the rock carried by the ice erodes the sides and floor of the valley

Wind is an important agent of erosion in desert areas and its effects can be enormous. These 'towers' in Monument Valley, Arizona, appeared when the sandy areas between them were eroded by wind

Deposition

When winds, rivers and waves can no longer carry weathered material, it is deposited as **sediment**. The heaviest material is deposited first and the lightest last. You can see a gradation in particle size from coarse gravels to sands and then silts in the sediment. Soils are formed as these new sediments are subjected to further weathering processes – poor soils from gravels and sands, better soils from the silts and clays.

QUESTIONS

1 How does mechanical weathering make chemical weathering more effective?
2 Heat speeds up chemical reactions. So why does chemical weathering take place slowly in hot deserts?
3 Why is wind erosion relatively more important in arid (dry) regions of the world than in humid (wet) ones?
4 Describe two ways in which waves cause erosion. What typical landforms of erosion are seen along the coastline?

Figure 1
Formation of igneous, sedimentary and metamorphic rock

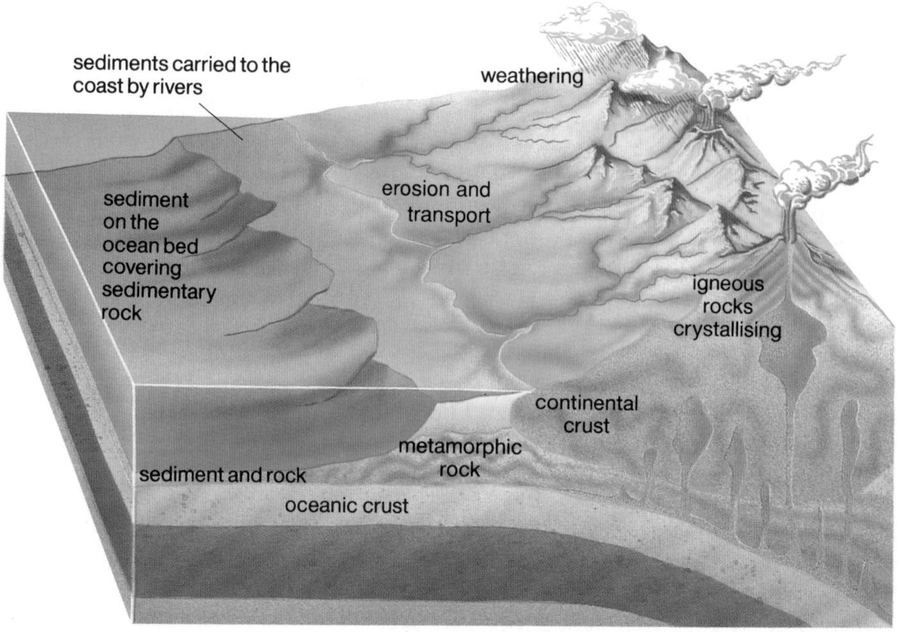

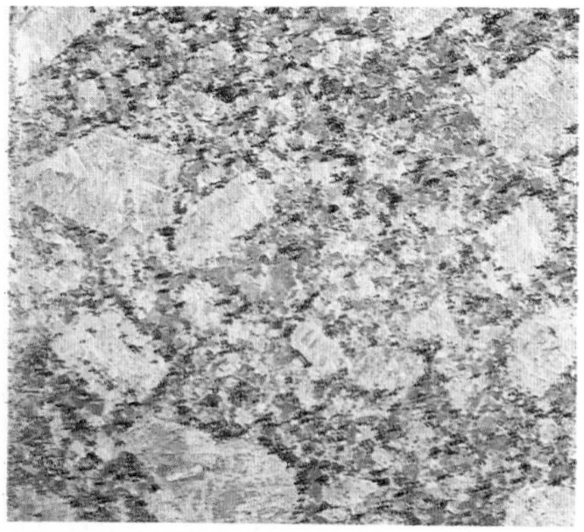

This is an example of an igneous rock. 'Igneous' means 'formed by fire' because these rocks result from the cooling of molten magma on, or within, the Earth. The size of the crystals depend on how quickly the magma cools. This rock has cooled slowly so its crystals are large.

When the Earth first cooled, its molten crust solidified to form **igneous rock**. Igneous means 'formed by fire'. Initially there was no other type of rock. Over thousands and millions of years, two other types of rock were created. These were **sedimentary rocks** and **metamorphic rocks**. Figure 1 shows how these three types of rock are forming today.

Igneous rocks

Igneous rocks originate from the liquid rock, or magma, below the Earth's crust. **Lava** from volcanoes is magma pouring out from cracks in the Earth's crust. When magma or lava cools and crystallises, igneous rock is formed (Figure 1). The size of crystals in the igneous rock depends on the rate of cooling. Lava loses its heat very rapidly to the air, so above the Earth's surface it quickly becomes a solid. The lava crystallises in days or months depending on its thickness. This process produces rocks, such as **basalt**, with small crystals. Magma which is cooling *below* the Earth's surface retains its heat far longer. It may not cool to a solid for centuries so the crystals have a very long time in which to grow. This produces rocks like **granite** with large coarse-grained crystals.

This is an example of a sedimentary rock, formed from the breakdown products of older rocks

Sedimentary rocks

When igneous rocks are exposed at the surface of the Earth, they undergo weathering and erosion. These processes form sediments which are carried from their original location and deposited elsewhere. Most sediment is deposited on the shallow **continental shelves** surrounding land masses (Figure 1). As the weathered material is brought down from the eroding land masses, successive beds, or **strata** form. In some areas, chalk sediments have formed from the calcium carbonate present in the shells of sea animals. As the beds become buried deeper, they are compacted under the weight of the layers above. This converts the soft sediments to harder sedimentary limestone rocks.

Metamorphic rocks

In some cases, sedimentary rock has been buried to great depths and been changed by enormous pressure and high temperatures. This has produced **metamorphic rocks** (Figure 1). There are various types of metamorphic rock depending upon the type of sediment from which they originate. The minerals in the parent sediment are sometimes changed under the stresses of heat and pressure. In some cases, they are changed to chemically similar, but harder and more stable metamorphic rocks. This results in a series of physically different metamorphic rocks. Deep down inside the Earth, the base of the metamorphic rock may melt, creating magma. This will eventually solidify as igneous rock, beginning the rock cycle again.

Figure 2 shows the various stages of the rock cycle. The complete rock cycle, lasting hundreds of millions of years, does not always take place exactly as described. Sometimes, igneous rock, instead of weathering at the surface, may be subjected to heat and pressure and become metamorphic rock. Metamorphic and sedimentary rocks may also weather and erode to form new sedimentary rocks. Not all old rocks are turned into sediment and recycled. Some of them have been covered and protected by successive layers of younger rocks since the Earth was first formed 4600 million years ago.

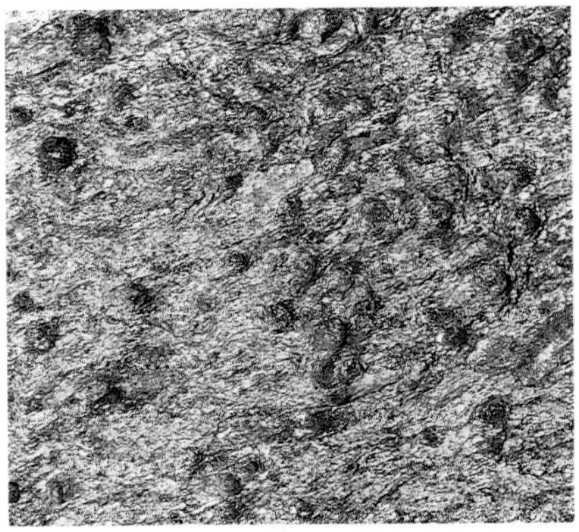

This is an example of a metamorphic rock, formed from sedimentary rock under high temperature and pressure. The minerals in metamorphic rocks can help geologists to trace their origin

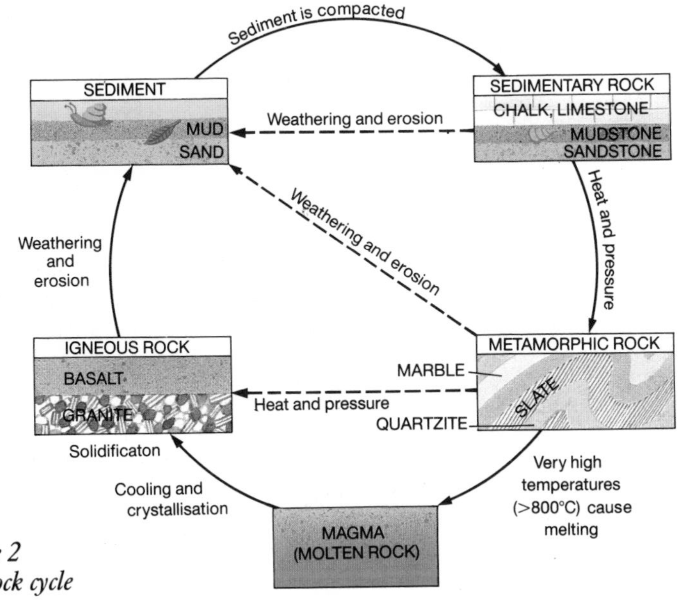

Figure 2
The rock cycle

QUESTIONS

1 Using the rock cycle in Figure 2, explain the following statement: 'one rock is the raw material for another'.

2 When the Earth first formed, there were only igneous rocks.
(a) Why was this?
(b) How did sedimentary and metamorphic rocks come into existence?

3 (a) Describe how sediments may form on land and in the sea.
(b) How are they turned into sedimentary rocks?

4 You are given three pieces of rock by a friend who wants to know more about them.

Rock A is made of rounded pebbles held together by a hard, sandy layer.

Rock B is white in colour and contains several small fossils which look like shells you have seen at the seaside.

Rock C is very hard and is made of large crystals which you can see quite clearly.

What could you tell your friend about the rocks?

5 THE STRUCTURE OF THE EARTH

Internal structure of the Earth

The Earth is shaped like an orange, spherical but slightly flattened at the poles. Its structure and surface is like a badly cracked egg. The 'cracked shell' is the very thin **crust**, the 'white' is the **mantle** and the 'yolk' is the **core** (Figure 1). These concentric layers increase in thickness, density and temperature towards the centre. Evidence for the Earth's structure comes from various sources. Information about the crust comes from studying mines, volcanoes and earthquakes. Evidence for the mantle and the core comes from the study of earthquakes, deep volcanoes, meteorites and the Earth's magnetic field.

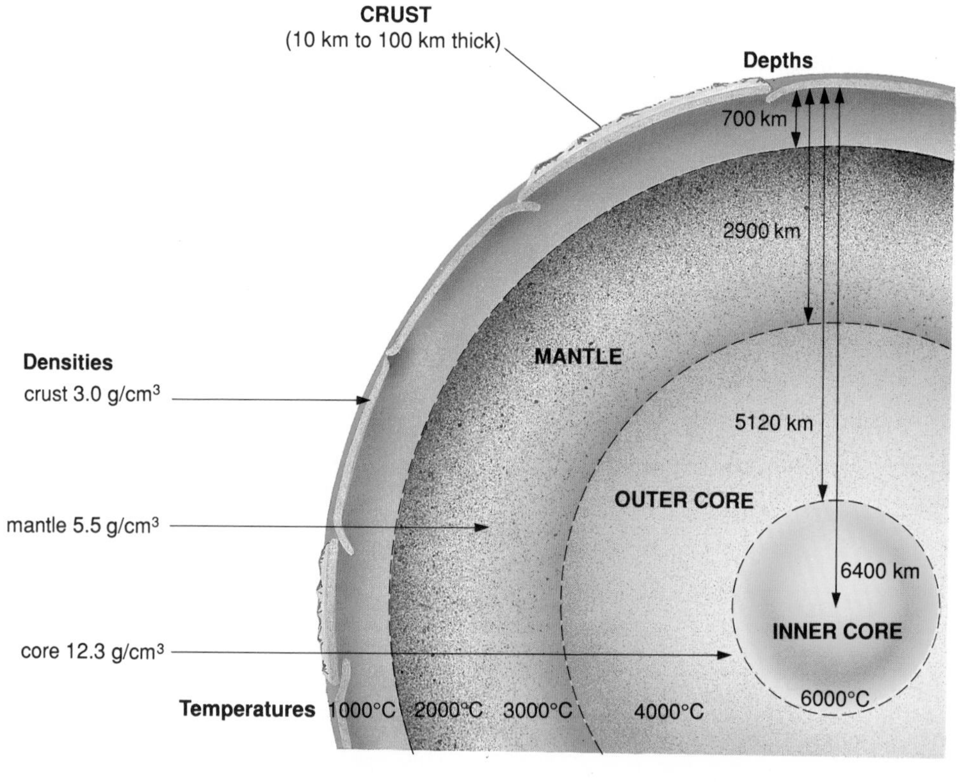

Figure 1
Cross-section of the Earth to show its internal structure

Figure 1 shows a cross-section of the Earth. The Earth's crust is composed of igneous, metamorphic and sedimentary rocks. The crust is very thin and is broken into large and small parts, called **plates**. These float on the denser mantle, parts of which are molten. Over long periods of geological time, changes occur in the crust. Immense internal forces cause the plates to move resulting in faults, rift valleys, volcanoes and earthquakes. In some parts, the crust is being stretched and is under tension. Deep cracks appear in the crust, rocks break and one side slips down filling the space created. This forms a **fault**. **Rift valleys** form where two normal faults lie alongside each other (Figure 2). Deep cracks in the Earth's surface can also lead to volcanoes (Unit 6). In other parts of the Earth, the crust is being pushed together and compressed (Figure 2(b) and 2(c)). This causes layers of the crust to ride over each other and fold, resulting in earthquakes (Figure 2(d)).

The mantle and the core are so deep in the Earth that they are normally inaccessible but, in The Lizard in Cornwall, mantle rock has been pushed to the surface and can be seen. Mantle rocks can also be studied from **xenoliths** ('strange rocks') brought up in volcanoes.

Evidence for the Earth's structure

Much greater detail of the Earth's structure comes from the study of earthquakes and magnetic fields. During an earthquake, the crust ruptures. Energy generated at the focus of the earthquake creates a train of shock **(seismic)** waves. These extend outwards through the Earth and their pathways are shown in Figure 3. Earthquake energy creates three main types of waves: **P** or **primary waves; S, secondary** or **shear waves; L, longitudinal or surface waves.** L waves are slow moving and travel through the crust. P and S waves move faster and travel through the deeper layers of the Earth (Figure 3). P waves travel at 5 km per second and pass through liquids and solids. S waves travel at 3 km per second and only pass through solids. They cannot pass through the Earth's liquid outer core. From the movement of these waves we can obtain evidence for the Earth's internal structure. Surface waves roll invisibly through the Earth's crust. Around the **epicentre** of an earthquake (right above the focus) these waves cause the most damage to buildings.

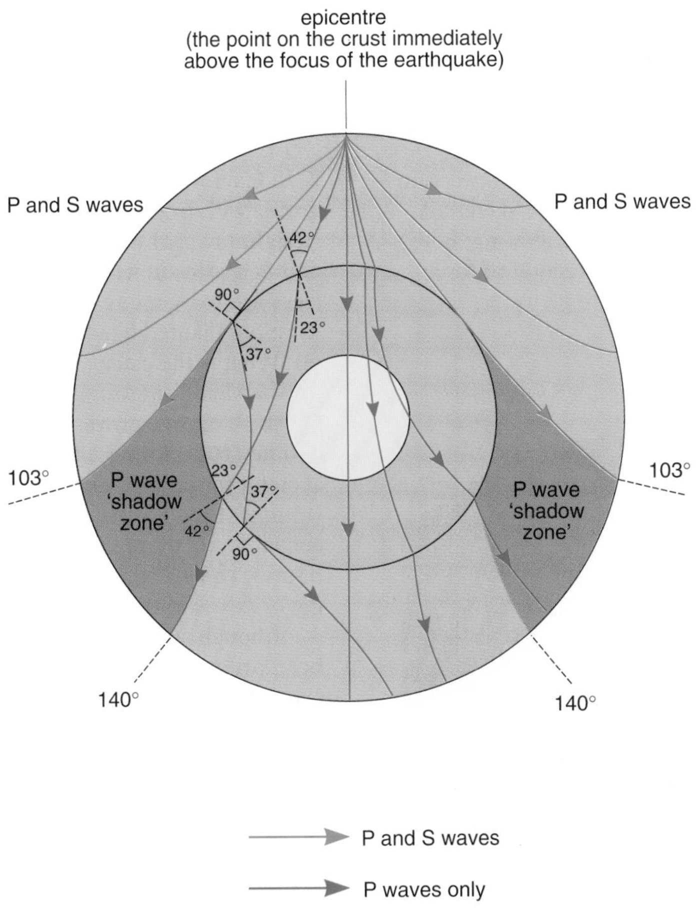

Figure 3
Cross-section of the Earth showing the paths of seismic waves from an earthquake. Both P and S waves spread out to angles up to 103° from the epicentre, but S waves cannot travel through the liquid core. P waves entering the liquid core slow down and are focused (like light waves through a lens) into the region at the bottom of the diagram (angles >140°). No P wave travelling through the liquid outer core reaches the 'P wave shadow zone'. However, weak P waves are detected in that zone, which have been speeded up, and therefore refracted, by the solid inner core

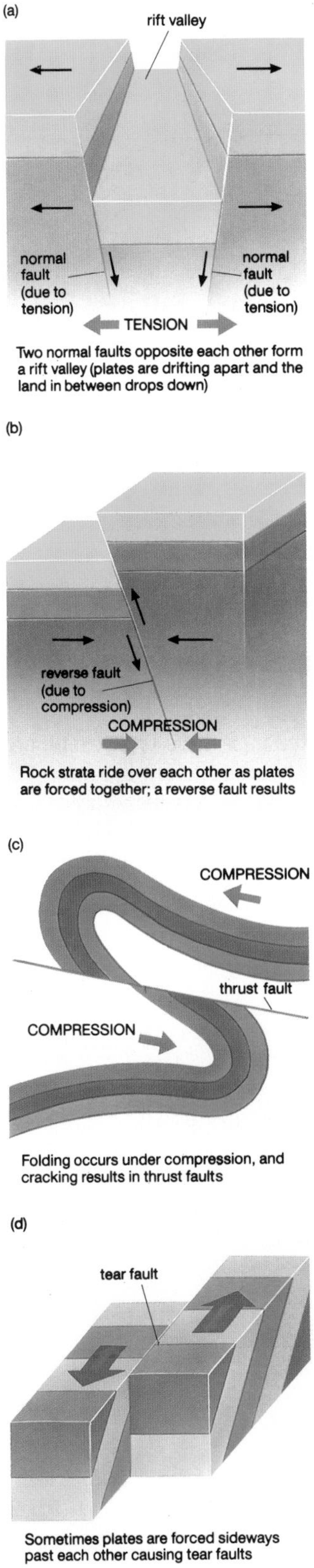

Figure 2
(a) Normal faults and a rift-valley
(b) Reverse fault
(c) Thrust fault (d) Tear fault

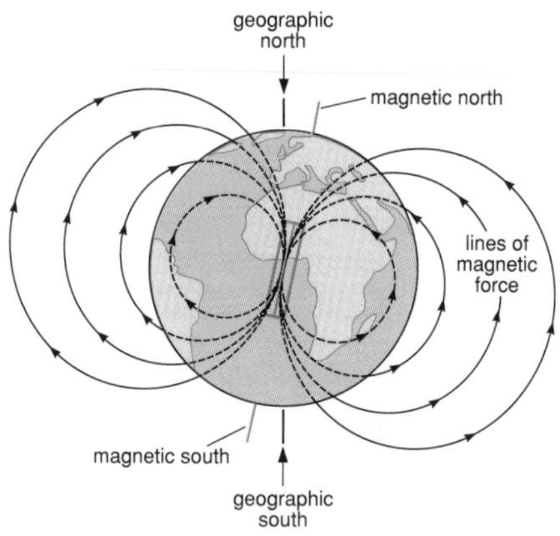

S waves do not register on earthquake recording equipment at an angle greater than 105° from the **epicentre** of the earthquake. P waves are also cut out after 105° but reappear again at 140° from the epicentre. Between 105° and 140°, a **shadow zone** exists where no earthquake information can be traced. This suggests that S waves cannot pass through the liquid outer core. P waves, are, however, refracted through the core, reappearing at 140°.

The Earth has a strong magnetic field which is produced as the Earth spins on its axis. The fluid outer core allows the mantle and the crust to rotate faster than the solid inner core. This produces a magnetic field similar to an enormous bar magnet (Figure 4). The magnetism exists because the Earth has an iron-rich core. Rocks in the Earth's crust which contain iron also take on this magnetism as they solidify. Magnetic particles in the crust show that the crust must have moved over geological time.

Figure 4
The Earth's magnetic field

QUESTIONS

1 The students in the photograph are drawing a field sketch of the fold.
(a) If the camera was facing east, which directions do you think the forces which created the fold came from?
(b) Was the force one of compression or tension? Draw a simple sketch of the fold and label the following: beds, upfold, downfold, direction of force, type of force.

As continents gradually move together and the oceans between them become smaller, the rocks of the Earth's crust on the ocean bed are pushed up into folds. These small folds at Lulworth in Dorset are an example

2 In California, you can see orange groves with trees growing in perfect, straight lines. However, there are some orange groves in which the lines of trees are for some reason kinked. They were not planted like that.
(a) What is the cause?
(b) In what interesting geological area are these orange groves?

3 The crust of the Earth is thought to account for only 1% of the total volume of the Earth. The mantle accounts for about 82%. Using these figures, construct a pie chart for the major layers of the Earth.

4 Although some very deep holes have been drilled into the Earth, none has ever reached the mantle. We have no direct evidence of the nature of the interior of the Earth. Our knowledge is based on indirect evidence. What is this evidence and what has it told us about the Earth's interior?

5 Imagine that a major earthquake happens somewhere in the British Isles. Seismological (earthquake) stations in Paris, New York, Peking and Nairobi all know that an earthquake has occurred. In Auckland, where they often monitor major earthquakes, only weak P waves are detected. Explain this.

6 PLATE TECTONICS

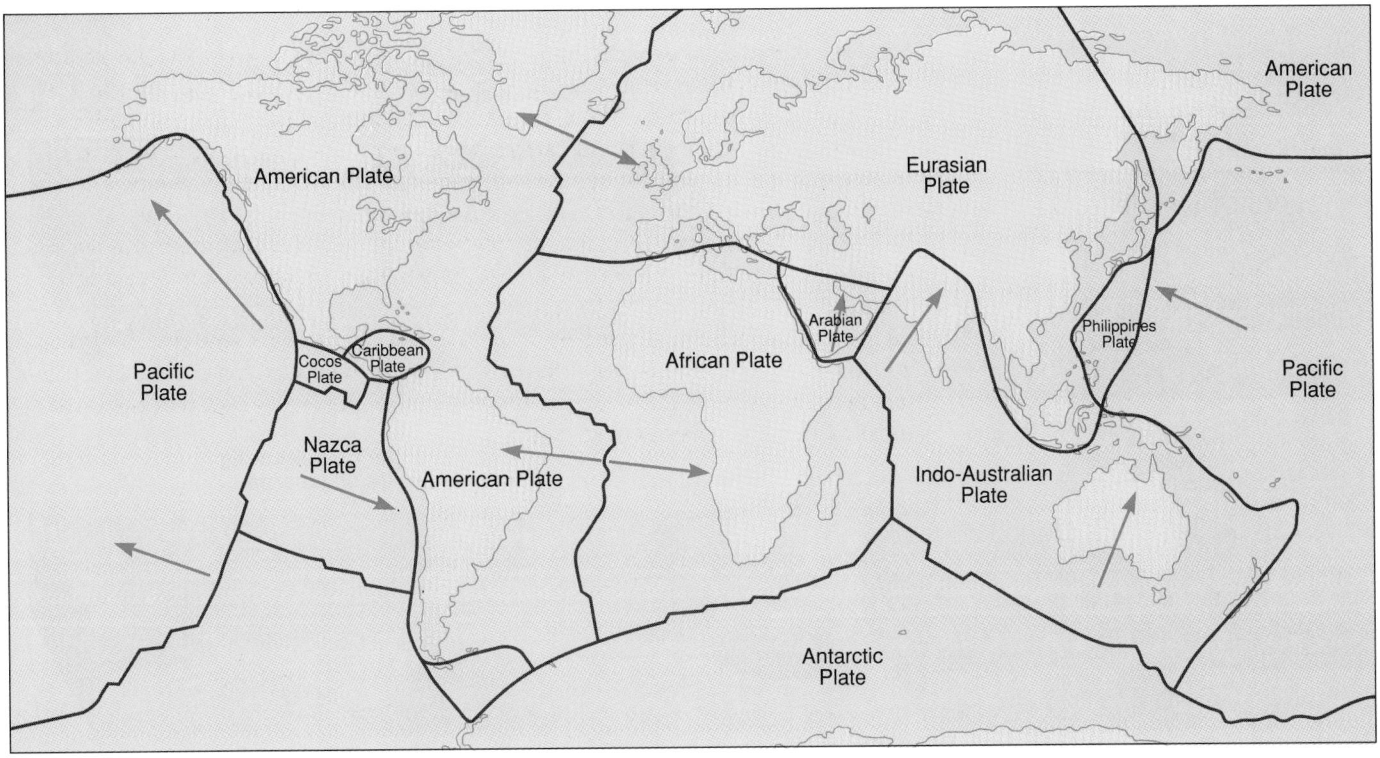

Figure 1
World map showing the main plates and their directions of movement

The structure and physical properties of the Earth are the key to understanding **plate tectonics**. The Earth's core is as hot as the surface of the Sun. This causes slow convection currents in the liquid mantle which result in slow movements in the plates of the Earth's crust. This is the start of the **tectonic cycle**. Figure 1 shows how these plates are moving.

Many years ago, Africa and Arabia used to be joined together. They are now growing apart as the Red Sea gradually widens. This sea may eventually become as large as the Atlantic Ocean

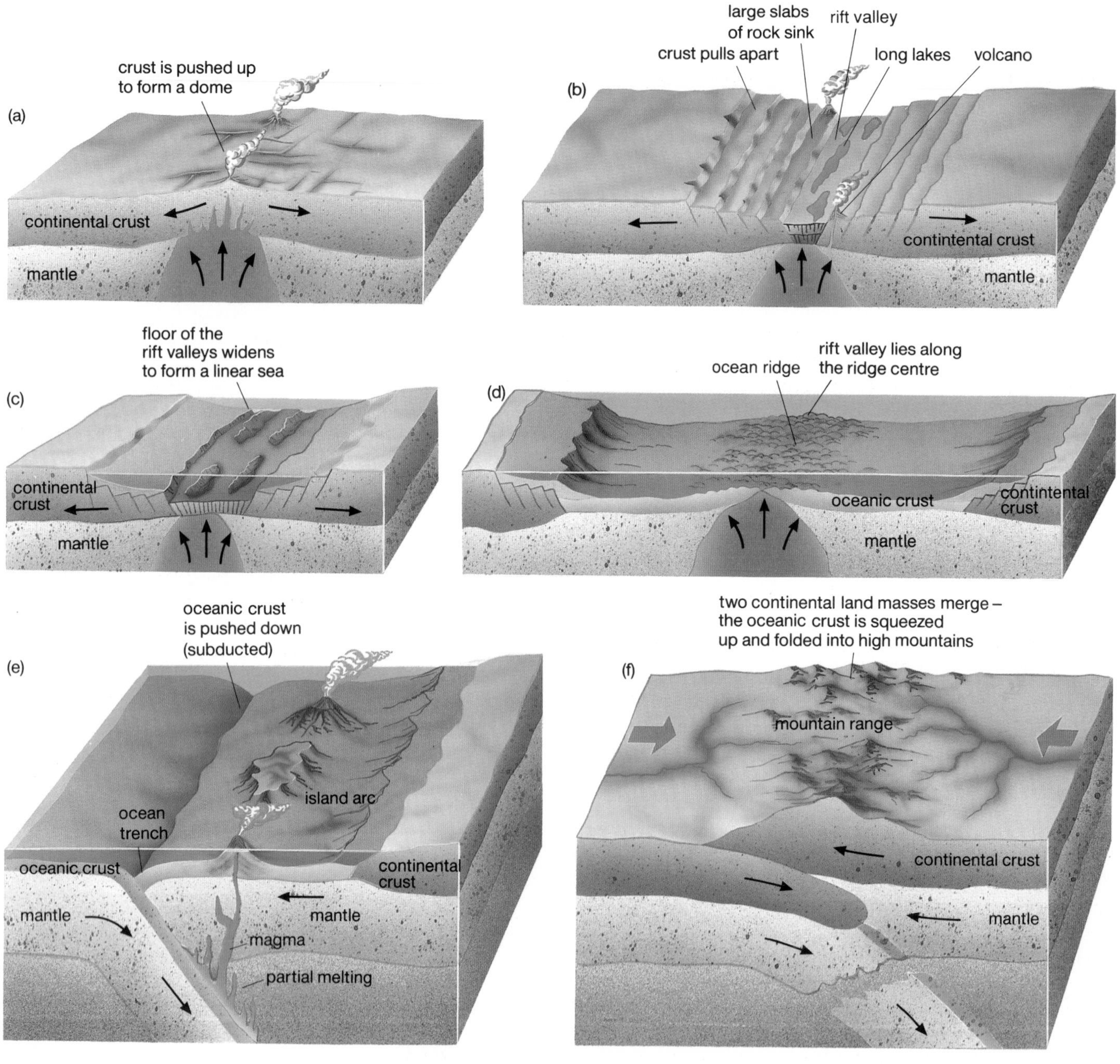

Figure 2
(a) The continental crust is pushed up into a dome
(b) Rift valleys form with volcanoes
(c) A linear sea develops
(d) An ocean ridge forms and the ocean widens
(e) An ocean trench forms with island arcs; the oceanic crust is subducted (pushed down)
(f) As continents collide, the ocean closes and a larger land mass with high mountains is formed

The convection currents circulating in the liquid mantle may take millions of years to rise to the surface. If currents of hot, molten rock rise under the thin oceanic crust they form '**hot spots**' of intense volcanic activity. The Hawaiian islands lie above one of these hot spots. On these islands, magma rises to the crust surface forming '**shield volcanoes**'; the 'runny' basalt lava flows quickly so shield volcanoes have gentle slopes. Under the thicker continental crust, rising convection currents push the crust up into a dome, causing tension and cracking (Figure 2(a)). As the crust is pulled apart, large slabs of rock sink and rift valleys form. Volcanoes appear where the magma escapes from cracks in the rift valley (Figure 2(b)). This is typical of the rift valleys in East Africa.

The Red Sea and Gulf of Aden illustrate the next stage of the tectonic cycle when the floor of the rift valley can widen to form a **linear sea** (Figure 2(c)).

This region will gradually form a new ocean, separating Africa from Arabia. The Atlantic is an example of an expanding **young ocean**. It is widening as fast as your fingernails grow. A **submerged ocean ridge** has formed down the centre of the Atlantic from the Arctic to the Antarctic (Figure 2(d)). This submerged ridge follows the boundary between the American Plate and the African and Eurasian Plates (Figure 1). A rift valley lying along the centre of this ridge, slowly generates new oceanic crust.

As the Atlantic Ocean widens, the Pacific Ocean is slowly closing. The Pacific is a much older ocean than the Atlantic. Over millions and millions of years, thick deposits of sediment have collected on the ocean floor. The layers of sediment are particularly thick near continents where rivers have carried silt into the ocean. As the oceanic crust is thin, the weight of sediment makes it sag. More sediment collects, and eventually the crust breaks. As new crust is being created at the ocean ridge, the old oceanic crust, near to the continent, is pushed down or **subducted** (Figure 2(e)).

The deep depression in the crust where subduction occurs is called an **ocean trench**. Subduction of the ocean crust and wet sediment into the mantle creates magma which rises up to the surface. Some of this escapes as thick, explosive lava, forming volcanic islands in arcs which fringe the trench (Figure 2(e)). These **island arcs** extend from the Aleutian Trench in the north-west Pacific to the Tonga Trench in the south-west Pacific. Some islands, with metamorphosed sediments, form larger island groups like Japan.

As the Atlantic grows, the American continents are moving westwards. Along the western edge of South America, the Nazca Plate is being pushed against the American Plate (Figure 1) Here the ocean crust is being forced under the advancing landmass. This is pushing up marine sediments and creating the Andes mountains.

A tectonic cycle ends when two continental land masses converge and the ocean between them disappears. The layers of oceanic crust are squeezed into tight folds forming high mountains (Figure 2(f)). This is what happened when India moved north to collide with Asia. The ancient Tethys Ocean disappeared and the Himalayan mountains were formed. The crust here is so thick that volcanic activity has stopped, although earthquakes are common.

Plenty of geological evidence supports the theory of plate tectonics. The latest evidence comes from lasers mounted on satellites which can measure oceanic growth of only 2–3 centimetres a year.

These mountains were formed from horizontal beds of sediment at the bottom of an ancient ocean. As the continents moved together and the ocean closed up, this sediment was pushed up into a fold mountain range

If the Earth was filmed from outer space using time lapse photography over a period of 1000 million years, you would be able to see the tectonic cycle clearly: continents moving apart, oceans waxing and waning, new continents joining and mountain chains forming.

QUESTIONS

1 Using an atlas, find named examples of volcanoes for each of the following stages in the tectonic cycle (the areas in brackets will help you):
(a) continental rifting (East African Rift valley),
(b) linear sea spreading (Red Sea and Gulf of Aden),
(c) growth of ocean (islands on the Atlantic ridge),
(d) subduction of ocean floor far from continents, (island arcs near deep trenches),
(e) subduction of ocean floor near to continents (the Andes and the Rockies).

Are there any volcanoes in the final stage of the cycle, (e.g. the Himalayas)? Are there any other locations where volcanoes can be found? If so, why are they found there?

2 What do we mean by the 'jig-saw fit' of continents?

SECTION L: THINGS TO DO

Study questions

1 Make a simple copy of the map below. On it label the winds with the direction from which they come. For each wind, use an atlas to find out what areas they blow over and from which air mass they originate. Explain why the winds are labelled 'good', 'fairly good', 'fairly poor' and 'very bad'.

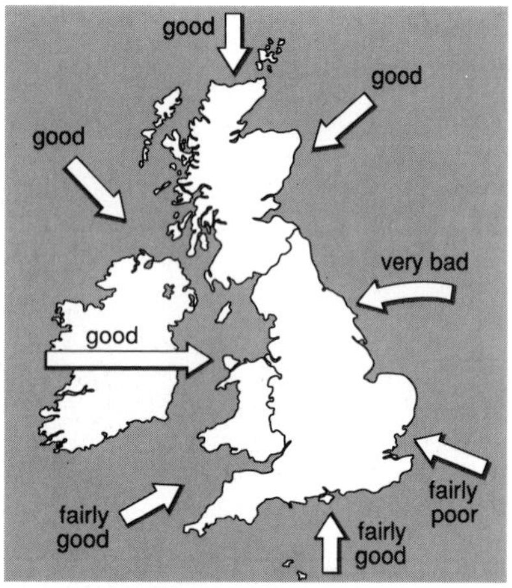

2 What are the main sources of carbon dioxide in the atmosphere? What is the role of carbon dioxide in the atmosphere? What could happen if atmospheric levels of carbon dioxide changed significantly?

3 What is the significance of ozone in the atmosphere? What is causing the amount of ozone to change? What may be the long term consequences of increasing or decreasing the amount of ozone in the atmosphere?

4 The table below shows how incoming radiation is absorbed or reflected by the Earth and its atmosphere. The values are percentages of the total radiation at the outside boundary of the Earth's atmosphere.

Absorption/reflection of radiation	Percentage of total radiation
Reflections from clouds to space	21%
Absorption by clouds	3%
Diffuse reflection to space by dust etc.	5%
Absorption by air molecules (CO_2, O_3, H_2O) and dust	15%
Reflection to space from the Earth's surface	6%

(a) What is the total percentage of incoming radiation reflected by the atmosphere?
(b) What is the total percentage of incoming radiation absorbed by the atmosphere?
(c) What percentage of incoming radiation is transmitted to the Earth's surface?
(d) What percentage of incoming radiation is actually absorbed by the Earth's surface?
(e) What percentage of incoming radiation is lost back to space by reflection by the Earth *and* the atmosphere?

5 The table below shows the major *natural* reservoirs for water on the Earth. The volume of water in each of these reservoirs is given in cubic metres.

Natural reservoir	Volume of water held in reservoir/m^3
Oceans	1350×10^{15} m^3
Glaciers and polar ice	29×10^{15} m^3
Underground water in rocks and natural wells	8.4×10^{15} m^3
Lakes and rivers	0.2×10^{15} m^3
The atmosphere	0.013×10^{15} m^3
The biosphere	0.0006×10^{15} m^3

(a) Calculate the amounts of water held by the oceans as a percentage of the total water in natural reservoirs.
(b) What percentage of the total in natural reservoirs is found as glaciers and polar ice?
(c) How much water is held in the atmosphere as a percentage of all *non-ocean* water?

6 An **aquifier** is a rock which holds water. An **aquiclude** is a rock which will not hold water. An **artesian basin** is a region where water is trapped in an aquifer, often far below the surface. London had a plentiful supply of underground water before it had such a large population. Using the geological cross-section of the London Basin, explain why London had such a good water supply.

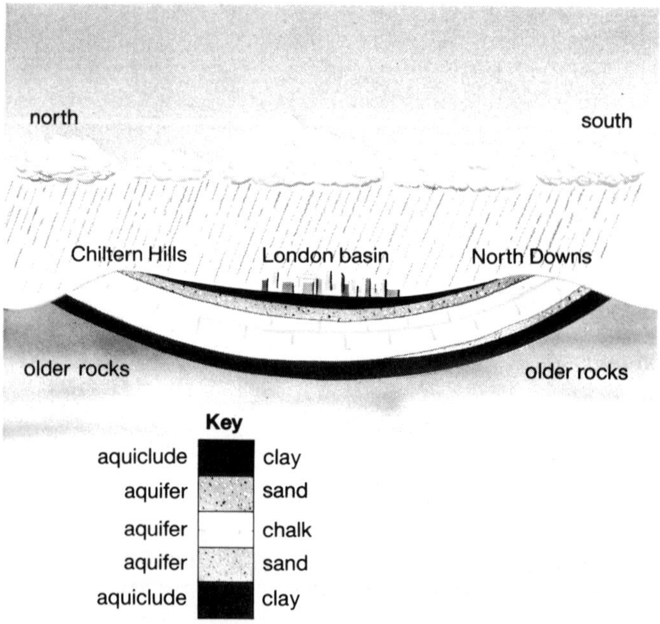

7 Households today use on average 70% more water than households 30 years ago. This is mainly because we now have more household appliances that use water. The table below shows the average amounts of water used by a household.

Use the figures in the table to answer the following questions.

(a) Estimate your own daily consumption.

(b) Estimate your own weekly consumption.

(c) Estimate your own annual consumption.

(d) Now answer (a), (b) and (c) for the other members of your family.

(e) Construct a bar graph to show the amounts of water you use for each purpose.

(f) How could you reduce the amount of water you use?

Reason for water consumption	**Average amount of water used at one time/litres**
PERSONAL WASHING	
Wash basin	1–3
Bath	70
Shower	25
Power shower	50 (per minute)
LAVATORY	
Old lavatory (before 1989)	7.6
New lavatory (1989 onwards)	6.8
DRINKING AND COOKING	2.5 (per day)
WASHING CLOTHES	
Hand washing	40
Machine washing	80
DISH WASHING	
Hand washing	5
Dishwasher	30
CAR WASHING	
Power spray	40
Bucket	7
WATERING GARDEN	
Sprinkler	800 (per hour)
Watering can	7
LEAKING TAP	
Fast drip (1 drip/second)	100 (per day)
Slow drip (1 drip/10 seconds)	10 (per day)

8 The diagram below shows an alpine area. Make a simple copy of the diagram. By each label line, put the appropriate letter(s) as follows: 'W' for weathering, 'E' for erosion, 'T' for transport, 'D' for deposition. If you know the names of the features shown, you can add these to your diagram too.

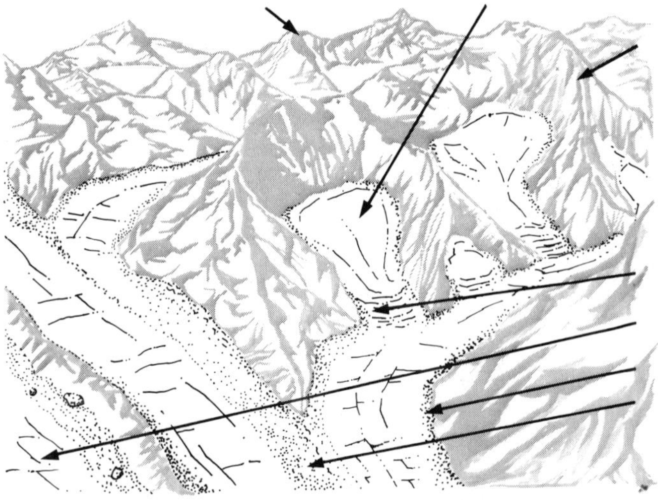

9 The following diagram shows three samples of rock material. They are the same rock type and have an equal volume. Calculate the total surface area for each sample. Which sample would be most susceptible to weathering processes? How will their rates of weathering compare?

1

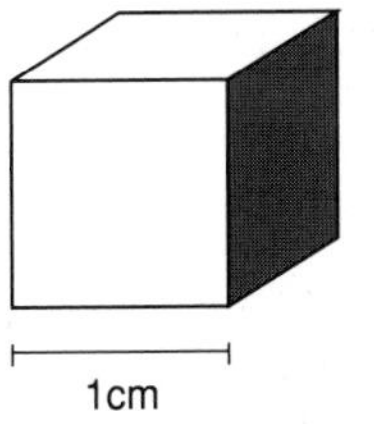

2

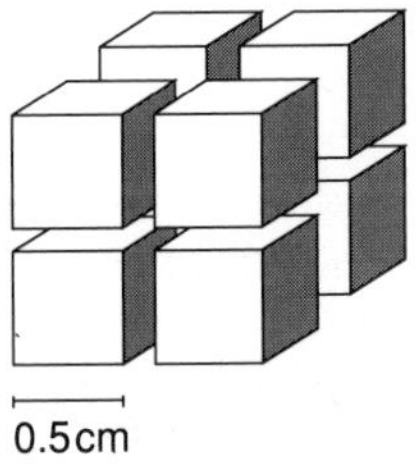

3

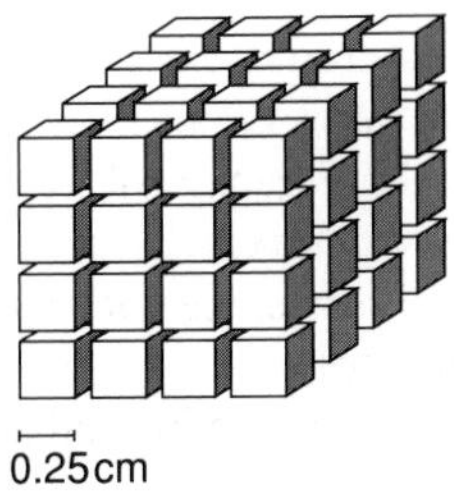

10 The table shows data obtained by school children doing a weathering project in a graveyard.

Dates on gravestones	Type of rock		
	Sandstone	Marble	Igneous
1720–1770	extremely badly weathered	very badly weathered	moderately weathered
1771–1819	very badly weathered	badly weathered	moderately weathered
1820–1870	badly weathered	moderately weathered	slightly weathered
1871–1919	moderately weathered	slightly weathered	unweathered
1920–1970	slightly weathered	unweathered	unweathered
1971–1991	unweathered	unweathered	unweathered

(a) Why do you think they chose a graveyard for this study?
(b) What do the results show?

11 The graph below shows the velocity of flow of a stream of water needed to pick up and *transport* particles of different sizes (Curve A). Curve B of the graph shows the velocity of flow at which particles of different sizes come out of suspension and are *deposited*. Use the graph to answer the following questions.

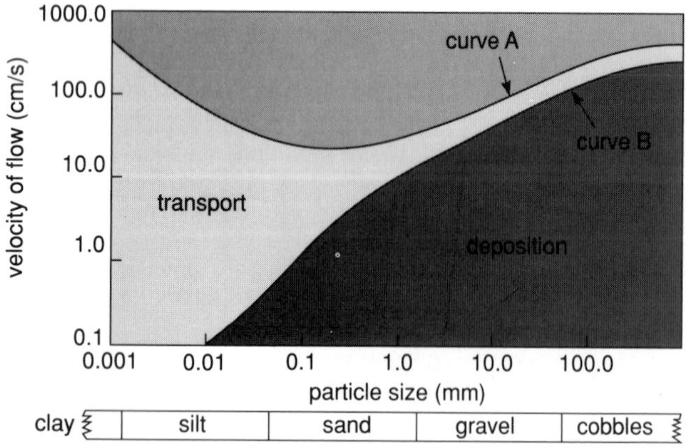

(a) Which type of material is most easily transported?
(b) At approximately what velocity will clay particles be picked up?
(c) At approximately what velocity will coarse gravel be picked up?
(d) In a stream flowing at a velocity below 50 cm per second, what type of material is most likely to be picked up?
(e) Explain the shape of Curve B.

12 The map below shows the location of three seismological stations which record earthquakes. The time that each station receives information on its equipment depends upon its distance from the epicentre of the earthquake. P waves arrive at stations faster than S waves; the time between the arrival of the two types of wave can help the stations to locate the epicentre. The table shows how far each station is from the epicentre.

Continent	Distance of station from epicentre/km	
	Quake 1	Quake 2
A	2500	1800
B	2000	1500
C	1100	3200

(a) Trace the map accurately and, using a compass, construct arcs to locate each epicentre.
(b) Why is it essential to know three distances?
(c) Why do stations receive little data for some earthquakes

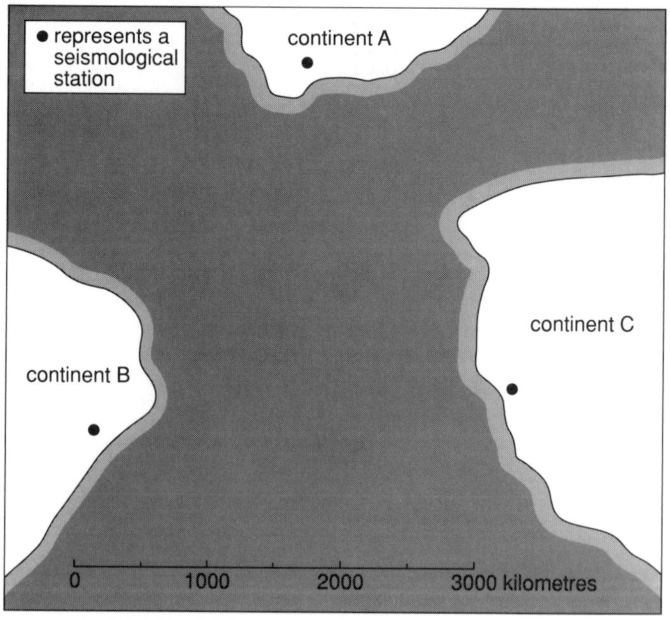

13 The map shows how geologists believe the world might look in eight million years time. What major changes have taken place? What evidence has led geologists to predict these changes?

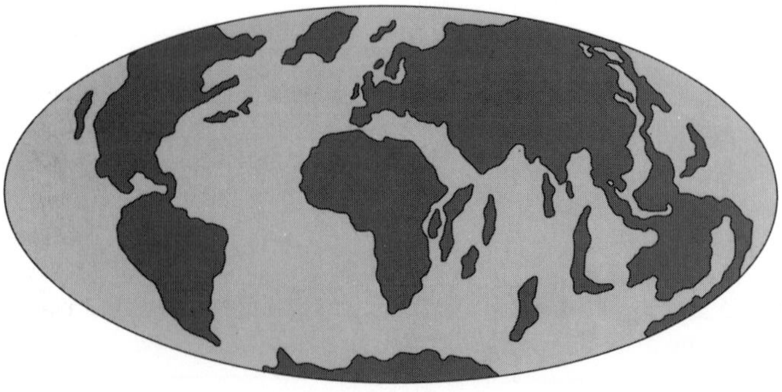

RELATIVE ATOMIC MASSES

Element	Symbol	A_r	Element	Symbol	A_r
Aluminium	Al	26.9	Magnesium	Mg	24.3
Antimony	Sb	121.8	Manganese	Mn	54.9
Argon	Ar	39.9	Mercury	Hg	200.6
Arsenic	As	74.9	Molybdenum	Mo	95.9
Barium	Ba	137.3	Neon	Ne	20.2
Beryllium	Be	9.0	Nickel	Ni	58.7
Bismuth	Bi	209.0	Nitrogen	N	14.0
Boron	B	10.8	Oxygen	O	16.0
Bromine	Br	79.9	Phosphorus	P	31.0
Cadmium	Cd	112.4	Platinum	Pt	195.1
Caesium	Cs	132.9	Potassium	K	39.1
Calcium	Ca	40.1	Rubidium	Rb	85.5
Carbon	C	12.0	Scandium	Sc	45.0
Chlorine	Cl	35.5	Selenium	Se	79.0
Chromium	Cr	52.0	Silicon	Si	28.1
Cobalt	Co	58.9	Silver	Ag	107.9
Copper	Cu	63.5	Sodium	Na	23.0
Fluorine	F	19.0	Strontium	Sr	87.6
Gallium	Ga	69.7	Sulphur	S	32.1
Germanium	Ge	72.6	Tellurium	Te	127.6
Gold	Au	197.0	Thorium	Th	232.0
Helium	He	4.0	Tin	Sn	118.7
Hydrogen	H	1.0	Titanium	Ti	47.9
Iodine	I	126.9	Tungsten	W	183.9
Iridium	Ir	192.2	Uranium	U	238.0
Iron	Fe	55.8	Vanadium	V	50.9
Krypton	Kr	83.8	Xenon	Xe	131.3
Lead	Pb	207.2	Zinc	Zn	65.4
Lithium	Li	6.9			

Index

Acknowledgements

We are grateful to the following companies, institutions and individuals who have given permission to reproduce photographs in this book. Every effort has been made to trace and acknowledge ownership of copyright. The publishers will be glad to make suitable arrangements with any copyright holder whom it has not been possible to contact.

Action Plus (178, 194 top); D. Allan BAS (40 top); J. Allen Cash (27, 28 bottom, 166 right, 195, 199); Banking Information Service (70 bottom); Professor J. Barbar, Imperial College, London (237); BASF (218 all 3); BICC (93); Biophoto Associates (132 right, 222); Syd Bishop and Sons Ltd (210 top left); Paul Brierly (85, 87 right); Brighton Borough Council (75 top); Britain on View (156 right, 157 top, 158 bottom); British Alcan (87 left, 92); British and European Sales Ltd (2 top); British Geological Survey (154, 157 bottom); BP Oil Ltd (75 bottom, 201); British Rail Ltd (213 bottom); British Steel (44, 111 left, 131); Burmah-Castrol (UK) Ltd (43 top); Calor Gas Ltd (198 both); Cambridge University Library (229, 230); Camera Press (2 bottom, 46 top, 246 bottom); Canadian Embassy (193 left); Cavendish Laboratory, Cambridge University (232); CEGB (37, 81, 84 bottom, 189, 193 right); Chloride Automotive Batteries Ltd (150); Corning (210 bottom left); Sean Crampton (30 top); Marie Curie Memorial Foundation (9 left); Farmer's Weekly (91, 221 top, 223 top); Fire Research Station (119 bottom); Ford Motor Company Ltd (94 left, 211 left); Friends of the Earth (3 top); GSF Picture Library (132 left); Geological Museum (164 right); Sally and Richard Greenhill (158 all 3); Greenpeace (25, 38, 135, 146 bottom left, 181, 207 bottom left, 244); Robert Harding Picture Library (6 top left H 8); John Hillelson (17); Hulton Picture Library (32 bottom, 171 right); ICI (136, 144, 175, 220, 221 left); IMI Range Ltd (105 bottom); Japanese Ministry of Foreign Affairs (35 left); Rosalind Jones (H6); Martin Leach (36); Lee Valley Park (141 bottom); Lever Brothers (152); Manchester City Council (51); Alfred McAlpine plc (59); The Meteorological Office/Institute of Geological Science (H1, H2, H3); Museo Prado (32 top); NHPA (12, 16 top, 39 right, 41, 188, 223 left/Stephen Kraseman (H7); National Gallery (238 top); National Power (207 bottom right); Natural History Museum (H9); Norsk Hydro Polymers (137 top); NASA (124 top); Nuffield-Chelsea Curriculum trust (106 top); Ohmeda (213 top); Oxford Scientific Films (28 top); David Phillips (177 top); Polysius Ltd (77); Popperfoto (9 bottom right, 22 bottom, 101 top 3, 137 bottom, 139, 245 top); Rex Features (236 bottom); Rio Tinto Zinc (234); Royal Society (6 bottom, 90 top); Slater Clark Associates (81 bottom); Jerry Sampson (35 right); Science Museum (9 right, 66 bottom, 67, 74, 124 bottom, 233); Science Photo Library (8, 26 top, 30 bottom, 53, 60, 62, 66 right, 106 left, 114 top, 121, 129 left, 164 left, 165 both, 168 bottom left, 242 right/David Vaughan H5/Dr Morely Read H4); Scot rail (43 bottom, 47); Scottish Tourist Board (110 left); Dr Seakins (21); Shell (1, 192, 196); Sulphur Institute (145); Syndication International (34 top, 39 left); Tate and Lyle (12); Colin Taylor Productions (4 top, 5 top, 3, 24 top right, 26 bottom, 35 middle and bottom, 48, 52 centre, 56 top right, 73, 76 top, 79 bottom, 82 top, 96, 100 right, 102 right, 113 top, 120 top, 123, 130, 149, 163, 166 left, 216 both); John Topham (42 top, 141 top, 179, 182 top left and right, 183, 185 both, 236, 247 left); UKAEA (9 centre, 128, 227, 235 top, 242 left, 243 right, 243 left, 247 right); Van de Burghs and Jergens (203); Wessex Water (40 bottom); Woodmansterne (10 bottom, 79 top, 174 top); Zefa (76 right).

British Library Cataloguing in Publication Data

Hill, Graham
Materials. – (Science Matters Series)
I. Title II. Series
620.1

ISBN 0-340-55845-8

First published 1993
Impression number 10 9 8 7 6 5 4 3 2 1
Year 1998 1997 1996 1995 1994 1993

Typeset by Keyset Composition, Colchester, Essex CO1 2LP.
Printed in Great Britain for the educational publishing division of Hodder & Stoughton Ltd, Mill Road, Dunton Green, Sevenoaks, Kent TN13 2YA by New Interlitho SPA.